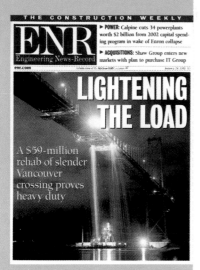

Student

Please enter my subscription for **Engineering News-Record**.

6 months ☐ $29.50 (Domestic)

Name

Address

City   State   Zip

☐ Payment enclosed   ☐ Bill me later   McGraw_Hill CONSTRUCTION ENR

5EN2DMHE

Friendly

Please enter my subscription for **Architectural Record**.

6 months ☐ $19.50 (Domestic)

Name

Address

City   State   Zip

☐ Payment enclosed   ☐ Bill me later   McGraw_Hill CONSTRUCTION Architectural Record 5AR2DMHE

Savings

Please enter my subscription for **Aviation Week & Space Technology**.

6 months ☐ $29.95 (Domestic)

Name

Address

City   State   Zip

☐ Payment enclosed   ☐ Bill me later

CAW34EDU

The Complete Technical Illustrator

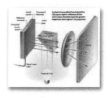

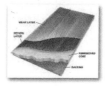

The Complete Technical Illustrator

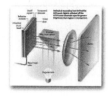

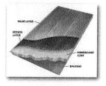

Jon Duff
Arizona State University

Greg Maxson

Mc Graw Hill **Higher Education**

Boston   Burr Ridge, IL   Dubuque, IA   Madison, WI   New York   San Francisco   St. Louis
Bangkok   Bogotá   Caracas   Kuala Lumpur   Lisbon   London   Madrid   Mexico City
Milan   Montreal   New Delhi   Santiago   Seoul   Singapore   Sydney   Taipei   Toronto

Higher Education

THE COMPLETE TECHNICAL ILLUSTRATOR

Published by McGraw-Hill, a business unit of The McGraw-Hill Companies, Inc., 1221 Avenue of the Americas, New York, NY 10020. Copyright © 2004 by The McGraw-Hill Companies, Inc. All rights reserved. No part of this publication may be reproduced or distributed in any form or by any means, or stored in a database or retrieval system, without the prior written consent of The McGraw-Hill Companies, Inc., including, but not limited to, in any network or other electronic storage or transmission, or broadcast for distance learning.

Some ancillaries, including electronic and print components, may not be available to customers outside the United States.

This book is printed on acid-free paper.

1 2 3 4 5 6 7 8 9 0 VNH/VNH 0 9 8 7 6 5 4 3

ISBN 0–07–252996–2

Publisher: *Elizabeth A. Jones*
Senior sponsoring editor: *Suzanne Jeans*
Developmental editor: *Melinda Dougharty*
Marketing manager: *Dawn R. Bercier*
Lead project manager: *Jill R. Peter*
Production supervisor: *Sherry L. Kane*
Lead media project manager: *Audrey A. Reiter*
Senior media technology producer: *Phillip Meek*
Senior coordinator of freelance design: *Michelle D. Whitaker*
Cover designer: *Rokusek Design*
Interior book design and composition: *The West Highland Press*
Senior photo research coordinator: *Lori Hancock*
Typeface: *11/13.5 Caslon*
Printer: *Von Hoffmann Corporation*

Cover images: Gutter Image: *Reprinted with permission from The Family Handyman magazine, ©2002 Home Service Publications, Inc., an affiliate of The Reader's Digest Association, Inc., Suite 700, 2915 Commers Drive, Eagan, MN 55121. All rights reserved.*
All other cover images: *Maxson Technical Illustration*

Library of Congress Cataloging-in-Publication Data

Duff, Jon M., 1948–
   The complete technical illustrator / Jon Duff, Greg Maxson. — 1st ed.
      p.   cm. — (McGraw-Hill series in computer science)
   Includes bibliographical references and index.
   ISBN 0–07–252996–2
   1. Mechanical drawing. I. Maxson, Greg. II. Title. III. Series.

T353.D83   2004
604.2—dc21   2003044181
                                        CIP

www.mhhe.com

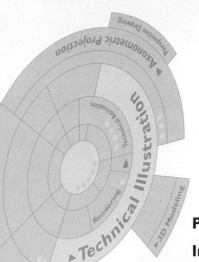

Table of Contents

▼ Part Two
Technical Illustration Layout and Construction

Chapter 3 — Orthogonal Layout 41

▼ Part Three
Technical Illustration Rendering

▼ Part Four

Modeling, Animation, and Technical Illustration

Chapter 20 — Animation and Technical Illustration 457

Perspective Drawing

Axonometric Projection

Technical Animation

Technical Illustration

Rendering

3D Modeling

Technical Illustration

Preface

Technical illustration is a field of graphic practice that uses static and dynamic images to explain the nature of technical relationships. There was a time when the people who made technical illustrations were referred to in their job titles as "technical illustrators." As such, they used a unique set of tools and techniques to produce their work. They were different from graphic artists, commercial artists, technical photographers, drafters, or videographers because a completely different tool set was necessary to do their jobs.

The advent of flexible digital graphic tools changed all that. The same program used by an engineer to design a piping installation can be used to create dimensionally accurate axonometric or perspective pictorials, render realistic views, or animate operations. This means that technical illustrations can be easily made by engineers, technologists, technicians, and graphic artists—and even by managers and administrative assistants.

Just because technical illustrations can be made by individuals not trained as technical illustrators doesn't mean that this diverse group doesn't need greater knowledge of technical illustration considerations. That is, of course, the reason for this book.

So if you are a technical illustrator by training, this book will expand your understanding of the subject's subtle nuances. If you are not specifically prepared as such, but find yourself creating, using, or evaluating technical illustrations, this book provides the information you missed.

Besides being digital artists, the authors are also classically trained technical illustrators. What does that mean to you, the reader? It means that in the pages of this book you'll find more than simple software tricks. You'll acquire the in-depth knowledge that in time will prepare you to discover your own

xix

tricks. You'll learn how to represent *any* geometry, in *any* view, using *any* tool, rendered using *any* technique, for *any* reproduction method, and when you do, you'll be the *complete technical illustrator*.

There isn't just one technical illustration tool because illustrations are made from such a wide range of data—sketches, photographs, engineering drawings, CAD data, raster scans, real objects that you measure, and your boss's verbal descriptions. Add to that the fact that technical illustrations appear in manuals and books, are displayed in Powerpoint presentations, are included in CD-ROM training, and are passed around on company Intranets. You have to match the tool to the data and its intended use.

We have based this book on the tools that we, the authors, use every day in our work. The actual software programs are less important than the methods of planning, executing, and evaluating digital technical illustrations. Those of you who use different software, or continue to use manual methods, should look past specific software references. If you *understand* raster, vector, and modeling methods, you'll have the flexibility to *apply* those methods as tools change.

The Complete Technical Illustrator really took 30 years to write. We hope it becomes the authority for the study, teaching, and practice of presenting technical information in a visual form. The authors would like to thank the many designers, engineers, architects, and technical illustrators who have graciously provided examples to supplement Jon's and Greg's own work. We hope this book becomes your dog-eared technical illustration companion.

Many people contributed to this project. Kelly Lowery and Melinda Dougharty of McGraw-Hill provided the direction and encouragement to keep the project on track. The manuscript was reviewed by Kurt Becker (Utah State University), Tom Bledsaw (ITT Technical Institute), Michael R. Aehle (St. Louis Community College at Meramec), Rick Vertolli (Cal State–Chico), Adam Smith (Rhode Island School of Design), Ernest B. Ezell, Jr. (Bowling Green State University), and William A. Ross (Purdue University). Their careful reviews were greatly appreciated, and many of their suggestions are reflected in the pages of this book. We would also like to thank those illustrators who contributed to and whose work appears in the case studies. Also, we thank the Microelectronics Modeling Group comprised of Nate Gelber, Mike Kelley, Ryan Graham, and Jay Hibler whose work appears throughout this text. Your willingness to share your illustrations and insight has contributed greatly to the practicable nature of the finished product. Special thanks go to George Ladas, K. Daniel Clark, Kevin Hulsey, Andre Cantarel, Richard Tsai, and Miller Visualization, Knott Laboratory, Daniels & Daniels, and Vector Scientific for their case study contributions. Finally, the authors would like to thank fellow and former students, illustrators-in-arms, and their clients for support and encouragement.

Introduction

The increased availability of digital graphics tools has inexorably changed the way technical illustrations are made. In fact, the very definition of a technical illustration has changed. Traditionally, technical illustrations were static graphic images. With the advent of three-dimensional CAD, modeling, rendering, and animation programs such as AutoCAD, ProEngineer, 3d studio max, Maya, and Lightwave, technical illustrators now can model and animate their solutions. It's a brave new world of technical illustration.

Although many technical two-dimensional illustrators have traded their Rapidograph pens, airbrushes, and Zip-a-Tone for Illustrator and Photoshop, the knowledge that an illustrator brings to the job really hasn't changed all that much over the past 40 years. Though the tools have changed dramatically, the body of knowledge has remained the same.

When Dr. Duff wrote his first book on the subject *Industrial Technical Illustration*, in 1982, he had been working in a traditional illustration environment for the Jeffrey Mining Machinery Division of Dresser Industries. He would take engineering drawings and sketches and turn them into line, and occasionally tone, illustrations for training, product development, and marketing. When he updated the book in 1994 as *Technical Illustration with Computer Applications*, the switch from analog to digital graphics was just gaining momentum.

This totally new textbook, with the addition of Greg Maxson as coauthor, completes the move to digitally created and printed technical illustrations. Greg's extensive experience over the past 15 years providing high-quality digital illustrations for publishing and magazine industries makes this text the authoritative source for education, training, and reference.

The authors acknowledge that some technical illustrations continue to be produced by traditional manual means. In fact, technical sketching using pencils and felt-tip pens continues to be the primary planning medium even though the final solution may be digital. The complete technical illustrator uses sketching to test ideas, record technical information, and communicate with clients and other illustrators. It's a lot less expensive and time-consuming to sketch your ideas rather than use computer tools. Plus, every computer program has a limited tool set. If you use a computer program to plan your illustrations, your final products will all look like.

▼ Quotable

"If the only tool you have is a hammer, every fastener will look like a nail to you."

This text is intended to supplement documentation that comes with your digital tools. We would have preferred to have covered the entire gamut of raster, vector, and modeling-animation programs but that would have resulted in a book as thick as it is tall (and way too expensive). So we have settled on four of the most common programs currently used to make technical illustrations: Adobe's Illustrator and Photoshop, and Autodesk's AutoCAD and 3ds max. If you use CorelDRAW, PhotoPAINT, and Lightwave, you can use our examples and our processes. With a little translation, the methods described on the following pages can apply to whatever tools are available.

Modern Technical Illustration

For the purposes of this textbook, a technical illustration is considered as:

any visual presentation which has the purpose of communicating technical information intended to aid in the design, manufacture, assembly, storage, distribution, use, disassembly, or disposal of a product or process.

Exempted from this definition are visual presentations whose primary purpose might be to entertain (though a technical illustration may entertain as a secondary function); visual artwork whose primary purpose is to express a statement by the artist; or other graphic communications without specific technical purpose. So you can see that a technical illustration is a special kind of image, even a special kind of technical image.

Technical illustrations are created from many different sources of technical data. As a general description, this data can be either analog or digital. Analog data includes sketches, paper drawings, prints, designs, maps, photographs, diagrams, charts, actual parts, models, mock-ups, and prototypes. Each of the preceding sources can also be found in digital form, ready to be used by computer graphic programs to produce finished technical illustrations. In fact, it may be more efficient to convert analog data to digital data by scanning or vectorizing rather than starting from scratch on the computer.

You'll find that productive technical illustrators are able to solve illustration problems with whatever technical data is available. We call these people "complete technical illustrators," and it is for them that this book is written.

For example, if you are an illustrator who works only from existing engineering CAD geometry, and all the data available is photographic, you may be out of luck. Likewise, if the only way you can create pictorials is by tracing photographs, you'll be lost when looking at hundreds of engineering prints. This textbook covers almost every imaginable scenario in which a technical illustration can be made.

A technical illustrator needs knowledge across several technical areas: engineering and technology, printing and publishing, computer graphics, aesthetics, and visual psychology. In this textbook, these topics are presented using notes when appropriate. If you are new to technical illustration, please make consistent use of our extensive glossary.

What You Need to Know to Use This Book

This book is not intended to be a basic text for Adobe Illustrator or Photoshop, AutoCAD, or 3ds max You should have fundamental knowledge of these tools in order to apply them to technical illustration tasks. It would be unfair to think that each of you using this textbook has every scrap of knowledge necessary to become a productive technical illustrator. The broader your knowledge, however, the greater value you will be to your employer, or possibly to yourself. As a technical illustrator, you are shooting for:

▼ Sketching that communicates. You must be able to communicate using sketching. This means planning illustration strategies, communicating to your client, your boss, and fellow workers. Some technical illustrations can actually be sketches.

▼ Broad manufacturing and construction knowledge. You need to know how objects are made, assembled, fastened, and used. You should have a good understanding of materials and how they look after being formed, processed, cut, or finished.

▼ Broad knowledge of printing and publishing. Technical illustrations will be repurposed for training, marketing, and customer service in books, manuals, multimedia, and over the Internet. Each medium has strengths and weaknesses that determine both the nature of the visual as well as its ability to communicate technical information.

▼ Extensive knowledge of geometry. It's impossible to create effective technical illustrations without the ability to accurately evaluate both two- and three-dimensional geometry.

▼ Strong knowledge of computer graphics tools. The better you know a tool, the more you'll be in control of the illustration. Additionally, the more tools you know, the greater the chance you can pick the best tool for the job.

▼ A good understanding of color, light, and material. Critical decisions may be made based on subtle representations of a technical illustration. You may need to attend to the nuances of color, texture, surface finish, reflection and refraction, and environment.

▼ Knowing TGE. A technical illustration must be cost-effective. That means you need to know when to say "that's good enough." If it's too good, it means you probably haven't made any money on the project. If it's not good enough, you probably have reduced its effectiveness.

What This Book Contains

The chapters in this book contain two informative features: Tips and Case Studies. The Tips put into practice the points covered in the chapters. They give a hands-on explanation you can use right away. Case studies bring you real-world technical illustration examples. Many professional illustrators have graciously contributed their work, and they discuss their trials and tribulations in arriving at the solution. You'll also find an extensive glossary at the end of this book. Rather than define every term as it's being used, terms and abbreviations are defined in the glossary.

You already know this book doesn't cover everything you need to know to be a technical illustrator. What it does contain are strategies and procedures for applying your knowledge about geometry, assemblies, materials, and processes to communicate technical information. You'll find lists of text and Internet resources at the end of almost every chapter that will lead you to additional sources of information. This book is divided into four separate parts.

Part One: Technical Illustration Background

These first two chapters review computer graphics and image reproduction that the complete technical illustrator needs to keep in mind as jobs are analyzed, planned, and executed. They don't necessarily have to be read first; you may want to dive right into the subjects covered in Parts Two, Three, and Four. But before you tackle that first big technical illustration assignment, review these chapters. You'll find valuable information that will increase your effectiveness and efficiency.

Part Two: Technical Illustration Layout and Construction

Every effective technical illustration is based on accurate layout and construction. In these chapters, you'll learn just about every technique you'll need to turn drawings, sketches, photographs, and engineering CAD data in into technical illustrations.

Part Three: Technical Illustration Rendering

A great illustration needs to show the difference between glass and metal, wood and concrete, rubber and chrome. Knowing which technique to use requires both technical and artistic ability as well as an understanding of the eventual distribution medium. The chapters in this part cover the gamut of rendering techniques—from simple line rendering to photo-realism.

Part Four: Modeling, Animation, and Technical Illustration

This part contains methods for approaching technical illustration as a modeling as opposed to a drawing exercise. Technical illustrations are no longer simply two-dimensional line drawings. The same digital data used for two-dimensional illustrations can be the basis for three-dimensional models and animations. To reverse the process, three-dimensional models can be the basis for two-dimensional illustrations. If you have an accurate, well-rendered model, you have instant access to *every* possible pictorial view, something impossible with two-dimensional drawings.

On the CD-ROM

The first productivity tool included free on the accompanying disk is "AxonHelper," a calculator that can be kept on your desktop. With it, many of the calculations necessary to make accurate axonometric constructions are simplified. Professor Duff created this tool in Macromedia Flash, and it is distributed free of charge to purchasers of this book.

The authors have assembled examples of the most popular illustration tools on the CD-ROM included with this book. Each of the chapters that present specific instruction has corresponding files on the CD-ROM that can be used to follow along with the instruction. A full 15-week course including many of the exercises Professor Duff uses in his technical illustration classes is included with a web browser. Greg Maxson has generously included examples of textures and materials from his studio, as well as a gallery of professional illustrations to encourage you to do your best work.

Contacting Us

We enjoy hearing from the readers of our books, both constructive criticism as well as "attaboys." Every textbook is a compromise, but we would be more than willing to consider adding or subtracting material in future editions based on what you, our readers, feel.

Jon M. Duff (*jmduff@asu.edu*)

Greg Maxson (*gmaxti@shout.net*)

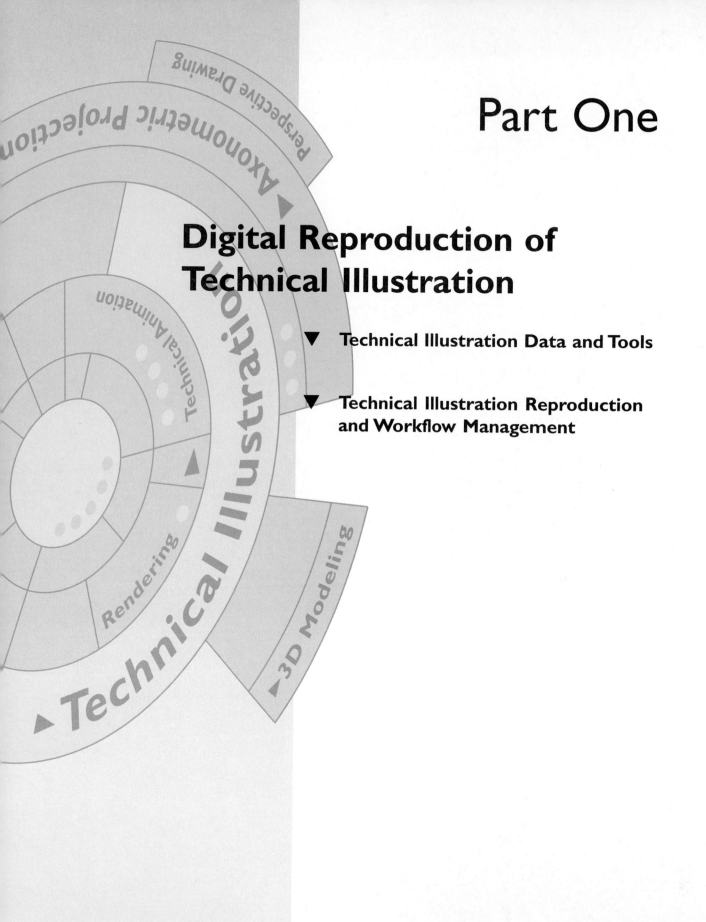

Part One

Digital Reproduction of Technical Illustration

▼ **Technical Illustration Data and Tools**

▼ **Technical Illustration Reproduction and Workflow Management**

Chapter 1

Technical Illustration Data and Tools

In This Chapter

Although technical illustration is characterized by product or output (the actual graphic), there is a background of helpful knowledge that can make your job easier. With that knowledge in hand you can more easily make effective and efficient decisions to solve particularly pesky problems. True, digital illustration (especially illustration from three-dimensional modeling) is more forgiving than are analog (manually drawn) techniques. Why is this so? In analog illustration, an inappropriate decision early in the illustration process could mean the loss of hundreds of hours of effort because if you're wrong, you usually have to start over. Digital techniques can be more forgiving, more flexible, and in the end, more powerful.

This text assumes that you have general knowledge of computer graphics: tools, techniques, file formats, and printing. This first chapter builds on this general knowledge and shows how it can be applied to modern technical illustration.

Chapter Objectives

In this chapter you will understand:

▼ The purposes of technical illustration

▼ The range of technical illustration tools

▼ How computer graphics processes two-and three-dimensional geometric data into a form compatible for technical illustration

▼ The forms of data relied on to create technical illustrations

▼ The technologies of raster and vector graphics

Technical Illustration Defined

A technical illustration is a graphical representation showing spatial, geometric, or conceptual relationships for the purpose of marketing, training, operation, assembly, maintenance, or disposal. It walks the thin line between engineering and art. It requires accurate geometry, position, materials, and order so that the *intent* can be accomplished. If the intent of the illustration is to facilitate the assembly of a product, but is ineffective because of unfaithful representation, parts may not be assembled in the correct order or orientation.

Sometimes the results of a substandard illustration are not very significant. But illustrations are often used in nuclear reactor operations, bomb disposal training, surgical operations, and pilot training. A mistake here can have tragic consequences. It is for this reason that technical illustration cannot, at its core, be art.

Everyone has heard of artistic license. This term describes the freedom an artist has in interpreting imagery through the his or her own experience. Technical illustrations exhibit technical license, similar to artistic license but because of the aforementioned consequences, of greater importance. It is not enough to faithfully capture all the spatial, geometric, and representational aspects of a subject. In fact, such faithfulness may increase the amount of visual information in the illustration and lead to confusion or information overload. So beyond being technically faithful, a technical illustrator must be prepared to *abstract* visual information so that a user can focus on important features without extraneous visual "noise."

▼ Quotable

"You must be able to leave out unimportant information without impacting the effectiveness of the illustration."

So a technical illustration must be:

▼ Geometrically, spatially, and visually faithful

▼ Carefully abstracted to accomplish the intent

In Figure 1.1 a photograph, a continuous-tone color illustration, and an abstracted line illustration of the same subject are shown for comparison. The photograph (Figure 1.1a) is faithful to the subject but is probably unusable (notice the drips, scratches, and splatters). The color illustration (Figure 1.1b) removes this extraneous and distracting information and better conveys assembly information. The line illustration (Figure 1.1c) further abstracts the subject so that parts can be more easily identified independent of color and background. This is a difficult but important function of a technical illustrator. This ability comes with experience, and research and by involving others with needed technical expertise.

The Complete Technical Illustrator

This book is called *The Complete Technical Illustrator* because its authors believe that to be a successful technical illustrator today, you must be able to manage data in almost any form using a variety of tools. There are several reasons for this.

First, an illustrator normally doesn't have control over the form of input information for technical illustration. Reference data may be sketches, photographs, engineering drawings, scale models, computer models, or simply a verbal description. If you require engineering drawings in order to complete accurate illustrations, and no engineering drawings are available, you are in serious trouble. Likewise, if you rely only on visual correctness, and a client wants dimensionally exact representations, you are equally limited.

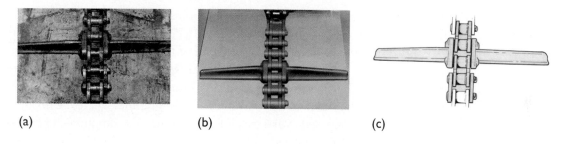

(a)   (b)   (c)

Figure 1.1   Abstracting an illustration may make information clearer.

On the output side, you may not immediately know in what form your technical illustration will be used. If you have only produced black-and-white illustrations because printing full color has been too expensive, and the client decides to use the Web to distribute the images, what do you do when color is as cheap as black and white?

The bottom line is that the complete technical illustrator takes information in any form and produces illustration in whatever output format is required.

Several industries still employ individuals they call technical illustrators. Government job classification codes include categories for illustrator-draftsman and branches of the military continue publishing manuals for the training of such occupational specialties. Aerospace and defense contractors make use of technical illustrators to produce the voluminous documentation required for aircraft, equipment, and defense systems.

The trend, however, has been for more and more technical illustration to be producd by individuals with job titles other than technical illustrator. This is a fundamental change in the technical illustration paradigm.

▼ Old Paradigm

Technical illustration is produced as the two-dimensional pictorial documentation of the completed design. It uses engineering drawings, photographs, models, and sketches, along with specialized geometric drawing methods, by individuals trained in those procedures.

▼ New Paradigm

Technical illustration is the by-product of the 3D process and may take the form of 2D drawings, 3D models, animations, or virtual experiences. It is the specialized view of the product database, used for specialized purposes, and can be created by anyone with access to that data.

Computer Graphics and Technical Illustration

This text is intended to show how computer graphics concepts can be used as the basis of modern technical illustration. If you desire a closer look at computer graphics itself, a list of resources is presented at the end of this and Chapter 2.

Computer graphics is the area of computer science that deals with the representation of information in digital graphic form. Technical illustration (when executed using computers) is then that part of computer graphics where the imagery is used for technical as opposed to artistic ends. Figure 1.2 is a computer-generated graphic. Is it a technical illustration?

The same software, hardware, utilities, and knowledge are used throughout computer graphics so it is not unusual to see creative animators used on technical projects or technical illustrators used in entertainment. The technical skills are easily transferable. However, the mind-sets of computer artists and technical illustrators are vastly different.

The technical illustrator works for a *client*. It is the client who sets deadlines and requirements such as size, color, and data format. The technical illustrator applies computer graphics knowledge and skills, along with technical product and process knowledge, to solve the problem within the client's specifications.

Technical Illustration Data

Technical illustration data can be divided into *input data* and *output data*. Input data is what you use to create a technical illustration. This includes engineering drawings, sketches, photographs, models, files in various computer-aided design (CAD) formats, art files in any of several raster and vector formats, and verbal descriptions. Output data describes the form in which a technical illustration may be produced: raster bit map (TIF), PostScript (EPS), stereo lithographic (STL), or one of several proprietary vector formats. Technical illustrators may be even called to submit data in virtual reality formats (VML) where the illustration becomes an interactive experience.

Figure 1.2 Many forms of computer graphics are not technical illustrations.

The intent of an illustration dictates to some degree the appropriateness of the possible input and output formats. Using six-decimal place engineering data for quick concept illustrations is an example of overkill. However, using photographs and approximate dimensions for an illustration that will be used in exacting multi-million dollar decisions is probably not a good idea either.

Data in computer format is referred to as *digital* data. The characteristics of digital raster and vector data are discussed throughout this text because not only are technical illustrations output in these two formats, but an increasing amount of input data may be found in raster and vector formats as well.

Raster and Vector Data

As an overview, *raster data* (also called bitmap data) is a pattern of dots at fixed resolution (Figure 1.3). In order to achieve high quality, raster images must have tens of millions of pieces (bits) of data; raster files can be quite large [1–50 megabytes (Mbyte)], requiring the most powerful computer you can afford. The most important raster file format for technical illustration is the *Tag Image File Format* (TIF) because it communicates its raster information to high-quality printers to produce optimized output. Other raster formats (GIF, JPG, BMP, PCX, DIB, TGA) do not have this desirable feature and are appropriate for other forms of output.

▼ Tech Note

When a bitmap is displayed on a monitor, each dot becomes a picture element or *pixel*. When the displayed bitmap is printed, each pixel is represented by a matrix of *printer dots*. The resolution of the bitmap as displayed on a monitor is expressed in dots per inch (dpi). The resolution of the printer is expressed in printer dots per inch (ppi). Therefore, a 1200-ppi printer represents each pixel of a 300-dpi bitmap with a 4 x 4 matrix of printer dots (1200/300 = 4). The larger this matrix (1200/150 = 8), the greater the number of values each raster dot can assume and the smoother the tones in the illustration.

Vector data uses mathematical descriptions for lines, text, shapes, and fills (Figure 1.4). Vector data is *resolution independent*, an important characteristic. The same vector data can be sent to an 800-ppi printer for inexpensive distribution and to a 2000-ppi printer for poster-quality reproduction.

Vector files require considerably less processing and storage resources because vector graphics are not stored as individual bits. The most important vector file format for technical illustration is the *PostScript* (PS) format or its cousin *Encapsulated PostScript* (EPS). As you might expect, robust PostScript files make use of the library of fills, blends, and operations available in PostScript printers. Other vector file formats (DWG, PRT, CGM, HPGL, PICT, WMF) may be converted to PostScript but fail to make use of the full range of PostScript routines. Very high resolution raster graphics (1200 dpi and above) and PostScript graphics may appear identical. But a close inspection of the vector graphic reveals an absence of pixels; shapes are described by vectors (curves), while tones are resolution-independent mathematical fills (Figure 1.4).

If you compare the wheels in Figures 1.3 and 1.4, you might be inclined to think they contain the same data. But a closer inspection reveals that the raster illustration in Figure 1.3 owes its smoothness to the density of bits, while the vector illustration in Figure 1.4 makes use of mathematical curves and fills.

If you would like additional information on the differences between raster and vector data formats, consult *The Encyclopedia of Graphic File Formats* by Murray and vanRyper.

Analog Data

Data not in electronic format is referred to as *analog data*. The information in analog data is continuous. Examples of analog data include paper drawings, blue prints, sketches, and photographs. Traditional technical illustrations made with pens, pencils, brushes, films, and tapes are referred to as *analog illustrations*.

To be used with computers this analog data must be converted (digitized) into electronic format. *Scanners* sample analog images and convert values (gray scale) and hues (color) into bitmap, or raster, data. For example, photographic prints can be converted into TIF format by scanning.

Figure 1.3   High-resolution raster images are comprised of millions of individual picture elements (pixels).

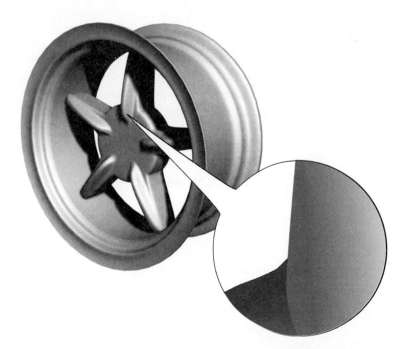

Figure 1.4   Vector graphics are comprised of mathematical descriptions of strokes and fills.

Raster to Vector Conversion

Once in raster format, software can detect abrupt changes in value and convert raster data to vector data. This is called *raster to vector conversion*; in some software this is referred to as *autotracing*. Scanned engineering drawings can be converted to vectors, and at one time much effort was spent "digitizing" paper drawings. Current practice is to do this only on a critical case-by-case basis.

Although parameters for detecting value change can be adjusted to fine-tune the conversion, autotracing remains problematic unless the image being traced has adequate contrast. Without sharp delineation between edges and background, results are often unacceptable. In these cases it is often more productive to simply trace over the photograph by hand (see Chapter 11).

Raster to vector conversion can at least give you outlines that can be adjusted and edited to arrive at final vector paths. Figure 1.5 shows a raster scan after autotracing. The scanned photo is on the bottom layer, while the autotrace has its own layer. You can see that the tracing parameters need to be adjusted because too few control points have been used to define the curves.

Scanners can also convert pages of printed text (analog) into electronic text (vector) through the process of *optical character recognition* (OCR). Once in electronic format, the text can be incorporated into technical illustrations without retyping. When the text is formatted using a resolution-independent outline font (like PostScript or TrueType), it has essentially been vectorized.

▼ Tech Note

Raster to vector conversion doesn't actually *convert* the bitmap to vectors. Instead, vectors are overlayed following abrupt changes in bitmap values. The raster image can then be discarded. The vectors almost always need to be edited, so you need to carefully evaluate whether or not raster to vector conversion is worth the effort or if you should manually trace over the bitmap.

Vector to Raster Conversion

In the reverse process, vectors can be converted to a bitmap. This is called *rasterization*. Once in raster format, the data is fixed in resolution and all vector characteristics have been lost. Because of the resolution independence of vector graphics, there is little benefit in rasterizing technical illustrations for high-quality printing. However, vector illustrations must usually be rasterized for Internet and multimedia applications because these media don't (without special plug-ins) display vector graphics.

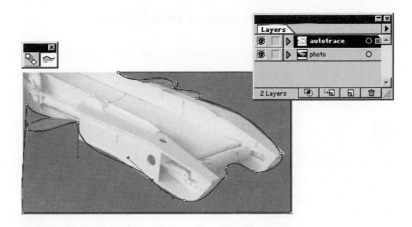

Figure 1.5   Raster images can be converted to vectors by detecting dramatic changes in value.

Additional Methods

Three-dimensional data (for computer modeling programs) can be derived from analog 2D sources in a *raster to model conversion*. Gray-scale information from photographs can be interpreted into depth (Z axis) information and when combined with the XY coordinates across the scan, interpreted into 3D models. For example, photographs from Mars can be interpreted into 3D terrain models. In Figure 1.6, a raster *displacement map* causes a flat plane to be deformed. White areas are raised, black areas are depressed, and middle gray areas provide a baseline elevation.

In an application of scanning, *3D digitizers* record the XYZ position of real world objects. This *tactile to model conversion* facilitates technical illustration of objects that already exist, but for which other sources of data are not available.

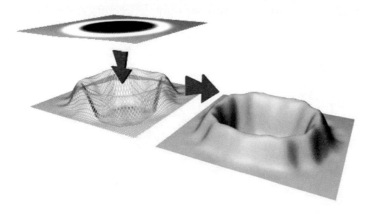

Figure 1.6   Brightness values of a raster bitmap can be interpreted into a 3D model.

Technical Illustration File Types

The complete technical illustrator is fully aware of the strengths and limitations of output data formats, matching the appropriate data to the intended use. A general rule governs almost all decisions as to output data format.

"Always capture or create your illustrations with the highest resolution and the greatest depth of geometric and color information."

What this means is that you can *sample the data down* when high resolution and great color depth are not required. Figure 1.7 shows the effect of reducing both resolution and color depth on a simple assembly saved in TIF format. Note that a 4 times reduction in resolution, from 300 to 72 dpi, results in a 15 times reduction in file size. Also notice how a severe reduction in color depth results in little reduction in file size due to the overhead of saving the data in TIF format. The same image, saved in 8-bit GIF format (appropriate for the Web, inappropriate for print) would result in a file only 2K in size. And as you can see at the bottom end, there is little savings with a great difference in quality.

Unfortunately, the converse isn't true. You can't get higher resolution or more color depth by sampling up. That's why you always want to start out with the greatest depth of information possible.

When a file is saved in a program's *native file format*, it contains information needed for the data to be worked on in that program. For example, Adobe Illustrator (AI) file format contains more information than is needed to be included in a technical document. By saving a copy of the file in EPS format you strip out Adobe Illustrator information and use only the vectors and calls to PostScript routines in the printer.

24-bit color at 300 dpi
File Size = 1145 K

24-bit color at 72 dpi
File Size = 75 K

8-bit color at 72 dpi
File Size = 62 K

Figure 1.7   The effect of sampling down high-resolution data.

Vector Illustration

Several vector formats are used in technical illustration but none have the power and flexibility of PostScript. This is because PostScript printers contain extensive libraries of resolution-independent effects. Non-PostScript vector formats (CGM and HPGL, for example) may produce resolution-independent vectors but without Bezier curves, screens, fills, and PostScript text. Encapsulated PostScript (EPS) provides device-independent and resolution-independent portability between illustration programs and page layout programs such as Quark, PageMaker, or InDesign. These page layout programs combine illustrations, photographs, and text to make technical documents.

Page layout programs can also be used simply to print EPS illustrations because they may offer more control over the printing process than illustration programs themselves. Encapsulated PostScript can be considered a container for vectors, bitmaps, and text described as outline printer fonts. This used to be called a *metafile*, but because most robust graphic files are now metafiles, its significance is diminished.

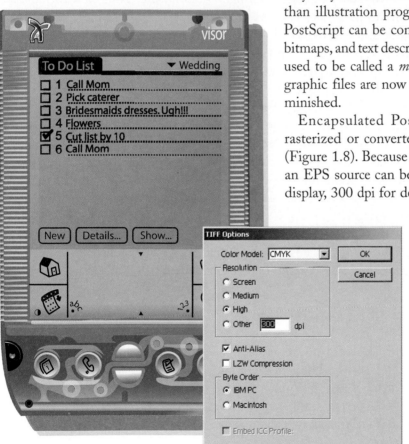

Encapsulated PostScript illustrations can be rasterized or converted to bitmap format, if needed (Figure 1.8). Because EPS is resolution independent, an EPS source can be rasterized at 72 dpi for screen display, 300 dpi for desktop printing, or 2000 dpi for high-quality printing, and in the raster format appropriate for the application. This would mean GIF, JPG, and PNG for the Web; BMP, GIF, and JPG for multimedia; and TIF for print publications.

Figure 1.8   PostScript vector graphics can be rasterized at any resolution (Genevive Christensen).

▼ Tech Note

Although EPS is resolution independent, it can't add resolution to fixed-resolution data. This means that any raster information included in the EPS description must be at the desired output resolution when incorporated into the file. Encapsulated PostScript can't magically transform low-quality data into a higher resolution.

Raster Illustration

Scanners, digital cameras, digital video cameras, and animation programs create images in raster format. As mentioned previously, an image in raster format is fixed resolution. This is not a bad thing, especially if the image is high resolution (1200 dpi and above) and deep in color information (24 bits, resulting in a 16.7 million color palette). However, such files are very large, requiring significant computing resources. Were a low-resolution, low-color palette raster image to be sampled up, the sampled information does not become finer nor the colors smoother. You only have more data describing the same coarseness.

The TIF format is appropriate as the basis of raster technical illustration for several reasons. It allows a large $[(2^{32} - 1) \times (2^{32} - 1)]$ physical size raster image with a deep color palette (24-bit). This is 16.7 million colors and a 4 billion pixel square canvas! Additionally, TIF format contains instructions (tags) that communicate with high-quality PostScript printers so that the raster bitmap is matched to the printer's characteristics (resolution, color depth, screen frequency, and dot shape). Created in high-resolution TIF format, the raster can be easily sampled down for various applications.

Encapsulated PostScript provides an envelope so that TIF formatted raster images can be combined with PostScript vectors. You get the benefits of resolution-independent text and vector graphics with the flexibility of the TIF format.

Technical Illustration Tools

There remain pockets of traditional manual noncomputer technical illustration. But for the most part, technical illustration is executed using computer hardware and software. The manual technique that we still employ is that of sketching. But of course, sketches can be scanned.

Technical illustration tools parallel input and output data formats: raster, vector, and 3D modeling.

▼ **3D Modeling**. To make use of existing 3D data and to create realistic technical representations you'll need a program that models and renders. Output: CMYK TIF raster files for still frames; vector files for use in 2D PostScript drawing programs (EPS, CGM, DWG, IGES); TGA, AVI, or MOV animation files.

▼ Illustration in Industry

George Ladas — Base 24 Design
Roselle, New Jersey

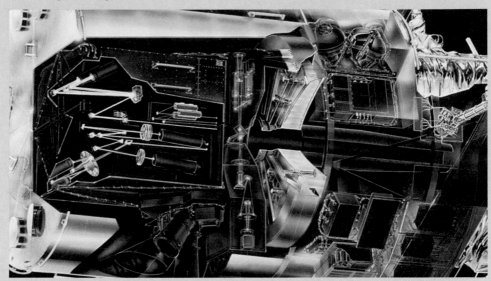

Image Copyright Base 24 Design.

Software
Generic CAD 3D

Hardware
Windows PC
Digital Plotter

Traditional
Airbrush

The challenge of this assignment was to design and illustrate an educational poster with a cutaway view of the Hubble Telescope showing all the important components and light paths to the various cameras. The National Aeronautics and Space Administration (NASA) provided several hundred engineering drawings from a number of agencies that had contracted for different components. A combination of accurate 3D CAD constructions and manual airbrush illustration techniques were used.

NASA required finished artwork as an oversize 50 x 48 in painting that was to be photographed and digitized into a final layout for the 33 x 26 in poster. Data from the engineering drawings was entered onto separate layers for each component. The 3D CAD files were then rotated and locked in the chosen view and plotted to the scale of the finished artwork. The plots were analyzed and drawn individually into one single finished line drawing on frosted Mylar which was then printed on Mylar and dry mounted on foam board, ready to be hand airbrushed to produce the finished art. NASA changed the specifications of the poster during the preliminary studies and insisted on accuracy of dimensions. Some of the detail was simplified to clarify the light paths and fulfill the requirements of an educational poster. The CAD drawing printout gave a very accurate and clean line that could be adjusted or altered as needed and provided a backup transparent Mylar drawing.

▼ **Terms**

A complete glossary of terms can be found in the glossary at the end of the book.

▼ **Black-and-White Proofing**. Early proofs don't need color, and many publications are designed in black and white anyway. Tabloid (11 x 17) capabilities are nice but an 8.5 x 11, 800 ppi PostScript printer usually suffices.

▼ **Color Proofing**. A high-quality (1440 ppi) color inkjet printer is essential for proofing. Though you can get away with an 8.5 x 11 format, 11 x 17 allows for bleeds and crops. Printers are fairly inexpensive. Consumables (ink and paper) are not.

▼ **Digital Camera**. To capture reference photographs, textures, or scenes for photo tracing, a digital camera with a minimum of 2 megapixel resolution is necessary. Output: RGB color raster files in TIF, JPG, or GIF format.

▼ **Digital Scanner**. To use paper drawings and photographs you need a scanner. Consider 1440 x 800 resolution and 24 bits of color depth. Output: red, green, and blue (RGB) or cyan, magenta, yellow, and black (CMYK) color; gray-scale or black-and-white raster files in JPG format for tracing; and TIF format for printing and sampling down for Web and multimedia.

▼ **Graphics Workstation**. You'll need a computer capable of storing, processing, and displaying graphic files twice the complexity you anticipate working on. In other words, if 128 MByte of memory seems to be enough for your needs right now, you will want to equip your computer with 512 MByte. Whether you utilize a Macintosh or Windows system, always buy more hard disk, more memory, a faster processor, and the biggest monitor you can afford.

▼ **PostScript Drawing Tool**. To create publication-quality resolution independent 2D graphics, you need a drawing tool that creates compliant PostScript files. Output: CMYK color vector files in PS or EPS format; GIF or JPG raster files in RGB color space for Internet publication.

▼ **Raster Editor**. To edit scans and other raster images, you need software that lets you work on bitmap data. Output: CMYK color, gray-scale, or black-and-white raster files in TIF format for print publication; RGB JPG, PNG, and GIF for Internet publication; TGA, JPG, and BMP in RGB color space for animation frames.

▼ **Storage**. Technical illustrations can quickly fill up a hard disk. It may be helpful to connect a large, fast secondary disk that you use to hold active projects. Archive illustrations that have been completed to CD or DVD ROM.

Raster and Vector Products Used in This Book

There are a number of very good software tools available for making quality technical illustrations; it is not the purpose of this text to pit one tool against another. Likewise, it would be infeasible to cover multiple applications without a book over a thousand pages in length. The authors have highlighted products that are generally the most accepted in the industry. If you pick up the classified section of any major city newspaper you will consistently see four products mentioned: Adobe Illustrator, Adobe Photoshop, AutoCAD, and 3ds max. *The Complete Technical Illustrator* features these products.

Review

Technical illustrations combine many of the elements of computer graphics and computer art with aspects of computer-aided design. Digital tools such as scanners, cameras, digitizers, and raster and vector programs form the core of modern illustration practice. Because information can be found in both analog and digital forms, the complete technical illustrator must have the flexibility to complete a task no matter the input and no matter the output formats. Whether for print, Internet, or multimedia, today's technical illustrator must be cognizant of the strengths and limitations of raster and vector file formats and be prepared to utilize each when appropriate.

Text Resources

Information for printing and data formats can be found in several print resources.

Cost, Frank. *Pocket Guide to Digital Printing*. Delmar/Thompson. Albany. 1997.

Miano, John. *Compressed Image File Formats: JPEG, PNG, GIF, XBM, BMP*. ACM Press. Washington, D.C. 1999.

Murray, James D., vanRyper, William, and Russell, Deborah (Ed.). *Encyclopedia of Graphic File Formats with CD-ROM*. O'Reilley. Sebastopol, California. 1996.

Ramano, Frank. *Delmar's Dictionary of Digital Printing and Publishing*. Delmar/Thompson. Albany. 1997.

Schildgen, Thomas. *Pocket Guide to Color with Digital Applications*. Delmar/Thompson. Albany. 1998.

Internet Resources

Resources for computer graphics, technical illustration data, and production tools can be found on the Web.

Autodesk Resource Site. Retrieved from: *http://pointa.autodesk.com/portal/nav/index.htm*. May 2002.

CADKEY. Retrieved from: *http://www.cadkey.com/resources/docs/index.asp*. May 2002.

Micro Station. Retrieved from: *http://www.bentley.com/vizcenter/links.htm*. May 2002.

Portsort professional illustrators' portfolios. Retrieved from: *http://www.portsort.com*. May 2002.

Solid Edge. Retrieved from: *http://www.solid-edge.com/media/media_center.htm*. May 2002.

Solid Works. Retrieved from: *http://www.solidworks.com/html*. May 2002.

3D Cafe. Retrieved from: *http://www.3dcafe.com*. May 2002.

Chapter 2

Technical Illustration Reproduction and Workflow Management

In This Chapter

The complete technical illustrator must be intimately knowledgeable about how illustrations are made, reproduced, transmitted, and stored. This is because technical illustrations by and large are not artwork; not simply hung on walls, they find their way into technical manuals, parts books, training classes, product reviews, maintenance manuals, CD-ROM, and onto the Internet. Reproduction quality is important because the information contained in a technical illustration must be clear and unambiguous. Substandard printing or display can render an effective illustration worthless. The more a technical illustrator knows about reproduction processes, the more successful the final product.

Of course, technical illustrators are not printers, or at least they weren't until the advent of digital printing. Analog printing (negatives, stripping, plates, ink, solvents and rags, etc.) requires a completely different set of skills and tools as compared to illustration. For a technical illustrator to get the desired results, a close working relationship with the printer had to be formed.

Take the printer out of the mix—substitute a digital device that directly reproduces the illustration file—and suddenly the illustrator has to be cognizant of a myriad of issues that before were the domain of the printer. There is no doubt that traditional analog printing will continue to be a dominant technology but eventually it will be replaced by direct digital reproduction. This means that future illustrators will become more and more involved in the entire technical publication process: from illustration to finished technical document.

Chapter Objectives

In this chapter you will understand:

- ▼ The strengths and limitations of printing processes
- ▼ How technical manuals are executed
- ▼ The characteristics of spot and process color
- ▼ The characteristics of analog and digital printing
- ▼ The characteristics of digital displays
- ▼ The preparation of technical illustrations for printing from the Web
- ▼ How networks impact the illustration process
- ▼ File compression uses and abuses
- ▼ Methods of job control and tracking
- ▼ Reproduction of technical illustration

Analog Reproduction

Analog offset lithography accounts for the greatest percentage of technical documents printed in industry. Factors that must be considered when preparing technical illustrations for analog offset lithographic reproduction include:

- ▼ Analog printing is always a trade-off between quality and cost.
- ▼ Analog offset lithography remains the most cost-effective method for making a moderate number (1000–50000) of copies.
- ▼ For analog reproduction, digital illustrations must be output to film or printed and a negative made of the print. The process is quite involved: A paste-up of text and graphics becomes the page that is photographed to make the negative that is used to make the printing plate that is used to make the ink impression that is finally transferred to the sheet of paper. You can see that if you start digital, you should print digital. If you start analog, scanning the illustration into digital format is the first step in increasing eventual usability.

▼ Analog reproduction requires a darkroom, cameras, stripped flats, negatives, printing plates, presses, and inks. Each element introduces a measure of variability in the eventual reproduction and, once the job is submitted to the printer, it is out of the control of the illustrator.

▼ Line illustrations can be reproduced using the natural distribution of silver particles found in fine-grain photographic film. Still, under extreme reductions, fine detail can be lost (drop out) or muddied (ink gain).

▼ Continuous-tone illustrations must be broken into a pattern of dots, called a *halftone*. The halftone negative must then be combined with other page elements (and aligned) on a stripped flat. The finer the pattern of dots (called the screen pitch), the more difficult the image is to print accurately. The coarser the screen, the easier it is to print but the less detail shown.

▼ Color increases costs in analog offset lithography because individual plates must be made for each color. Full color requires four plates: cyan, magenta, yellow, and black. You *can* print four colors on a one-color press but this would necessitate reloading printed sheets and reimaging, mounting, and inking the additional printing plates, one by one.
Anything other than one color (usually black) increases costs.

▼ Note

To bridge the gap between analog and digital methods, digital files can be directly imaged to plates. This is called *digital-to-plate* or DTP.   By using DTP manual stripping is avoided while printers continue to use analog presses with digital input.

Electrostatic Copying

For very short reproduction runs, copiers can produce cost-effective results with reduced but oftentimes acceptable quality. The same paste-up used for longer-run offset lithography can be used for copying. You should consider the following:

▼ Use electrostatic copying when you have flat art paste-ups.

▼ This type of reproduction is appropriate for print-on-demand applications and reduces the costs associated with printing, warehousing, and waste when documents are updated and obsolete versions are disposed of.

▼ Copier optics are considerably less accurate than camera optics. Copiers that use lenses can be susceptible to distortion when copying at the extremes of the copy field.

▼ Digital copiers may be more accurate across the entire copy field but are limited to the sampling frequency of the *charge-coupled device* (CCD).

▼ Electrostatic copying uses dry toner that is fused to the paper; the toner doesn't penetrate the surface. Copied sheets are susceptible to cracking and peeling and do not stand up to heavy use as well as analog printed pieces do where ink penetrates the paper.

▼ The range of colors available from color copiers is much smaller than that of offset lithography. Color copying is more appropriate for one color in addition to black (spot color) or where color fidelity is not critical.

Digital Reproduction

Directly printing digital data on digital devices removes many of the sources of variability found in analog offset printing. Issues such as color registration are automated. The *raster image processor* (RIP) translates the digital file into a format understood by the printing device. Raster and vector data in your digital illustration file are ripped (translated into printer dots) for the target printer. Color can be reproduced on digital printing devices with only a portion of the overhead cost (labor, materials) found with analog methods. Currently, digital reproduction is appropriate for print-on-demand, low-production runs, and for variable data.

▼ Low-volume digital printing can be accomplished with laser printers, digital copiers, and thermal and piezoelectric ink-jet printers.

▼ Digital dry toner four-color copying (Agfa ChromaPress, for example) combines qualities of analog process color with digital electrostatic imaging. Quality is appropriate for spot color and noncritical continuous color fidelity.

▼ Digital ink printing (Indigo E-Print Pro, Turbostream, Ultrastream) combines the qualities of ink-jet printers with the flexibility of a high-volume printing press. This technology can even print on flexible media like plastic or foam.

▼ Direct digital offset (Heidelberg Quickmaster DI, Speedmaster DI) removes manually intensive prepress (make ready) and images the digital file directly on the press. There are no negatives, no stripping, and no hand mounting of printing plates.

The Characteristics of Spot and Process Color

Process color uses four or more inks in combination to represent the full range of colors that humans are able to see. Spot color (Figure 2.1) uses one or more colors to highlight important features of an illustration. In Figure 2.1 a spot color from the *Pantone Matching System* (PMS) color palette is used to highlight an area of an illustration. In Figure 2.2 a custom spot color, made from CMYK inks, is used. In both cases, the use of a single color in combination with black-and-white line art effectively draws your attention to the desired features. Yet most people would tell you without hesitation: "color is better." But a perusal of research on color perception reveals some disheartening data on the predictable impact of color (Horton, 1991).

▼   There is wide variability in how humans *perceive* color. These differences can be physiological. But other differences are environmental, social, and cultural. Different viewers will not perceive a color identically, even if that color is specified correctly.

▼   Human physiology is designed to discriminate changes in brightness to a greater extent than changes in hue. We look for differences in value and overlook differences in color.

▼ **Tech Tip**

A CMYK spot color is really the result of process inks arranged in a halftone and not a single color.

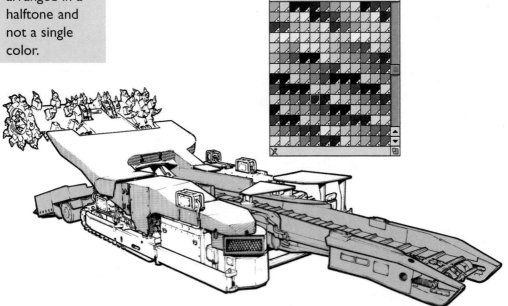

Figure 2.1   Spot color using the Pantone Matching System.

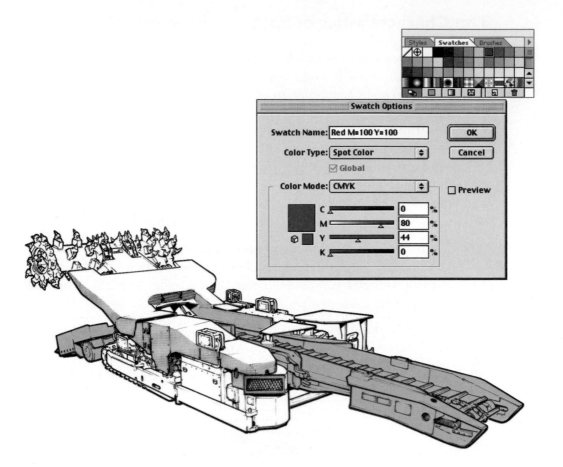

Figure 2.2   Spot color using CMYK inks.

- ▼ Fully 10 percent of males have divergent color vision, most commonly confusing red and green.

- ▼ Of the 16.7 million colors possible in the 24-bit color palette, only 7.5 million are actually measurable.

- ▼ Trained colorists (such as people who work for paint companies) can distinguish only 1 million colors even under ideal conditions.

- ▼ People without extensive color training can distinguish only seven colors within their field of view (without scanning around an image).

- ▼ A single color has the highest reliability in target recognition (locating something on an image).

The reason these points are important is because with digital graphics it's often just as easy to create images in 24-bit color as not. If images are going on the Web, there is no added expense to color and the human condition being what it is (more is better; bigger and more is even better!), the temptation is to throw color around without any thought.

The bottom line is that while color can be effective and useful, images containing many colors may present too much or confusing visual information. The color requirements of a technical illustration must be analyzed early and color used intelligently.

For that reason, the use of a highlight, or *spot color,* is still important in making effective illustrations. In analog printing spot color is not made from component CMYK inks. Instead, individual inks (specified in some system such as PMS color) are used. In digital printing, spot colors may be made from CMYK components when the press has only four-color capability. When more colors are available (as in a six-color press), specialty inks, in addition to CMYK, can be used for spot color applications.

Specifying Color

The entire range of colors a particular device is capable of recording or displaying is called its *gamut.* Printing presses, scanners, cameras, and the human eye all have different gamuts; each has its own technology for reproducing color (Figure 2.3). It would be impossible for this text to adequately cover the strengths and weaknesses of the various systems. Several excellent resources have been listed at the end of this chapter should you desire additional information. But you need to accept that every device has a unique gamut. Passing color information between devices is a matter of mapping one gamut to another.

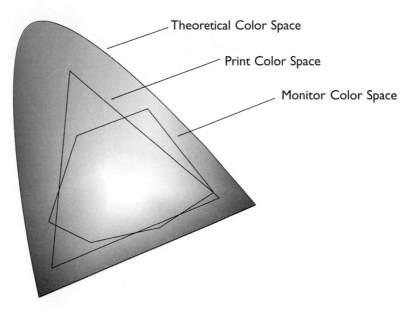

Figure 2.3   The gamut for a particular device will not include all colors in a particular color space.

Printing presses understand color specified in cyan, magenta, yellow, and black (CMYK); monitors understand red, green, and blue (RGB). But illustrators don't think in CMYK or RGB. Illustrators generally think in terms of color, color intensity, and brightness. This may be referred to as hue, light, saturation (HLS) or hue, saturation, brightness (HSB).

An example of this is working with highlight, midtone, and shadow of a particular color. By adjusting the brightness value, the same color appears in bright sunlight, shade, and shadow (Figure 2.4). The car fender in this example begins as a gradient mesh in Adobe Illustrator. The fender's overall color is specified by the HSB value of 192, 50, 100 (a medium blue). By adjusting the saturation at the top of the fender to 10, you arrive at the highlight color. By adjusting the brightness to 50, you arrive at the shade color. The hue stays at 192, so the fender has consistent coloration. Trying this in CMYK or RGB requires mental gymnastics most people

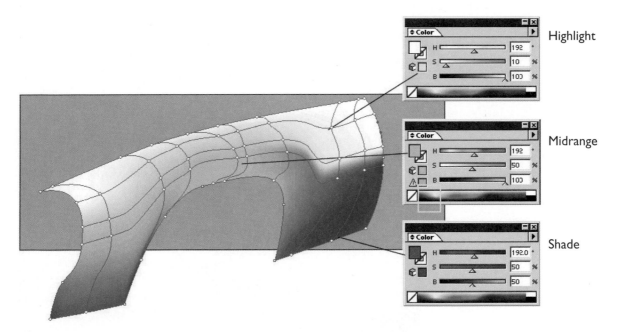

Figure 2.4   Changing a base color (hue 192) to highlight, shade, and shadow is intuitive in HLS or HSB color space.

Most illustration programs provide flexible color pickers that let you work in HLS or HSB while displaying both CMYK and RGB values (or even hexadecimal color values, but that's an entirely different method). In this manner, you can determine if a color is outside the gamut for a particular output device and match colors specified in the various systems (Figure 2.5).

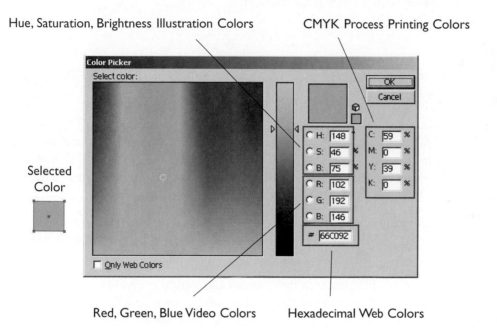

Hue, Saturation, Brightness Illustration Colors

CMYK Process Printing Colors

Selected Color

Red, Green, Blue Video Colors

Hexadecimal Web Colors

Figure 2.5   Color can be specified in any of several color systems on Illustrator's Color Picker.

Characteristics of Digital Displays

More and more technical illustrations are never printed on paper. Instead, they are displayed on monitors as part of on-line documentation, CD-ROM training, or digitally archived documents. When preparing illustrations for digital display, realize that:

▼ You probably will have no control over the capabilities of the computer equipment used to display your work.

▼ Digital displays don't naturally display vectors. Vector files must be rasterized to be displayed as pixels on the screen. If you provide vector data, you'll have no control over this rasterization process.

▼ Digital displays have fixed resolution. Zooming in on raster data increases the coarseness of the image. If closer inspection of parts in an illustration is necessary, specialized software is required to re-rasterize vectors for every zoom level.

▼ Digital displays only display colors as determined by the memory available on the graphics controller. Color depth is inversely related to display resolution. For example, assume you have a graphics controller with 4 MBytes of video memory. At 1024 x 768 resolution only 8-bit color may be available (256 colors), while at 640 x 480 you have access to the entire 24-bit palette (16.7 million colors).

▼ Note

Unless you have more than 16 MBytes of video memory (dedicated or shared), your monitor can display high resolution and fewer colors or lower resolution and more colors.

▼ Monitor display is susceptible to wide color variations. Some monitors are bluish, while some are greenish. Some get brownish with age. All monitors are influenced by ambient light. Monitor color is different in a room lit by incandescent light than one lit by fluorescent light.

▼ Operating systems describe color differently. The same illustration displayed on Macintosh and Windows computers will show subtle differences.

▼ There are differences in how various application programs describe color. Some programs may restrict display to 8- or 16-bit color, regardless of the capabilities of the graphics controller. Colors outside the palette are displayed by *dithering*, the practice of placing available color pixels in a pattern that approximates the unavailable color. Again, you will have no control over how accurately this is done.

Printing from the Web

It may be more advisable to display a facsimile or thumbnail of a technical document on the Web and provide a way for the client to view or download the actual document in a form that better preserves the detail of photographs and illustrations. Currently, the most flexible technology to accomplish this is with a *portable document*, of which there are several types currently in use. The most pervasive of these is Adobe's *Portable Document Format* (PDF). PDF documents are efficiently compressed for acceptable network downloads. When viewed in the *PDF Viewer*, technical documents become searchable, linkable, and enlargeable by zooming in on detail without loss of resolution. Because the documents are described in a variant of PostScript, they are resolution independent and can be printed at any desired level of quality. If you have downloaded tax forms, you have probably seen PDF files.

Programs such as Quark, PageMaker, InDesign, and FrameMaker are capable of producing PDF files. Even Microsoft Word can do it, though few would use Word for complex technical documents rich in technical illustrations.

Networks and Technical Illustration

The equipment on which technical illustrations are produced represent but a portion of the modern technical document production environment (Figure 2.6). As described earlier, technical illustrations are best produced on dedicated workstations as opposed to general-purpose computers. Even when working freelance, you should be aware of the architecture of the production facility used to print (or process for display) your illustrations.

The prototypical environment shown in Figure 2.6 demonstrates the diverse methods used in technical publications containing illustrations. Follow the path from an illustration workstation to the *direct digital press*. Then follow the path of analog content to the analog press. The following are several scenarios in publishing technical illustrations. While reading them, follow Figure 2.6.

Method A
Digital output from in-house illustrator to digital press

This is the optimum case where all text, graphics, and composed pages start and stay digital.

▼ Illustrator creates digital artwork and stores the files on an open press interface (OPI) server.

▼ Page layout combines art (linked) and edited text from the server into a composed technical document and stores the pages on the server.

▼ RIP station pre-flights and converts the pages to an image file.

▼ Digital press accepts the ripped pages from the RIP, combines the composed pages with linked art from the OPI server, and prints the document.

Method B
Analog output from in-house illustrator to analog press

This is not a desirable procedure but at times digital artwork must be combined with analog art and text.

▼ Illustrator creates digital artwork and stores the files on sa erver.

▼ Files are proofed and output at desired resolution on a digital printer. A negative is made of the art print.

▼ Or, digital file is sent to the image file setter where a negative is produced.

▼ Negative is stripped into flat.

▼ Flat is proofed. Plates are made from flats and mounted on an analog press.

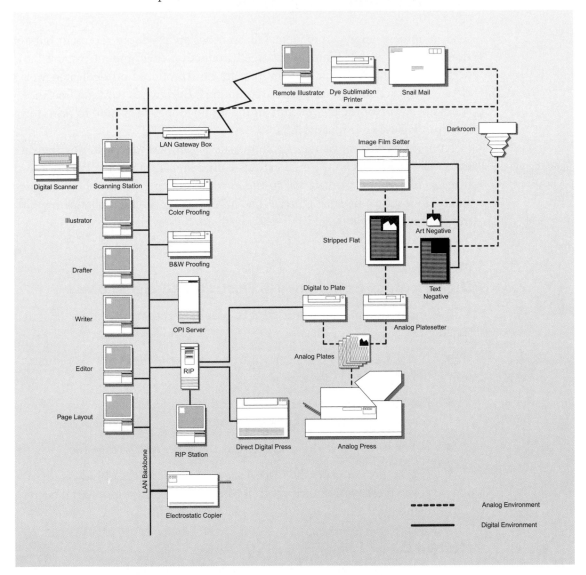

Figure 2.6   A technical illustration production network includes input, storage, and output devices.

Method C
Digital output from in-house illustrator to analog press

This is common in situations where companies have large investments in analog printing equipment.

- ▼ Illustrator creates artwork and stores the files on an OPI server.
- ▼ After digital proof, the illustration is combined with the text in page layout.
- ▼ Composed pages are sent to image film setter to create a negative.
- ▼ Negative is stripped into flat and proofed.
- ▼ Plates are made from flats and mounted on an analog press.
- ▼ Or, RIP station preflights pages and then creates an image file for DTP plate setting.
- ▼ Plates are mounted on the analog press.

Method D
Digital output from remote illustrator to digital press

This is the best possible situation for a subcontract or freelance illustrator.

- ▼ Remote illustrator creates digital artwork and stores it locally.
- ▼ File is sent electronically [file transfer protocol (FTP), WhamNet, email attachment] to the local area network (LAN) server through the LAN gateway.
- ▼ Combine with method A.

Method E
Analog output from remote illustrator to digital press

- ▼ Combine methods B and D.

Method F
Pre-digital age output from remote illustrator to analog press

This is still done although it is rapidly disappearing.

- ▼ Remote illustrator creates artwork and stores it locally. Manual art is mounted and covered with tissue and a flap. The digital file is printed at highest resolution and deepest color palette.
- ▼ Printed art is put in envelope and mailed to art director. See second half of method B.

Most production environments make use of an OPI server. This allows servers to store high-resolution files and send low-resolution placeholders while pages are being composed. The high-resolution images are substituted at print time.

▼ Illustration in Industry

Miller Visualization
Howell, Michigan

Images Copyright Miller Visualization.

Software

3ds max
PhotoShop
Premier
After Effects

Hardware

Windows PC
Faro Digitizer
Perception Board

These graphics were produced to reconstruct an unfortunate head-on collision for a court case. We inspected the destroyed vehicles, and using a Faro digitizing arm, created 3D models of both vehicles in their postaccident state. These 3D models allowed the vehicles to be precisely aligned as they interacted throughout the collision.

We worked closely with an accident reconstruction expert to create the interaction of the two vehicles during the collision. The vehicle models needed to correctly pass over all the physical evidence surveyed at the accident scene, including tire marks and gouges in the roadway. Attention to motion curves provided within 3ds max made sure that the motions of the vehicles adhered to the laws of physics. We assured, for example, that the vehicles behaved in accordance with physics while airborne. Much of the animating was actually done by manipulating the motion curves rather than moving the models in the viewports. The detail in the vehicles, their surroundings, and the dust and debris assured that jury members felt that they had seen the actual accident when they entered the jury room.

File Compression

File compression is a technique whereby redundant data is represented in a way that reduces the overall file size. Smaller files require less storage and are transmitted more quickly over a network. The *compression ratio* refers to the ration of the file size of the uncompressed data to the compressed data. A 500K file compressed 10:1 will result in a file 50K in size. A compression ratio of 10:1 is 5 times more efficient than 2:1.

Data can be compressed *internally* to the file, as is the case with TIF, BMP, JPG, and GIF file formats. By opening the file, the data is uncompressed or *extracted*. Additionally, data can be compressed externally to the file, as is the case with ZIP (PK Zip) and SIT (Stuffit) file formats.

File compression can be *symmetrical* or *asymmetrical*. Symmetrical compression compresses and extracts the data in roughly the same time. This would be appropriate for data that needs to be compressed and then read on the fly. Asymmetrical compression is more appropriate for backup or storage. You may tolerate very slow compression on data that you want extracted repeatedly. Conversely, data that may only be occasionally used can be compressed rapidly and extracted more slowly.

Files that have long strings of identical data can be efficiently compressed. For example, a black-and-white raster line illustration may be compressed efficiently because the background can be described by long strings of white data. Continuous-tone raster illustrations (gray scale or color) have fewer patterns of repeating data, so as you would imagine, different compression methods must be used. Vector data, because objects are described mathematically, is a poorer candidate for compression. Because vector files start out smaller, compression is not as critical.

Matching the data type to the appropriate method is very important. Choosing an inappropriate compression method can actually result in a larger file than you started with (such as using GIF for a continuous tone image). Data compression methods (algorithms) can be *lossless* or *lossy*.

Lossless Data Compression

Using lossless techniques, a file is compressed and uncompressed (extracted) without the loss of any data. The extracted file is identical to the original. Lossless methods include *run-length encoding* (RLE), *Huffman* encoding, and *Lempel-Ziv-Welch* (LZW) compression. File formats include TIF, BMP, and GIF.

Lossy Data Compression

Lossy techniques achieve reduced file size by removing certain information about the data. Generally, lossy methods make use of our inability to discriminate fine variations in color. The greater the compression ratio, the more data is removed. When the file is extracted, the missing data is *interpolated* by various means to fill in the missing pieces. Depending on the method, the extracted data can be virtually indistinguishable from the original or full of anomalies such as *artifacts*. Lossy methods include JPG, MPG, and AVI. The solid area of white background makes the image in Figure 2.7a a poor candidate for lossy compression. Though the compression of 5:1 is substantial, the artifacts introduced when the file is extracted (Figure 2.7b) may make this method unacceptable for the data.

(a)   (b)

Figure 2.7   The greater the lossy compression, the more artifacts are introduced into the extracted image.

Vector format technical illustration files are, by design, smaller than corresponding raster files. Compare the values in Table 2.1 for the rock hauler illustration. External compression (as in PK Zip) will result in savings depending on the level of redundant data. As shown in the table, an uncompressed EPS file is 521K in size. When zipped, the file is reduced to 98K (much of this is the compression of the 8-bit TIF image header—the picture of the EPS file you get to see when you place it on the page).

On the bitmap side, a 150-dpi, 24-bit TIF raster file of the rock hauler weighs in at a chunky 2.5 MBytes. The EPS file from which the TIF file was rasterized is a svelte 512K (with the added benefit of being resolution independent!). The 24-bit TIF file is reduced to 61K when medium-quality JPG lossy compression is applied. An uncompressed 8-bit BMP file is reduced to 96K when GIF lossless compression is applied.

.tif	.tif (LZW)	.bmp	.jpg (jpeg)	.gif (LZW)	.eps	.zip
2.5 MBytes	257K	629K	61K	96K	521K	98K

Table 2.1   Raster and vector file compression (Sam Steinborn).

Methods of Job Control and Tracking

As you might imagine, small illustration projects are easy to manage. But a large technical publication may have thousands of illustrations in several different formats and hundreds of pages of text—all being worked on by dozens of individuals. Managing a project such as this requires certain control mechanisms just to keep everything in order.

The Beauty of Regular Backups

Without regular secure backups any project, large or small, is at risk. Regular means at least once a day. Having two copies of a file on the same device is not considered secure backup. Secure also means that the backup and the active file are not stored at the same location.

The Job Jacket

This is a term that refers to the organization of everything necessary for a job. Job specifications, communication with the client and vendor, quotes, roughs, approvals, original art, galley proofs—anything and everything necessary to control the job, including the *job ticket*. In analog illustration this is actually a real jacket, or folder, that holds all pertinent materials. In electronic illustration, this is the organization or directory structure in which all electronic materials are stored for the job (Figure 2.8).

The Job Ticket

This control document is a record of the activities required to accomplish the job (Figure 2.9). For example, the approval of sketches, proofs, and final design would be recorded here; the specification of sizes, copies, colors, etc., would also appear on the job ticket. The job ticket is the ultimate source of information about the job. A *job control number* is assigned to the job as it comes in; this number should be easily interpreted and yield important information at a glance.

Other control documents help manage an illustration project. If billing is done by the hour, a *time sheet* is invaluable. Even when billing is done by the job, recording time spent on each job gives you information so that realistic (and profitable) estimates can be made on an *estimate request*. A *final quotation* uses information from the estimate and previous time sheets to assure that the illustration can be delivered on time, at the required quality, and for the price agreed upon. Sample control documents can be found on the accompanying CD-ROM.

Figure 2.8   An electronic job jacket. Note the organization of all pertinent information.

Review

Because the same digital technologies are used to create technical illustrations, create technical documents, and display or print the results, technical illustrators must be more aware of reproduction processes than ever before. Likewise, knowledge about computer technologies such as monitors, the Internet, digital printing, and file compression may impact the effectiveness of your solutions. Even the way you organize your work, name your files, and back up your data contributes to your ultimate success.

Text Resources

Adams, J. Michael, and Dolin, Penny Ann. *Printing Technology.* Delmar. Albany. 2002.

Field, Gary. G. *Color and Its Reproduction.* GATF Press. Pittsburgh. 1999.

Technical Illustration
Job Ticket

Job Number

☐☐☐☐☐☐ – ☐☐

Customer Approval

Layout ☐   Proof ☐
Rough ☐   Ship ☐
Color Comp ☐

	Due In	Preflight Due	Layout Due	Rough Due	Color Comp Due	Final Proof Due	Shipping Due
Sales Rep							
Bill To							

Attn

Job Specification

City	Type of Illustration
State Zip:	End Use
Contact	Projection System
Phone Fax Email	Color Requirements
Illustrators Dates	Data Formats
Text Specification	Delivered Format Media
Call Outs	Illustration Size
Notes	**Equipment Required**
Figure Legends	Computer
Body Copy	Storage
Fonts	Storage

Reference Materials

Engineering Drawings		Electronic Files		Photos	
Title	Number	Name	Format	Log Number	Subject

Figure 2.9   The job ticket is a complete record of technical illustration activity (adapted from Adams and Dolin, 2002).

Green, Phil. *Understanding Digital Color*. Graphic Arts Technical Foundation. Pittsburgh. 1995.

Hunt, R.W.G. *The Reproduction of Colour*. Fountain Press. Tolworth, England.1987.

Horton, William. *Illustrating Computer Information*. John Wiley & Sons. New York. 1991.

Murray, James D., and VanRyper, William. *Graphic File Formats*. O'Reilly. Sebastepol, California. 1996.

Internet Resources

Resources for print and electronic reproduction can be found on the Web.

Adobe Acrobat. Retrieved from: *http://www.adobe.com/acrofamily/main.html*. May 2002.

Agfa Graphic Arts. Retrieved from: *http://www.agfa.com/graphics/*. May 2002.

Heidelberg Web Site. Retrieved from: *http://www.heidelberg.com/ frm_prepage.asp*. May 2002.

Indigo digital offset. Retrieved from: *http://www.indigonet.com*. May 2002.

Planet PDF. Retrieved from: *http://www.planetpdf.com/*. May 2002.

On the CD-ROM

Look in *exercises/ch2* for the following files:

estimate.pdf   A sample job estimation sheet.

job_tix.pdf   A sample job ticket you may want to use or modify.

quotation.pdf   A sample final quotation sheet.

time_sheet.pdf   A sample time sheet for estimating and billing.

Part Two

Technical Illustration Layout and Construction

▼ **Orthogonal Layout**

▼ **Axonometric Views**

▼ **Axonometric Circles**

▼ **Axonometric Scale Construction**

▼ **Axonometric Projection**

▼ **Axonometric Shearing**

▼ **Perspective Techniques**

Technical Animation

Technical Illustration

Rendering

3D Modeling

▶ Technical Illustration

Chapter 3

Orthographic Layout

In This Chapter

Before you begin the study of axonometric and perspective illustration, you should be well grounded in the theory and practices of orthographic projection. This chapter is *not* meant to substitute for a traditional course in 2D technical drawing. It does, however, discuss how knowledge of technical and engineering drawing is applied in illustration tools—both 2D and 3D.

You may find occasions where an orthographic view is more illustrative than a pictorial view. In these cases, the geometric foundation of the illustration must be dimensionally accurate, something that is not particularly natural in Adobe Illustrator (no trim-to extension tool, for example). Accurate geometry can, of course, be brought into Illustrator from AutoCAD or 3ds max, but there are serious limitations in usability. See Chapter 17 where using technical drawing data in Adobe Illustrator is discussed in detail. Still, most 2D CAD constructions are directly applicable in Illustrator and can, with some twists and turns, be effectively replicated.

Chapter Objectives

In this chapter you will understand:

▼ The relationship of world axes to orthographic views

▼ Terminology used to describe orthographic relationships

▼ Technical scale in Adobe Illustrator

▼ Orthographic methods in Adobe Illustrator

▼ Using grids, snap, and guides

▼ Orthogonal geometry as a basis for 2D operations

▼ Orthogonal geometry as a basis for 3Dl operations

▼ Orthographic views in 3ds max

World Axes and Orthographic Views

Three mutually exclusive axes, referred to as the X, Y, and Z coordinate axes define world coordinate space. These world axes are fixed in space, so when a viewing direction is taken parallel to one axis, the two other axes appear of normal, or of true length. This practice of viewing geometry parallel to one of the world axes is the basis for standard orthographic views.

Objects, then, should be placed in logical relationships to these axes so that the views generated have some meaning. Figure 3.1 shows a technical part, a connecting link, aligned with these world coordinate axes. You can see in each *principal view* that two of the three axes are true, or normal and the third is seen in point view. So between two *adjacent* views (separated by 90 degrees of viewing rotation), all three of the spatial dimensions are seen true and can be measured, and the part is completely described.

The position of geometry in space and the assigning of view names (front, side, top, and so forth) are totally arbitrary. Use the following guidelines in planning view-axis-object orientations:

▼ The front view is usuallythe most descriptive view. That is, the view that would allow someone to pick this object out of a group of similar objects.

▼ The front view is normally the view at which a person usually encounters the object.

▼ The front view traditionally aligns the longest dimension (height, width, or depth) with the front view's X axis.

▼ If depth is vertical (as with drilled holes), align depth to the Z axis. If depth is horizontal and front-to-back, align depth to the Y axis.

▼ Choose a side view that shows the greatest amount of visible features. Avoid a side view that hides the majority of geometric features.

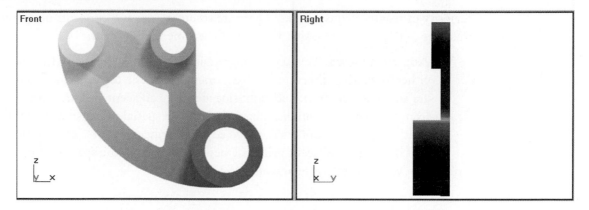

Figure 3.1   Geometry is aligned with the world coordinate axes.

Figure 3.2 shows the connecting link rotated about the Z axis. Note that only the height (Z dimension) of the link remains true. Width (X dimension) and depth (Y dimension) have been *foreshortened* and are not true length. Although either view in Figure 3.2 does a better job of communicating the overall geometry than do the principal views in Figure 3.1, the information necessary to produce the part is not available.

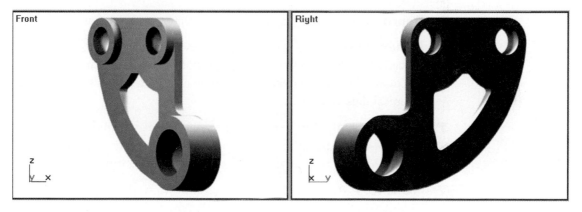

Figure 3.2   Object geometry rotated out of alignment with world coordinate axes.

What has this taught us? First, there are no right or wrong views, only views that are more or less descriptive and informative. It has taught us that the most important consideration is to align descriptive features—height, width, depth, and centerlines—with the axes. You may then choose to call the views what you will.

Orthographic Terminology

Orthographic literally means straight drawing. It is not the perspective manner in which humans view the world. It is, however, the way designers, architects, and engineers view their creations and is the projection method used for the vast majority of technical illustrations. There are several terms and concepts unique to this type of drawing system that you need to become familiar with.

▼ **Adjacent views**. Any two orthographic views separated by 90 degrees, either in viewing direction or object orientation. The top and front views are adjacent. The front and right side views are adjacent (Figure 3.3). The front and rear views are *not* adjacent. All three spatial dimensions (height, width, depth) are contained in two adjacent principal views. This means that if you have adjacent principal views, you have all the spatial information you need to create any other view.

Front View   Right Side View

Figure 3.3   Adjacent orthographic views.

▼ **Depth**. A horizontal dimension perpendicular to height and width.

▼ **Dimensioned views**. Orthographic views, drawn either to scale or not to scale, have true sizes that have been noted using standard dimensioning practices. These numerical values can be used to create scale views and technical illustrations.

▼ **Drawn to scale**. Orthographic views with sizes that reflect the true dimensional relationships of the object can be directly compared. You can use measurement of these views for technical illustrations.

▼ **Front view**. Also called the front elevation view, the view seen when looking parallel to the horizontal depth axis. This may be designated as either the Z or Y axis. (I prefer depth to be designated as the Z axis). The rear view is the reverse of the front view.

▼ **Height**. A vertical dimension perpendicular to width and depth.

▼ **In projection**. The position of views such that shared dimensions between adjacent views are in alignment; for example, height aligns between the front and side views. In an *alternate position*, the side view may be aligned with the top view such that depth measurements are in projection. The views in Figure 3.4 are in projection because height measurements in the front view and side views are perpendicular to the projectors.

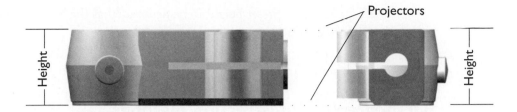

Figure 3.4   Projectors locate the same features in adjacent views.

> ▼ **Length**. A nonorthogonal dimension not parallel to height, width, or depth.

> ▼ **Not to scale**. Orthographic views without proportional relationships cannot be compared. You *cannot* directly measure these views for technical illustrations.

> ▼ **Orthographic view**. Any view taken from an infinite distance. This renders parallel features to be parallel no matter the point from which the view is taken. For example, Figure 3.5 shows a number of views that are orthographic (though not principal). You can tell they are orthographic because parallel edges remain visually parallel.

Figure 3.5   A number of orthographic views.

> ▼ **Principal orthographic view**. Any orthographic view that is taken parallel to the X, Y, or Z world axes. There are six principal views: top, front, right side, left side, back, and bottom. Figure 3.1 displayed two principal orthographic views.

> ▼ **Projectors**. Parallel eye sights used for locating features and aligning adjacent orthographic views (Figure 3.4).

> ▼ **Side views**. Also called profile views or side elevations, views seen when looking parallel to the horizontal width axis (X axis).

> ▼ **Top view**. Also called the plan view, the view seen when looking down on a vertical axis from above. This vertical axis may be designated as either

the Y or Z axis. (I prefer height to be designated as the Y axis.) The bottom view is the reverse of the top view.

▼ **Width**. A horizontal dimension perpendicular to height and depth.

Orthographic Construction Methods

It is helpful to have some experience in CAD methods before attempting accurate scale construction with a PostScript tool such as Adobe Illustrator. While other products, such as Macromedia Freehand and CorelDraw, have varying toolsets that aid in orthographic constructions, none are as robust as even the most modest CAD product. So up front, you know you are going to use work-arounds and make productivity concessions for the publication-quality output that PostScript tools produce.

There are also a number of plug-in products that automate certain operations and extend toolsets. Sources of these additions are listed at the end of this chapter. We will cover the tools you will encounter right out of the box in Illustrator.

Drawing Sheet Size

Because most illustrations will be placed into a page composition program for final output (Quark, FrameMaker, InDesign, PageMaker), you can make the drawing sheet in Adobe Illustrator as large as you want during the layout portion of illustrating (**File|Document Setup**). Because Illustrator takes its printable area from the active printer's characteristics, enlarging a sheet to a size not supported by your printer leaves the dotted print area outline offset to your layout. You can temporarily turn off this print frame by choosing **View|Hide Page Tiling**.

Illustrator still thinks it is printing to the active printer, so if you have a really big illustration, it will be hard to get quick proofs. Add to this Illustrator's inability to scale a drawing to the printer and you can have real problems. One solution is to create a printing frame before altering the document setup. That way you can copy the illustration to a printing frame layer and scale the graphics to fit. Another solution may be to copy and paste your illustration into a default file (which displays the printable area), group, and scale the illustration for printing.

▼ Tech Tip

Always work on a copy of your illustration when scaling and proofing. The last thing you want is to make your illustration smaller by some unknown amount and then be unable to correctly enlarge it later to match some other component.

Drawing Scale

Scale is not as important as it once was because resolution-independent PostScript graphics can be enlarged or reduced without effect on the quality of the final output. However, it is important to pick units that are appropriate for your construction. You can even mix systems of measurement. For example, if you are illustrating a golf course where distances are in integer yards, you could pick millimeters (**Edit|Preferences|Units & Undo**) to represent yards because numbers such as 100, 200, 300, etc., are reported in the ruler and dialog boxes. However, you will probably want to keep the *stroke* units in points because this is the common method of specifying line weights.

▼ **Note**

Adobe Illustrator has several display modes. You will be interested in **View|Preview** and **View|Outline**. This function is actually a throwback to the time when Illustrator required more computer than most people had. So rather than constantly display strokes, fills, and blends and wait 5 minutes for redisplay, you'd go to *outline* display mode. This is not currently the case with high processor speeds and lots of random access memory (RAM). You may find that you never need outline mode. It may, however, be more practicable to use the outline display mode in order to adjust and edit paths without having to look at fills and strokes.

Layers

You couldn't do CAD without layers, and the same is true with any PostScript tool. However, too many layers (and nested layers within layers) often confound the illustration process. You will need to develop a consistent system of layers so that it's easy to return to an illustration you completed 6 months ago. Always delete unused layers. Always give layers functional titles.

Grid

Grid divisions are set in **Edit|Preferences|Guides & Grid**. Illustrator displays subdivisions based on zoom level, so you can set subdivisions as small as you need. Setting units defines the type of grid you see (**View|Show Grid**).

▼ **Tech Tip**

Use the coarsest grid you can get away with. The finer the grid, the more difficult it becomes to discern construction geometry. Also, choose a grid color that contrasts in color and saturation with any strokes or fills in your illustration.

Snap to Grid

Once the units and grid are set, you can have the cursor snap to grid intersections by choosing **View|Snap to Grid**. Be advised that the grid snap overrides any other snap feature. Some objects adhere to grid divisions, and using a grid and snap is helpful. However, many objects don't adhere to consistent grid divisions and using a grid can be unproductive.

Ruler Guides

Ruler guides are like construction lines in CAD. They begin horizontal or vertical and are "pulled" from the rulers (**View|Show Rulers)**. These guides extend completely across the pasteboard. To operate on guides, they must be unlocked (**View|Guides|Lock Guides**). Once unlocked, they can be repositioned or rotated. It is helpful if guides are on their own layer.

Ruler guides can be "released" (**View|Guides|Release Guides**) and made into editable geometry. When released, the guides take on the current stroke and fill characteristics.

▼ Tech Note

Released guides still extend to the edges of the pasteboard. This can present problems when some guides are not stroked and included in a group. You end up with a much larger object than you expected because the hidden guides are included. Also, these ruler guides are included when EPS files are created, so turn them off before saving in EPS format.

The Zero Well

Located in the upper left corner at the intersection of the rulers, this X 0, Y 0 (origin) point can be relocated to any place on the pasteboard. When repositioned, the new origin is reflected in each ruler. Likewise, the grid is repositioned to align with the origin and the rulers.

Snap Settings

Several snap settings increase the accuracy of your orthographic constructions. **View|Snap to Grid** allows the cursor and tools to snap to grid subdivisions and overrides all other snap options. Grid snap works with or without the grid being displayed. When you turn on *smart guides* (**View|Smart Guides**) Illustrator finds entity points it can address, such as anchors, and centers, points on a line, etc. The most important smart guide snap for technical drawing is *intersection;* only actual intersections are found, unfortunately, not projected intersections. Without smart guides turned on, **View|Snap to Point** still allows anchors and centers to be addressed. Watch for a small box to appear beside the pointer to signify an anchor has been found. The cursor then changes from solid pointer to open pointer to show that the target anchor has been found.

▼ Tech Note

When using smart guides, the area inside which Illustrator searches for snappable points can be changed by choosing **Edit|Preferences|Smart Guides** and setting the snapping tolerance to a larger number (1–10 points). Actually this is 1 to 10 *screen pixels,* so the snap tolerance is a function of zoom. Zoomed out, 10 pixels is a large amount. Zoomed in, it isn't.

Trimming Tools

You will hardly ever initially create geometry that is the final size or shape, so trimming (editing) will be a major stage of any orthographic drawing. The *scissors tool* is used to cut lines and closed shapes. The *knife tool* is used to cut across closed or unclosed shapes and connects the ends of the cut, creating bound shapes. For accuracy, use snap options during editing.

There are several other options for trimming using the **Pathfinder Dialog Box**. These are covered in Chapters 13 and 15.

▼ Tech Note

Trimming can be problematic in Illustrator. First, Illustrator is sensitive to stacking order (front to back) and layer order. Also, the scissors tool will not work on end points of paths but will work on shape anchors. The knife tool can cut an unselected shape but can't cut a shape if another is selected. The knife tool won't work on straight lines.

The Transform Dialog Box

Because illustrators often know the general size of geometry, it may be advisable to create the features of an object from lines, rectangles, and circles whose sizes are specified numerically. Choose **Window|Show Transform** to gain access to numerical fields where the exact size of objects can be entered.

With this dialog box you can do three things:

▼ Specify which anchor you want to operate on by clicking on it once.

▼ Locate the object by the selected anchor at any XY position. The current X0, Y0 position can be changed by dragging the origin from the intersection of the two rulers.

▼ Size the object (relative to the selected anchor). You can mix units, for example, entering "mm" after a number even though Illustrator expects inches (in) or the inch character ("), when in millimeters.

Example Using Illustrator's Construction Tools

Probably the best way to see how these various tools work is to apply them to a typical construction found in technical geometry: a rounded corner. Figure 3.6 shows a typical rounded corner. Follow these steps for accurate construction:

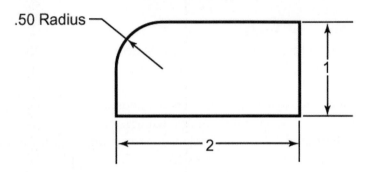

Figure 3.6   Typical rounded corner.

Step 1 Create a rectangle of approximately the correct size. Choose **Window|Show Transform,** and with the rectangle selected, enter correct width and height dimensions (Figure 3.7).

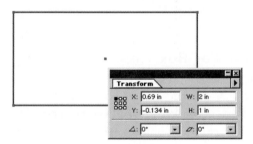

Figure 3.7   Numerical dimensions entered in the **Transform Dialog Box.**

Step 2 With **Snap to Point** active, relocate the origin to the upper left corner of the box (Figure 3.8).

Step 3 Create a circle and size it to a .5 in radius (1 width and height of 1) in the **Transform Dialog Box** (Figure 3.9).

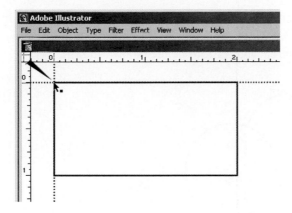

Figure 3.8   Origin relocated to corner of box.

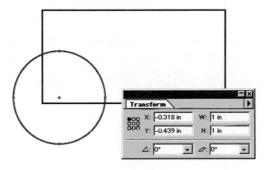

Figure 3.9   Circle sized correctly in **Transform Dialog Box**.

> **Step 4** Translate the circle by its upper left corner to X0, Y0
> (Figure 3.10). This places the circle perfectly tangent to the top and left
> sides of the box.

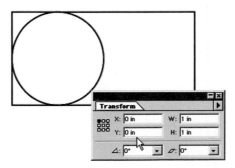

Figure 3.10   Circle positioned tangent to the box.

Step 5 Turn **Smart Guides** on. Trim the circle with the scissors tool at the tangent anchors (Figure 3.11). Use the open selection tool to select the portion of the circle you want to delete (shown in red).

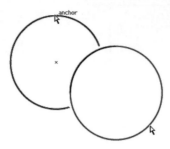

Figure 3.11   Circle trimmed to tangents.

Step 6 Bring the box to the front. Trim the box to the intersection with the rounded corner (Figure 3.12). Illustrator may identify the arc's anchors, but that's fine, because they *are* at the intersection. The portion of the box removed is shown in red.

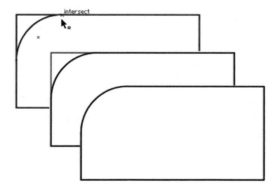

Figure 3.12   Box trimmed to tangents.

Step 7 Use the open selection tool and select the coincidental anchors at one of the tangent points (Figure 3.13). These two anchors will appear solid, while the others will appear open.

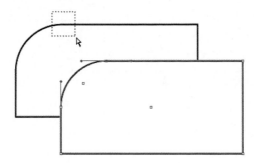

Figure 3.13   Select coincidental anchors.

Step 8 Choose **Object|Path|Join** and make the selection a smooth point. Repeat for the other tangent. You now have a contiguous bound shape, ready for illustration purposes.

Orthographic Construction Example

Figure 3.14 shows two scale views of an actuator handle. In these two views we can derive the necessary information for constructing accurate principal orthographic views. There are several considerations when the source information for an illustration is in paper form:

▼ When the views are drawn to scale, the sheet can be scanned, imported, and used as the basis for the technical views (see Chapter 16).

▼ For this to be practicable, you first *must* have scale views. Check for scale by locating an integer dimension and developing a scale to which other dimensions can be compared.

▼ Drawing over a scan will not be as accurate as drawing from scratch, but if five-digit accuracy isn't required, tracing can be more efficient.

▼ Scale views can be measured and those measurements used to construct accurate geometry that can be used as the basis for technical illustrations.

▼ If the views are not drawn to scale, dimensional information (dimensions) must be used to create accurate views.

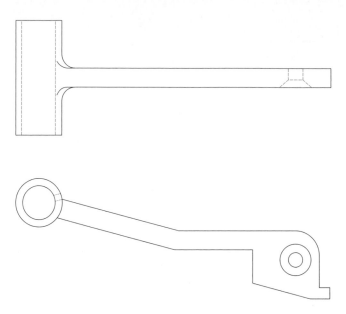

Figure 3.14   Scale views of the actuator handle.

Determine Appropriate Scale

Most manufactured consumer objects can be drawn at full scale, unless they are architectural, marine, automotive, etc. For example, the actuator handle in Figure 3.14 appears to be dimensioned in fractional inches. You'll want to set two preferences to help you in construction.

Step 1 Choose **Edit|Preferences|Units&Undo** and set **General Units** (Figure 3.15) to **Inches**.

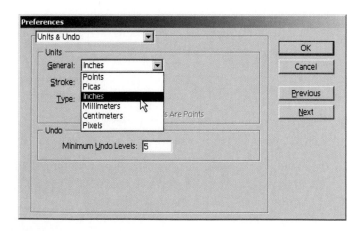

Figure 3.15   Units set to inches.

Step 2   Choose **Edit|Preferences|Guides & Grid** and set the grid line to every inch with 16 subdivisions (Figure 3.16). An inspection of the actuator handle shows 1/16 to be the smallest measurement.

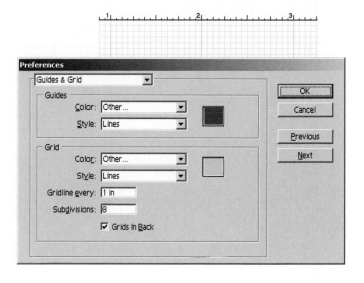

Figure 3.16   Grid set to 16 subdivisions.

Establish Datums That Control Features

If you carefully analyze the front view of the actuator handle, you'll see that several features control its shape: right and left edges, bottom and top of pivot end, centerlines of cylindrical and hole features. Choose **View|Snap to Grid**, and construct these datums over the grid by pulling ruler guides from the rulers. Select all the guides and choose **View|Guides|Release Guides**. Trim these guides to better reflect the handle's shape (Figure 3.17).

▼ Illustrator Tip

Tools in Illustrator are sensitive to how far the tip of the tool is away from the object's axis origin. Take rotation for example. The closer the tip of the rotation tool to the axis of rotation, the greater the amount of rotation caused by slight cursor movement. To get more control over transformations move the cursor to a distance further away from the axis origin. Then when you hold the mouse button down and move the cursor, you will create less transformation.

Control is even more of an issue when **Smart Guides** and **Snap to Point** are active. If you have a fairly dense drawing, you may need to selectively **Lock** objects or move some linework to a different **Layer** where they can be **Hidden**.

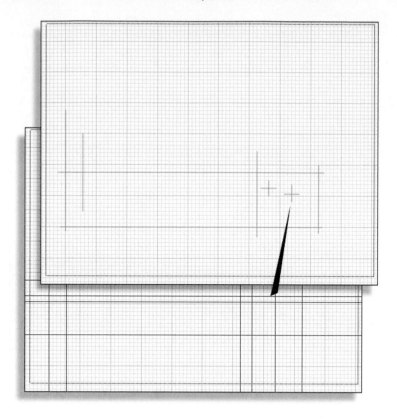

Figure 3.17   Datums control position of features.

▼ **Tech Note**

Drag the zero point from the intersection of the two rulers in the upper left corner of the screen to any place on the grid you want to measure from.

Construct Angular Features

The height of the handle at the left side isn't dimensioned, nor is the intersection of the angled and the horizontal sections. Begin with the start of the angular centerline (4 in from right), the true thickness (1/2 in), and rotate a horizontal piece to -15 degrees. Note that the perpendicular intersections *are not* on this 4 in line (see detail on Figure 3.18).

Complete the Front View

Figure 3.19 shows the complete front view after all the features have been added and line work trimmed correctly. A partial top view was constructed so that the depth of the countersink could be found (you will need that information in the future).

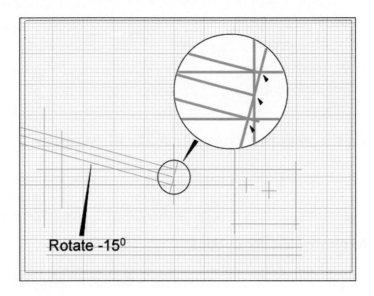

Figure 3.18   Angular handle begins in horizontal position.

▼ **Hint**

It's not unusual that some features must be found by combining information from other views. If you're stuck, it's probably because you're working in the wrong view. Look for the data in another view.

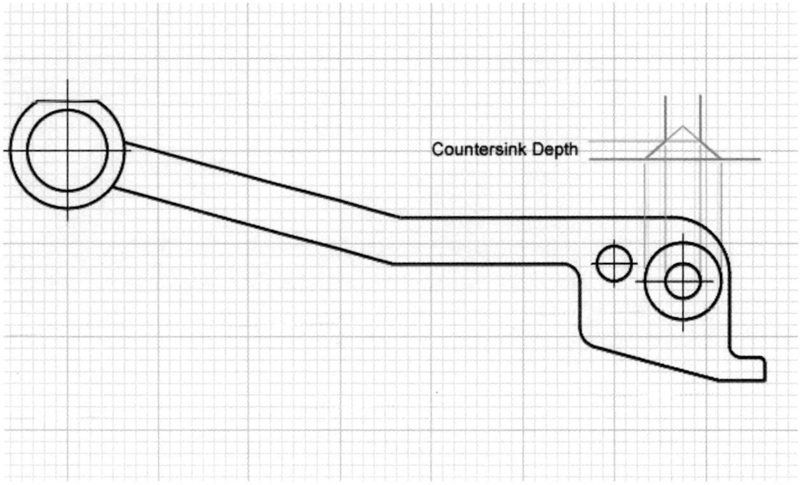

Countersink Depth

Figure 3.19   Completed front view with countersink detail.

What can be done with an accurate orthographic front view? Once the orthographic view is completed, it can be used for several illustration tasks.

▼   Create a rendered orthographic illustration (Figure 3.20).

▼   Shear the orthographic view into axonometric position and complete an axonometric illustration (Chapter 13).

▼ Place the view on a picture plane and complete a perspective illustration (Chapter 14).

▼ Bring the Illustrator vectors into 3ds max and complete a 3D model (Chapter 16).

Orthographic Views and 2D Illustration

Accurate orthographic views form the basis for a number of illustration techniques covered in subsequent chapters. By having these views, laborious constructions in axonometric or perspective can be avoided (and after all, time *is* money).

Axonometric Projection

Two-dimensional orthographic views can be arranged on an *axonometric diagram* and projected to form an axonometric projection (Figure 3.20). See Chapter 12.

Figure 3.20   Illustration from accurate orthographic front view.

Axonometric Shearing

Two-dimensional orthographic views can be scaled, rotated, and scaled again in an operation called *axonometric shearing* resulting in correctly scaled axonometric profiles (Figure 3.21). These profiles can then be extruded along axonometric axes to form an axonometric projection. See Chapter 13.

Perspective Projection

Two-dimensional orthographic views can be arranged on a *perspective diagram* and used to project a perspective illustration (Figure 3.22). See Chapter 14.

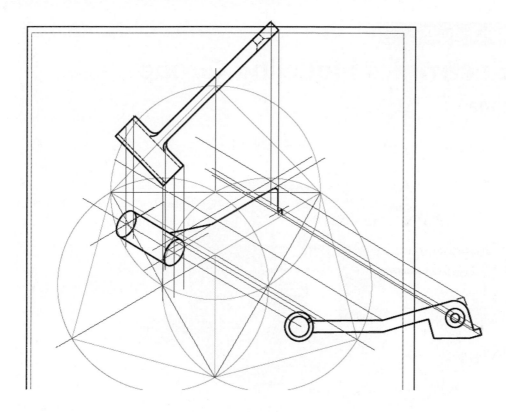

Figure 3.21 Axonometric projection from scale orthographic views.

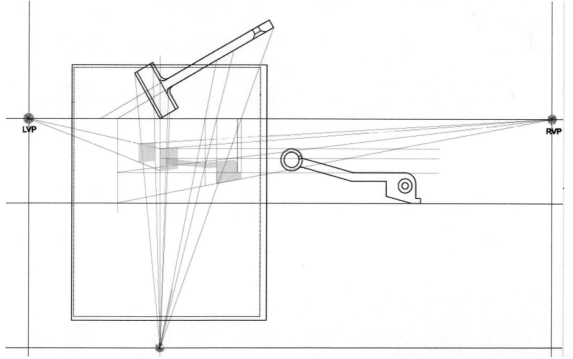

Figure 3.22 Perspective projection from scale orthographic views using a perspective diagram.

▼ **Illustration in Industry**

Micro Electronics Modeling Group

Mesa, Arizona

Image Copyright 2002 by The Micro Electronics Modeling Group.

The Micro Electronics Modeling Group was charged with creating the graphics to support training in a newly constructed fabrication facility. Although the group had access to the complete set of AutoCAD construction drawings from the contractor, much of the actual installation of utilities and the routing of ductwork was designed as the components were installed, and no drawings existed for these components.

The solution was to combine sketches and digital photographs with existing construction and installation drawings. The facility was modeled in 3ds max for several reasons. First and most importantly, a full range of static and motion technical illustrations including print pieces, Web graphics, and virtual multimedia training, as desired. A detailed 3D computer model would provide the basis for any type of technical illustration that might be required. The second reason for modeling was that an as-built document would provide a reference for maintenance, future expansion, or remodeling. So this was a case where a 3D technical illustration became an engineering document from which subsequent engineering documents could be made.

Software

3ds max
Photoshop
AutoCAD

Hardware

Windows PC
Digital cameras
CD RW drives

Orthographic Views and 3DIllustration

Orthographic geometry should be used as the basis for 3D illustration when available. Depending on your 3D tool, this may be problematic, however. Modeling programs usually require unambiguous geometry for successful lofting, extrusion, and sweeping. Depending on the condition of your 2D orthographic views, it may be more efficient to use dimensional data found on engineering drawings to create new geometry directly in the modeler. Even in the best of circumstances, a model made from orthographic views will have to be edited and combined with other elements to arrive at a final illustration. Figure 3.23a shows the shapes extruded from the Adobe Illustrator 2D profiles. Note the cone created and placed in correct position to form the countersink. Refer to Figure 3.19 where the countersink shape was constructed. The final result is shown in Figure 3.23b.

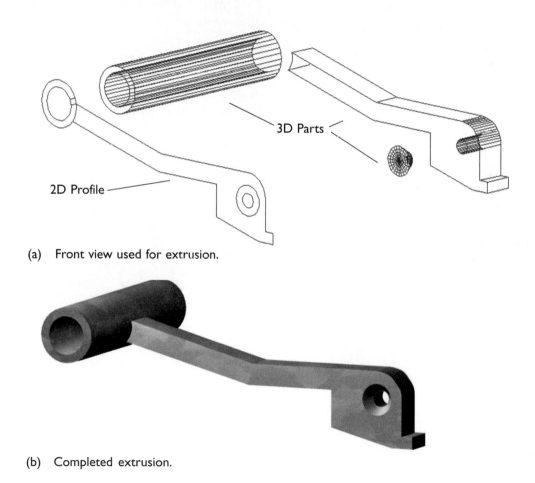

(a)   Front view used for extrusion.

(b)   Completed extrusion.

Figure 3.23   Two-dimensional front view (a) used to extrude a 3D model (b).

Orthographic Views in 3ds max

Orthographic views (top, front, left, right, bottom, rear) are set viewports in 3D illustration programs and other modelers, so achieving individual orthographic views presents little problem. Multiple composite views present a greater problem, however, because these programs do not scale or align views from one viewport to another. You may choose to duplicate the object and place instances in space so that when a single view is taken, multiple views are presented (Figure 3.24a and b).

A source of illumination appropriate for the front view may not yield desired results for other views. Each view may require its own source of illumination with other views excluded from the source Figure 3.24b.

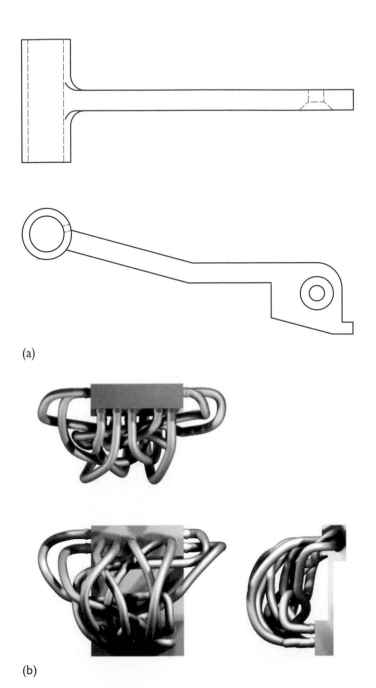

(a)

(b)

Figure 3.24   Multiple orthographic views achieved with instances (a). Multiple light sources assure each view is properly illuminated (b).

Additional Considerations with 3D Data

Although this is not a text on modeling as such, there are several considerations that are applicable when 3D data are used in technical illustration. Three-dimensional computer models can be *surface models* or *solid models*. Surface modelers are generally fine for technical illustration because we are usually interested in how objects look from the outside. Illustrators usually don't perform the engineering analyses that require solid models.

Modeling applications can be *parametric* or *nonparametric*. Parametric modelers establish complex relationships between features. For example, a part might have a hole that's a set diameter in a specific position relative to another set diameter hole—no matter how large the object. In a parametric application, the holes would remain in the same relationship and at the same diameter even if the object were enlarged. Parametric applications are not a necessity for technical illustration but offer great control and flexibility and are certainly worth a look.

Review

Two-dimensional orthographic views form an important part of technical illustration. When these views exist, illustrators should use them whenever possible. In other words, don't redraw views that are already accurate enough for illustration purposes. In the absence of orthographic views, it may be more efficient to draw them first and create axonometric or perspective illustrations by projection or shearing.

In 3D illustration, 2D views form the basis for extrusions, sweeps, lofts, and other operations that result in 3D geometry. You'll want to look at Chapter Eighteen, where these techniques are discussed in detail. Again, in the absence of accurate orthogonal geometry, it may be more efficient to first create 2D views. Models can then be built from the views.

Text Resources

The field of engineering drawing and CAD is based on orthographic views. You might look to the following sources for information on orthogonal geometry, its creation, and use.

Bertoline, G.R., Wiebe, E.N., and Miller, C.L. *Fundamentals of Graphics Communication.* WCB McGraw-Hill. New York. 2001.

Earl, J.H. *Graphics for Engineers.* Addison-Wesley. Upper Saddle River, New Jersey. 2000.

Frey, David. *AutoCAD2000: No Experience Required*. Sybex. Berkeley, California. 1999.

Leach, J.A. *AutoCAD Companion 2002*. McGraw-Hill. New York. 2001.

Mortenson, M.E. *Geometric Modeling*. John Wiley & Sons. New York. 1985.

Internet Resources

Resources for CAD and engineering drawing can be found on the Web.

Autodesk Resource Site. Retrieved from: *http://pointa.autodesk.com/portal/nav/index.htm*. June 2002.

CADKEY Resource Site. Retrieved from: *http://www.cadkey.com/resources/docs/index.asp*. June 2002.

Micro Station Resource Site. Retrieved from: *http://www.bentley.com/vizcenter/links.htm*. June 2002.

Solid Edge Resource Site. Retrieved from: *http://www.solid-edge.com/media/media_center.htm*. June 2002.

SolidWorks Resource Site. Retrieved from: *http://www.solidworks.com/html*. June 2002.

On the CD-ROM

For those of you who would like to try out the procedures described in this chapter, several dimensioned engineering drawings have been included on the CD-ROM. Look in *exercises/ch3* for the following files:

ch_03_01.pdf	ch_03_02.pdf	ch_03_03.pdf
ch_03_04.pdf	ch_03_05.pdf	ch_03_06.pdf

Perspective Drawing

Axonometric Projection

Technical Animation

Technical Illustration

Rendering

3D Modeling

Technical

Chapter 4

Introduction to Axonometric Views

In This Chapter

This chapter is an overview of Chapters 5–8. Each chapter deals with a different axonometric illustration or technique. Because axonometric or projection is the primary drawing technique employed in technical illustration, you need to fully understand its strengths and limitations.

The ability to create accurate axonometric drawings and projections is probably the one skill that separates a technical illustrator from a commercial artist. Axonometric views may appear at first to be extremely complicated. However, if you understand the relationship between orthographic views, you should see that axonometric views are just another type of orthographic view, governed by the same rules. While you achieve a principle orthographic view by looking parallel to one of the world axes, you achieve an axonometric view by looking at geometry at some other angle.

Chapter Objectives

In this chapter you will understand:

▼ The relationship of axonometric and orthographic views

▼ The difference between axonometric drawing and projection

▼ The types of axonometric views

- ▼ The use of axonometric axes and scales
- ▼ How to determine an axonometric scale graphically
- ▼ The standard views found in AxonHelper
- ▼ Methods for arriving at standard axonometric views in Adobe Illustrator
- ▼ Methods for arriving at standard axonometric views in AutoCAD and 3ds max

Axonometric Terminology

- ▼ **Axonometric axes**. The world axes in axonometric position along which true measurements can be made.

- ▼ **Axonometric diagram**. The coplanar arrangement of the axonometric plane and two or more orthographic planes rotated into the plane of the axonometric.

- ▼ **Axonometric drawing**. An axonometric view where one or more axis scales has been normalized to full scale.

- ▼ **Axonometric plane**. A plane not perpendicular to one of the world axes.

- ▼ **Axonometric projection**. An axonometric view formed by projecting features from orthographic views onto an axonometric plane.

- ▼ **Axonometric view**. A view whose direction of sight is not parallel to one of the world axes.

- ▼ **Dimetric view**. The axonometric view where the plane of projection is inclined equally to two of the three world axes.

- ▼ **Isometric view**. The axonometric view where the plane of projection is inclined equally to the three world axes.

- ▼ **Standard axonometric views**. Axonometric views for which information is known for any of the several methods of creation.

- ▼ **Trimetric view**. The axonometric view where the plane of projection is unequally inclined to the three world axes.

- ▼ **Rotation**. The method of creating an axonometric view in a principal view by rotating object geometry relative to the world axes.

- ▼ **Shearing**. The method of creating an axonometric view by scaling and rotating 2D orthographic views into correct axonometric projection.

- ▼ **Thrust axis**. The axis that remains perpendicular to a feature as it is revolved in space; the local Z-axis.

▼ **Viewpoint**. The method of creating an axonometric view by positioning the viewer's eye (the camera) at a specific location in space.

Axonometric and Orthogonal Projection

An axonometric view is an orthogonal (orthographic) view, so it's helpful to understand orthographic and non-orthographic views. Projections in orthographic views are made perpendicular to planes of projection. This places the observer at infinity (necessary to have parallel visual projections) and is directly opposed to the perspective method where projections converge at the viewer's eye.

Study Figure 4.1 where perspective and orthographic views are compared. Notice in Figure 4.1a that the *central projection* is, in fact, perpendicular to the plane of projection. All other projections are not orthogonal (perpendicular), so this is *not* an orthographic view. In Figure 4.1b, the projections are all perpendicular to the plane of projection. In other words, the observer has been removed to infinity. This *is* an orthographic view.

So in order to achieve an orthographic view, of which an axonometric view is a subset, all projections must be perpendicular to the plane on which it is drawn.

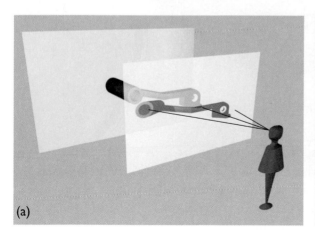
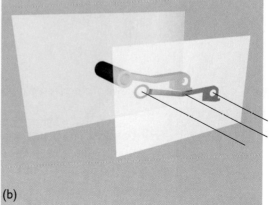

(a)   (b)

Figure 4.1   An orthographic view requires that all projections are perpendicular to the plane of projection.

Orthogonal views are characterized by parallel features such as edges, intersections, and centerlines will remain parallel no matter what view is taken. This is the critical strength of the axonometric in technique *technical* illustrations. In a commercial or creative illustration it's important that geometry *appears* correct; a technical illustration must *be correct* because technical decisions, often involving lives and considerable sums of money, will be made on the information contained in the illustration.

Principal views are projected onto mutually perpendicular planes. In Figure 4.2 this relationship is shown. Because the height, width, and depth of the handle are aligned with the planes of the views, special visualization skills are necessary to read the views. This principal alignment does not result in a pictorial view, necessary for many technical illustrations.

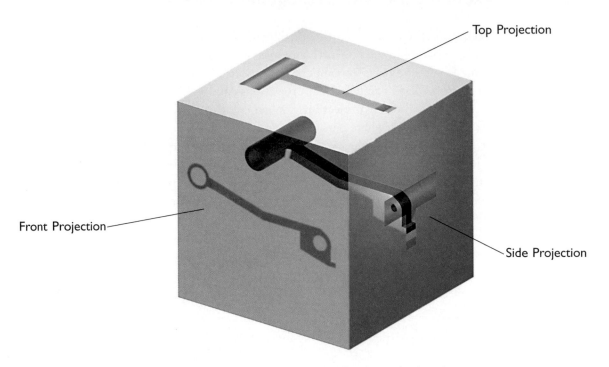

Top Projection

Front Projection

Side Projection

Figure 4.2   Principal views are projected onto mutually perpendicular planes.

Pictorials, Projections, and Drawings

A pictorial view is a view that shows an object's height (Y), width (X), and depth (Z) in the same view. There are an infinite number of possible pictorial views. An axonometric view is a pictorial because an axonometric view shows an angular view of all three axes; no axis appears normally or in point view. When geometry is projected onto a plane not perpendicular to the X, Y, or Z world axes (as the top, front, or side planes are), the resulting pictorial is called an *axonometric projection*. An *axonometric drawing* is a view constructed by measuring in axis directions and is not projected. Both drawing and projection result in identical axonometric pictorials, though at different scales.

Describing Axonometric Views

As with an orthographic view, an axonometric view is projected perpendicular to a plane of projection. Axonometric planes, however, are set at angles other than the six principal positions.

An *isometric view* results when the plane of projection is equally inclined (35.266 degrees) to the height, width, and depth axes (as well as the front, top, and profile planes). This plane connects the corners of a cube as shown in Figure 4.3. Measurements along the three axes are at the same scale. So you can always tell when you have an isometric view: the plane is an equilateral triangle and it makes equal angles with the three axes.

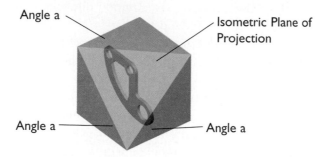

Figure 4.3   Isometric plane makes equal angles with world axes and principal planes.

A *dimetric view* results when the plane of projection is equally inclined to two of the three axes (Figure 4.4). Measurements along these two equally inclined axes are at the same scale. You can identify a dimetric plane in that it connects two corners of a cube.

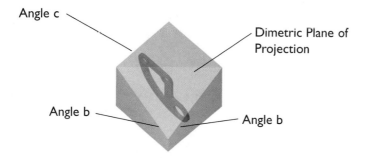

Figure 4.4   Dimetric plane is equally inclined to two of the three axes and two of the principal planes.

A *trimetric view* results when the plane of projection is inclined at a different angle to all three axes (Figure 4.5). Because each axis is inclined differently to the plane of projection, each axis is scaled differently. The trimetric plane's corners are at unequal distances along the three axes.

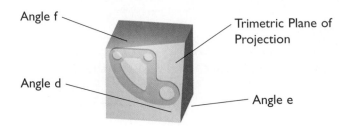

Figure 4.5   Trimetric plane is unequally inclined to the three axes and all principal planes.

Isometric views are the easiest to create, the most common, and the least realistic (because the isometric view is a special, perfect view), and cause the most visualization problems. Dimetric views are more realistic but can still allow symmetrical features to be visually aligned, making visualization difficult. Trimetric views are the most difficult to create but are the most realistic.

Though the axonometric approach provides ready access to measurements, its downside is that the view must be carefully matched to the geometry of the object and its intended use. Objects that are long and narrow are not appropriate for some axonometric views because from personal experience, we expect long objects to appear smaller as they get further away. Both isometric and dimetric axonometric views are symmetrical left and right, meaning that the viewer is rotated 45 degrees about the vertical axis. This presents a problem with features lining up one behind another. This can be avoided by changing to a trimetric view.

For example, study Figure 4.6 where storage tanks are aligned in a symmetrical grid. This design (fairly common in industry) causes the diagonal tanks to perfectly align in both isometric and dimetric views. However, when the view is changed to trimetric, the tanks are much more easily visualized.

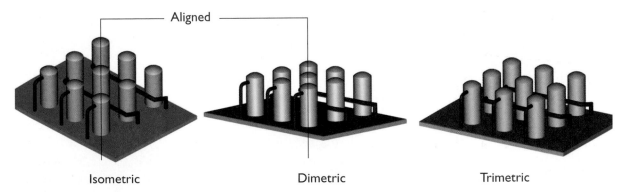

Isometric   Dimetric   Trimetric

Figure 4.6   Choose the axonometric view that promotes the greatest visualization.

In 3D applications, the natural perspective camera must be suppressed, by either orthogonalizing the pictorial view (using a user view) or by choosing a camera focal length of sufficient length that all perspective is removed. Focal lengths above 1000 mm generally result in near-axonometric views.

Axonometric Projection

In Chapter 3, Adobe Illustrator orthographic drawing techniques were discussed. When you assembled a front view from top and side views, all height and width dimensions in the front view appeared true size. This is because the top and side views are aligned normally to the X, Y, and Z-axes. This arrangement of three views on a single plane—like a piece of paper or a computer monitor—and identically scaled and aligned vertically and horizontally, can be referred to as an *orthographic diagram*.

Were the top and side geometry rotated into some other position, as shown in Figure 4.7, and the front view found by normal orthographic projection, a totally different front view would be formed—an axonometric view. So one way to arrive at an axonometric view is to rotate object geometry relative to the world axes and then project the result in one of the principal views, in this case the front view. If the top view is rotated 45 degrees about Y_W and the side view 35.266 degrees about X_W, the front view becomes an isometric view.

Consider only the front view in Figure 4.7. The top and side planes appear as edges and can be seen (when looking at the front view) only when revolved into the plane of the front view. At this point, all three views are *coplanar*, aligned, and identically scaled.

Because we have arrived at this axonometric view by projecting view information (top and side) perpendicular to a plane of projection (frontal plane), we refer to this method as an *orthographic projection*. Because the object has been rotated we see an *axonometric* view and all normal height, width, and depth measurements in the axonometric view are smaller than true size. This is called *foreshortening* and is a characteristic of axonometric pictorials. The percentage that each dimension is foreshortened allows scales to be developed so that distances can be compared and the view edited or added to.

Though this is a valid way of arriving at an axonometric view, it is not practicable because rotated views are not usually available in 2D orthographic drawings. It would be inefficient to redraw the views for an illustration so that they are in the correct tilted position. It would be much better to simply use the normal top, front, and side views that are usually available to project the axonometric view. Figure 4.8 shows how this is done.

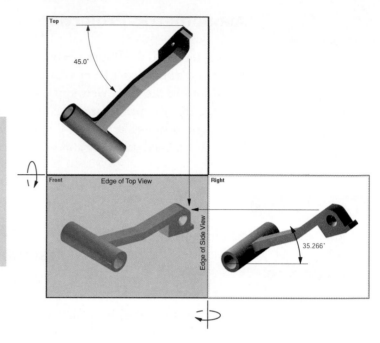

▼ Rotation

Make the first rotation about the vertical axis, in this case +45 degrees. Make the second rotation about the X-axis, in this case +35.266 degrees. These object rotations result in an isometric front view (orthographic diagram).

Figure 4.7   An orthographic projection diagram.

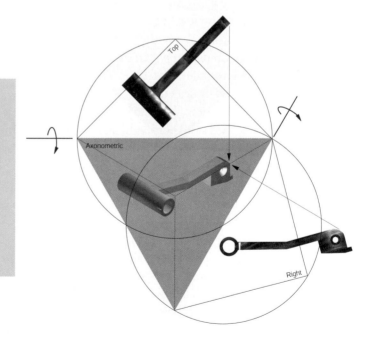

▼ Projection

Note how the orthogonal views maintain their alignment with the edges of the rotated principal planes. The axonometric view is created by projecting geometric features perpendicular to the axis of rotation between the normal principal view and the rotated axonometric plane (axonometric diagram).

Figure 4.8   An axonometric projection diagram.

Were the standard orthographic top and front views placed such that their projections create the correct foreshortening, the desired axonometric, correctly scaled, could be drawn. This arrangement of standard orthographic views for correct axonometric projection is referred to as an *axonometric diagram.*

The axonometric plane in Figure 4.8 is true shape. However, top and side planes are not true shape until they have been revolved and are coplanar with the axonometric plane. Chapter 7covers this technique in detail. It is important here to recognize the similarities between the top-front-side view relationships in Figure 4.7 and the top-axonometric-right view relationships in Figure 4.8. Consider these similarities between orthographic and axonometric diagrams:

▼ All views in both orthographic and axonometric diagrams lie in the plane of the paper.

▼ In both diagrams, geometry stays aligned to its principal plane as the view is rotated (the bottom of the object stays aligned with the bottom of the plane, for example).

▼ Because the views lie in the plane of the paper, they can be drawn as principal orthographic views.

▼ All projections are done perpendicular to the axis about which the view was rotated.

So the axonometric diagram is really just like the orthographic diagram, only the result is an axonometric view, and not a principal view.

Axonometric Drawing

An axonometric drawing is an axonometric view where one or more of the foreshortened axes have been normalized; in other words, the foreshortened measurement is enlarged as if it were full size. Why would this be an advantage?

Historically, when axonometrics were manually created with pencil and paper, special scales had to be developed because axonometrics had odd foreshortening (17.23 percent, 47.56 percent, 71.78 percent, and so on). There were even a few specialized commercial scales and tables to facilitate axonometric constructions. So if all three scales (isometric), two scales (dimetric), or one scale (trimetric) could be normalized, or brought up to full size, there would be fewer special scales to construct.

But because CAD and computer drawing makes scaling so easy, most of the concern about drawing versus projection has been eliminated. If you have a projection and want a drawing, scale it up. If you have a drawing and want a projection, scale it down. For example, Figure 4.9 demonstrates how an isometric projection is converted to an isometric drawing.

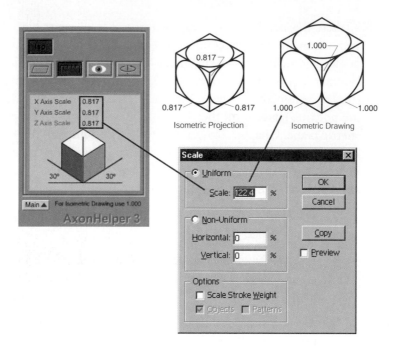

Figure 4.9   A projection can be scaled up to match a drawing.

In the Figure 4.9 example, an isometric projection (81.7 percent foreshortening in all three directions) is scaled up by 122.4 percent (1.00/0.817 = 1.224) to arrive at an isometric drawing.

Consider a dimetric projection where two of the axes are foreshortened 74.0 percent. These common axes can be scaled up by 135percent (1.00/0.740 = 1.35) to make two of the three axes full size. Note that this makes the third axis larger than full size (0.930 x 1.35 = 1.25).

A trimetric projection is more problematic. Any of the three axes (say, 0.860, 0.570, and 0.970) could be scaled up to full size. You would either adopt a convention—such as always scaling up the vertical in a trimetric drawing—or scaling the axis along which the most critical measurements are done. Using the 0.970 Z-axis as unitary creates a scale factor of 103 percent (1.00/0.970 = 1.03). This results in the axes shown in Table 4.1.

▼ Axonometric Projection to Drawing

	Projection	After Scaling 103%
X axis	0.860	0.886
Y axis	0.570	0.587
Z axis	0.970	1.000

Table 4.1   Converting a trimetric projection to a trimetric drawing results in one axis being unitary.

Drawing Axes

An axis shows a direction in space. World axes show the principal X, Y, and Z-axis directions of width, height, and depth. It is convenient to align as much of the object's geometry as possible with these three axes. There will be cases when certain features (such as angular brackets, pipes, or the like) are not aligned with the world axes. If you place major features where they can be easily constructed, it makes locating angular features easier. The actual length of an axis is variable. The direction and scale of the axis is, however, critical. In axonometric views, there are several axes used to define different directions.

Axonometric Axes

These axes are mutually perpendicular and generally meet at the X0, Y0, Z0 world origin. One axis represents X (width), one Y (height), and one Z (depth). These world directions can be designated $X_W Y_W Z_W$. Figure 4.10 shows an example of axonometric axes. An axonometric axis is defined by the following:

▼ The *angle* that the right or left axis is inclined to the horizontal expressed in degrees. The vertical axis remains vertical.

▼ The scale of the axis expressed as a percentage of full size.

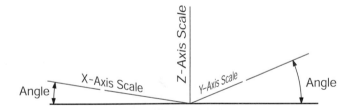

Figure 4.10   Axonometric axes.

You may choose to construct axonometric axes from a variety of vantage points. This is why a small number of standard views are usually sufficient for most illustration tasks. Figure 4.11 shows how one set of 15-55-30 trimetric axes can be rearranged to produce a variety of views.

▼ Tech Tip

The decision as to which orientation to choose is determined by the geometric features you need to represent and how the object is used in space. Orient the axes so that the most important features are in the most open or exposed face.

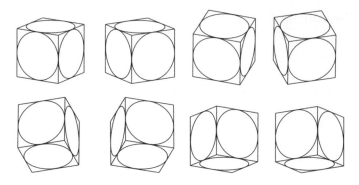

Figure 4.11   Axonometric axes can be rearranged to produce a number of different views.

Width (X) is universally a horizontal, left-to-right direction. Since architects are "top view centric," they consider vertical to be depth. Likewise, machine tools that hold parts horizontally (X and Y) consider up and down to be depth. However, if you think of Y as being vertical, world and view depth (Z_W and Z_V) are coincidental in the front view and fewer mistakes may be made. Many computer programs allow you, the user, to reassign directions to axes. Use a method that makes sense for the parts you are illustrating.

▼ Axonometric Tip

Technical illustrations having features *not* aligned with axonometric axes pose unique problems, as does the case of this skylight installation detail. Because the roof, skylight box, flashing, and open skylight frame are all at angles to axonometric axes, their sizes cannot be directly measured. Only the horizontal skylight edges and the horizontal shingle joints are parallel to an axis, making them measurable (Greg Maxson).

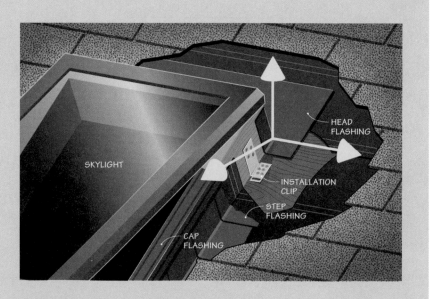

View Axes

Within a particular view or view port, up and down is Y, left and right is X, and in and out of the view is Z. We call these the *view axes*. They can be designated $X_V Y_V Z_V$.

For example, Figure 4.12 shows four view ports—top, front, side, and user (axonometric). All view ports consider up and down to be Y_V. This corresponds to Y_W in the front and side, though in the top it corresponds to Z_W. In the user view, because the view has been rotated axonometrically, Y_V doesn't correspond to any world axis direction. (Note: Although it appears that Y_V and Y_W are aligned, Z_W actually is tipped toward you.) You can see why it is important to know what axis system is active to make sure constructions and edits are done in the correct direction.

Local or User-Defined Axes

When a feature does not align with either world or view axes, it may be necessary to create an axis system aligned with that feature. This can be designated $X_U Y_U Z_U$. In Figure 4.13, a cylindrical feature has been rotated so that it no longer aligns with either the world or view axis systems. A local axis system has been applied to the feature, much as if the world axes have followed the feature as it was rotated. In this case, no direction of the local axis is aligned with either world or view axes.

Receding or Thrust Axis

This axis is perpendicular to a given feature (often called a face normal because normal refers to a perpendicular relationship). When the feature is frontal, horizontal, or profile, its thrust axis is an axonometric axis. When the feature is angular or skew (rotated about one or two world axes), its thrust axis remains perpendicular to the feature, only now it is a local axis. As the cylinder rotated in Figure 4.13, the thrust axis (Z_U) remained perpendicular to the circular end, and the minor axis of the ellipse remained coincidental with this thrust axis.

> ### ▼ Rule #1
>
> The minor axis of an axonometric ellipse *always* remains aligned with its thrust axis.

Axonometric Views

The three general classifications of axonometric views—isometric, dimetric, and trimetric—share many of the same properties.

▼ Their axes are mutually perpendicular and meet at the 0-0-0 world origin.

▼ Their axes are described by inclination above the horizontal and by a scaling percentage.

▼ Each view is named by the ellipses that correctly fit in the horizontal, left vertical, and right vertical faces. A 20-40-40 view would be dimetric; a 70-10-25 would be trimetric.

▼ Tip

It is convenient to construct axonometric objects about the world origin because dimensions are easier to express. You then can move the object anywhere.

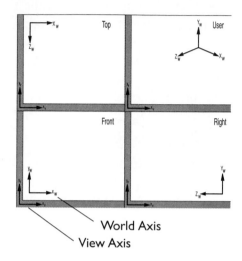

Figure 4.12   View axes are relative to the screen.

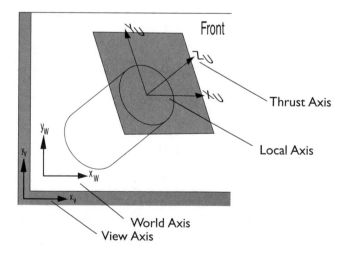

Figure 4.13   A local, or user, axis system applied to an angular feature.

Figure 4.14 displays the isometric cube, as well as one of the infinite dimetric and trimetric cubes. Study the common information provided to describe each view. Each has axis inclinations, axis scales, and ellipse exposures.

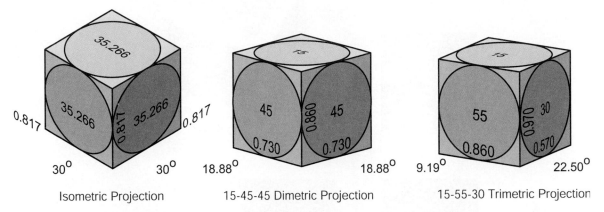

Figure 4.14   Information used to describe isometric, dimetric, and trimetric views.

These axonometric cubes are instrumental in creating accurate technical illustrations. See Chapter 6, for a step-by-step description of how these cubes are developed and used. Completed cubes are included on the CD-ROM in *tools/unit cubes*.

Standard View Methods

Because there are an infinite number of axonometric views, it is important to settle on a set number of views that can be used to show almost all possible geometry. Though there is no official set of standard views, most companies develop their own set of views so that illustrations done at different times by different individuals can be combined without excessive editing and redrawing. These standard views center on isometric views, and usually include several dimetric and trimetric views.

To automate the selection of a view and the necessary information to create that view by a number of means, a small utility, AxonHelper, has been provided on the CD-ROM for your use. This Macromedia Flash application can be copied to your computer and run as needed. AxonHelper contains approximate information to create views by four distinctly different methods.

Shearing Method

Shearing is a method that takes ordinary orthographic views, and by a process of rotation and scaling, twists or skews the view into correct axonometric position (Figure 4.15). Use this technique when you have scale views.

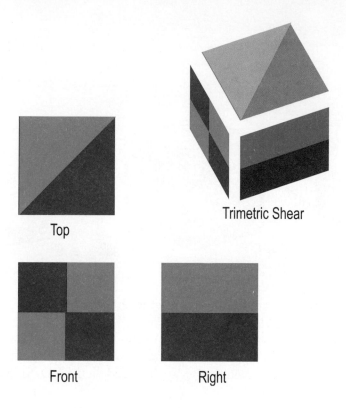

Figure 4.15   Shearing scales and rotating orthographic views into correct axonometric position.

After each orthographic view is sheared and correctly positioned on the axonometric axes, the faces must be extruded to assemble the final correct geometry. Figure 4.16 shows a sheared front view extruded along the depth (green) axis, using the sheared top view for correct positioning.

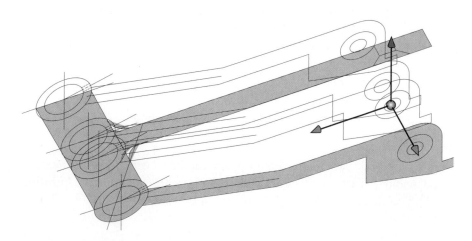

Figure 4.16   The sheared top and front views are extruded along the axonometric axes.

Scaling Method

Scaling a correct axonometric unitary cube provides axis angle, ellipse exposure, and scale information so that a view can be directly constructed (Figure 4.17). Use this technique when correct scale information exists but without scale views, or when you are illustrating from numerical data.

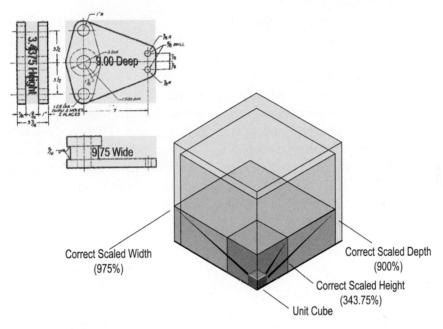

Figure 4.17   Scaling lets you directly construct an axonometric view from size information.

In Figure 4.17, a unitary axonometric cube has been scaled vertically and horizontally to match the measurements of the object shown in the sketch. The views of the crank Weldment are shown at reduced scale with correct height, width, and depth dimensional information shown superimposed; the views are not, however drawn at scale. By adding the dimensions, you can determine the overall width is 4.75, the overall height is 3.4375, and the overall depth is 9.00.

The unitary trimetric projection cube (available on the CD-ROM in the *tools/* directory) is scaled 475 percent to correspond to the width of 4.75; 343.75 percent to correspond to the height of 3.4375; 900 percent to correspond to the depth of 9.00.

The gray volume in Figure 4.17 is the isometric box that accurately describes the height, width, and depth represented by the gray rectangles superimposed over the views in the sketch. This is also referred to as the boxing-out method in many engineering design and drawing texts.

Viewpoint Method

The viewpoint method provides $X_W Y_W Z_W$ viewpoint (camera) information for achieving an axonometric view of 3D geometry. Figure 4.18 shows how a front viewpoint (camera at X0, Y0, Z1) is relocated to an isometric viewpoint (X1, Y1, Z1). The target of the viewpoint remains X0, Y0, Z0. Use this technique when you have 3D data that you want to form the basis of your illustration.

Only 3D programs can make use of axonometric viewpoints, and then only if object geometry is three-dimensional. Figure 4.18 shows the results of moving the viewpoint. A front view person, standing so that his or her eyes are at X0, Y0, and Z1 sees the front view view by looking down the Z-axis. Were the camera relocated to X1, Y1, Z1 the viewer would become an isometric view person and see an equally foreshortened view of the three axes: an isometric view.

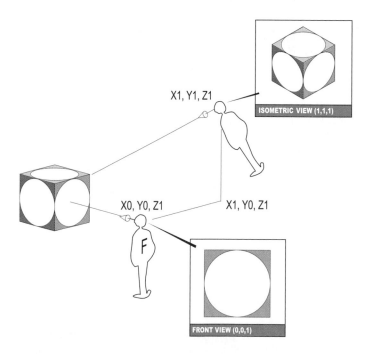

Figure 4.18   The viewpoint relocates the camera to a position that results in an axonometric view.

Object Rotation Method

Rotation yields the same results as does the viewpoint method, but with a stable front viewpoint and rotated geometry. Use this method when a program does not provide a method to locate a view by X, Y, Z camera position. This was done previously in Figure 4.7 where rotated top and side views resulted in an isometric front view.

Standard Views in AxonHelper

Once a dimetric or trimetric method has been selected in AxonHelper (shear, scale, viewpoint, or rotation), a number of standard views are presented (Figure 4.19). Note that a cube is associated with each view so that you can visualize the cube's position. The first view number refers to the exposure of the ellipse in the top face; the second number the exposure of the ellipse in the left vertical face; the third number the exposure of the ellipse in the right vertical face. (In subsequent chapters you will see how these cubes can be flipped around so that each standard view can result in many different orientations.) By selecting one of the standard views, the information needed to create a view by that method is displayed.

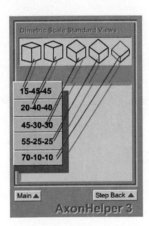

▼ **Tech Tip**

The exposure of a face relates to how much of the face can be seen. An orthographic view is fully exposed. A face with a seventy-degree ellipse is much more exposed than a face with a 10-degree ellipse. Given the option, put important features in more exposed faces.

Figure 4.19   AxonHelper's standard axonometric views.

Standard Views in Adobe Illustrator

Because Adobe Illustrator is a 2D PostScript drawing program, standard views are created by the following methods:

▼   Standard axonometric views are brought into Illustrator from 3D programs by **COPY-PASTE**, **OPEN**, or **PLACE**.

▼   Standard views are created in Illustrator by shearing scaled orthographic views.

▼   Standard views are created in Illustrator by direct construction using axis scale and inclination, and ellipse exposure information and by scaling correctly constructed unit cubes.

▼ Illustrator Tip

Illustrators may be called upon to represent wood-grained patterns. The goal is to create a pattern that looks natural and not mechanical. Begin by using the **Pen Tool** to create a basic wood grain (a). Then, convert the paths to outlines using **Object|Path|Outline Stroke**. This gives you shapes that can be further altered by repositioning individual anchor points. Always use shapes rather than lines because as shapes, the wood grain will always scale and distort proportionally (b). Fill the shapes with a natural wood color and **Mask** into the final axonometric or perspective shape (c).

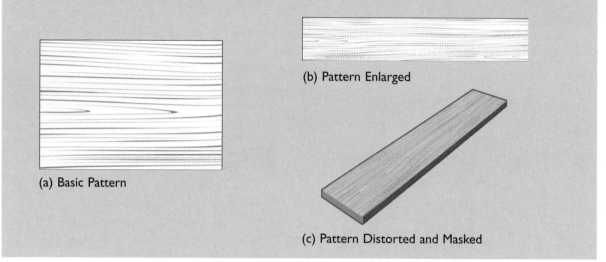

(b) Pattern Enlarged

(a) Basic Pattern

(c) Pattern Distorted and Masked

Standard Views in AutoCAD

Both shearing and scaling of 2D data can be used to create standard axonometrics in AutoCAD. However, AutoCAD is not a PostScript program and publication-quality resolution-independent illustrations would be impossible without subsequent use of a PostScript illustration tool. To create standard views using AutoCAD:

▼ Position the subject matter in the desired position relative to the world axes (usually centered about 0-0-0).

▼ Choose the eyeball (viewpoint) method in AxonHelper and select the desired standard view.

▼ Use the **VPOINT** command in AutoCAD to specify the camera position variables obtained from AxonHelper. Experiment with positive and negative values to arrive at the desired view.

▼ Choose **DISPLAY/HIDE** to remove hidden features.

▼ Use one of the methods described in detail in Chapter 17 to bring the view into Illustrator.

Standard Views in 3ds max

3ds max employs user-defined camera positions to create custom views (Figure 4.20). The view as seen by the camera is determined by the camera's position, the position of the camera's target, the focal length, and whether **Orthogonal View** is checked in the parameter roll-out window. This final option automatically creates an orthogonalized view. This procedure is covered in detail in Chapter 18. You can see from Figure 4.20 that we have an isometric view because the camera's X, Y, and Z-axis positions are the same absolute value.

Review

The heart of technical illustration is the axonometric pictorial. This type of orthographic view simultaneously displays three mutually perpendicular axes with the added benefit of consistent linear measurement in any axis direction. Computer-aided design (CAD) and modeling systems produce axonometric projections, views where axis measurements are foreshortened as necessary. Illustrations constructed in 2D are often axonometric drawings, where foreshortening has been normalized to a unitary scale.

Standard axonometric views are used to provide predictable and reproducible pictorial orientations. These standard views can be obtained in 2D with information from the AxonHelper application found on the CD-ROM. These standard views can also be made in CADD programs by viewpoint position or object rotation.

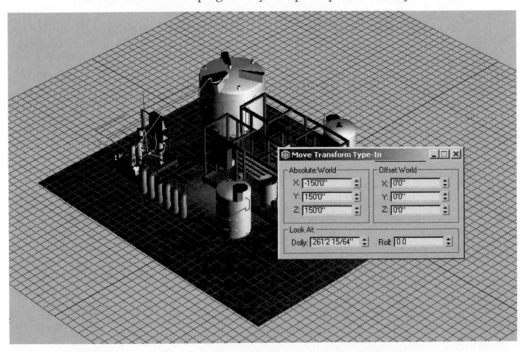

Figure 4.20   Axonometric camera position in a 3D program.

Text Resources

Bertoline, Gary R. and Wiebe, Eric N. "Axonometric and Oblique Drawings." Chapter 9 in *Technical Graphics Communication*. McGraw-Hill. New York. 2003.

Duff, Jon M. *Industrial Technical Illustration*. Brooks/Cole Engineering. Monterey, California. 1982.

Duff, Jon M. *Technical Illustration with Computer Applications*. Prentice Hall. Englewood Cliffs, New Jersey. 1994.

Duff, J.M. "Creating accurate standard axonometric projections by direct construction using two-dimensional PostScript drawing programs." *Engineering Design Graphics Journal*, Vol. 58, No. 1, pp. 4–10. 1994.

Gibby, Joseph C. *Technical Illustration*. American Technical Society. Chicago. 1962.

Hoelscher, Randolph P, Springer, Clifford H., and Pohle, Richard F. *Industrial Production Illustration*. McGraw Hill. New York. 1946.

Nicyper, Raymond. *Graphics Underlay Guides with Reference Data for Scale Drawing.* Graphicraft. Westport, Connecticut. 1973.

Thomas, T.A. *Technical Illustration*. McGraw-Hill. New York. 1978.

Treacy, John. *Production Illustration.* John Wiley and Sons. New York. 1945.

Uddin, Mohammed Saleh. *Axonometric and Oblique Drawing : A 3-D Construction, Rendering, and Design Guide.* McGraw-Hill. New York. 1997.

Internet Resources

An in-depth discussion of axonometric theory and the mathematics behind their creation. Retrieved from: *http://www.compuphase.com/axometr.htm*. August 2002.

Examples of axonometric drawings from landscape architecture. Retrieved from: *http://www.merilion.com/cgd/axons.htm*. August 2002.

Lecture video from Purdue's Computer Graphics Technology curriculum. Retrieved from: *http://www.tech.purdue.edu/cg/courses/tg217/lecture4/sld004.html*. August 2002.

A broad overview of axonometric and other projection methods. Retrieved from: *http://www.cs.brown.edu/stc/summer/viewing_history/viewing_history_1.html*. August 2002.

Chapter 5

Technical Animation

Technical Illustration

Rendering

3D Modeling

Technical

Axonometric Circles

In This Chapter

The accuracy of circular features impacts the effectiveness of technical illustrations probably more than any other geometric form. You should be able to see if an ellipse is not correct and be able to take corrective action. In this chapter you will develop an understanding of how circular features behave when viewed from different axonometric positions and how to represent these in technical illustrations

Chapter Objectives

In this chapter you will understand:

▼ How a circle becomes an ellipse and is constructed in Adobe Illustrator

▼ The relationship of an ellipse's major and minor axes

▼ How circular features such as cylinders, cones, spheres, tori, and other shapes with circular cross-sections are represented in axonometric

▼ The parallel relationship of an ellipse's minor axis to the receding perpendicular thrust

▼ The use of ellipses as measuring devices

▼ How sections through a sphere are represented as ellipses

▼ The use of a sphere as a measuring device

The Theory of Ellipse Representation

An ellipse, by definition, is formed when a plane cuts a right circular cone at an angle not perpendicular to the cone's axis, and not passing through the cone's base. But more common to technical illustrations, an ellipse is the *non-normal view of a circle.* If you view a circle straight on, your line of sight makes a right angle with the surface of the circle. If your line of sight makes any other angle, you see an ellipse. The only time a circle would appear circular and not elliptical is when the circle is truly perpendicular to your line of sight.

These ellipses must be represented correctly or the geometric form will appear twisted and unnatural. And because many human-made objects are cylindrical, spherical, or conical, technical illustrations contain numerous ellipses.

As a technical illustrator, you must be able to correctly choose the *exposure*, *scale*, and *tilt* of every ellipse. Failure to do so will result in the failure of the illustration to communicate its technical subject matter.

▼ Note

The discussion that follows concerns the correct construction, sizing, and alignment of ellipses in tow dimensions. In three dimensions, circular features are created as circles. As they are rotated in space, or as axonometric or perspective viewing positions are taken, correct elliptical foreshortening happens automatically.

How a Circle Becomes an Ellipse

A circle can appear in three ways: as a circle, as an ellipse, or as an edge the length of the circle's diameter. When your line of sight is perpendicular to the circle, you see the form normally (as a circle). When your line of sight is parallel to the circle, you see the circle on edge (as a line). Between these two positions are an infinite number of views where the circle appears elliptical.

Ellipse Exposure

An ellipse is described (named) by the acute angle formed between the line of sight and the plane of the circle (Figure 5.1). For example, a 20-degree ellipse is formed when the line of sight makes an acute angle of 20 degrees with the circle. This is called the *exposure* of the ellipse; greater exposure approaches a circle; less exposure approaches a line. So if you look at a circle in a perpendicular fashion you would see a 90-degree ellipse, in other words, a true circle. If you look at a circle parallel to its surface, you'd see a 0-degree ellipse, in other words, an edge view. All positions of the circle in between are ellipses designated by the acute angle between the line of sight and the edge of the circle.

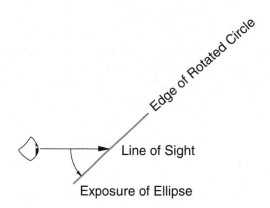

Figure 5.1   Ellipse exposure is determined by the acute angle between the circle and the line of sight.

Ellipse Geometry

Correct ellipse orientation depends on several factors. The first factor, exposure, was described in the previous section. Three other factors impact the correct placement of an ellipse.

Major Axis

The ellipse rotates through various angles about its major axis. The length of this axis remains constant and equal to the circle's diameter. In Figure 5.2 you can see that as the circle rotates from a circle through various ellipses, its major axis does not change. Therefore, a 4-in circle rotates through an infinite number of ellipses, each having a 4-in major axis, until it finally appears as a 4-in line.

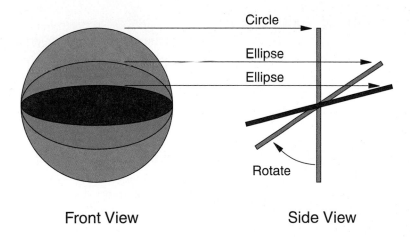

Figure 5.2   An ellipse is formed as the circle rotates about its major axis.

Minor Axis

Every ellipse has a major axis (the longest possible line through the center of the ellipse) and a minor axis (the shortest possible line through the center of the ellipse). The minor axis remains perpendicular to the major axis and becomes shorter as the angle between the line of sight and the circle becomes smaller. For an ellipse of given exposure, the minor axis can be expressed as a percentage of the major axis. For example, the minor axis of an isometric ellipse (35.266-degree exposure) is 57.6 percentof its major axis.

Because Adobe Illustrator knows nothing about ellipse exposure; ellipses must start out as circles. Using AxonHelper's ellipse function, the minor axis–to–major axis percentage can be determined. In Figure 5.3 a circle has been transformed into an isometric ellipse using the 57.6 percent minor axis scale. Note that the major axis (width or W) is unitary or 1 in. The minor axis (height or H) is 0.576 in. This produces a correct isometric projection ellipse.

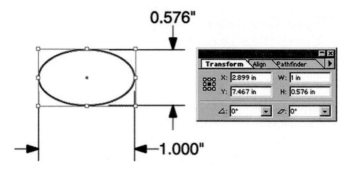

Figure 5.3   Minor axis expressed as a percentage of the major axis.

Thrust Axis

The thrust axis, also referred to as the receding axis, is perpendicular to the circle (as an axle is perpendicular to a wheel). When the circle becomes an ellipse, this axis remains coincidental with the ellipse's minor axis because the axis of ellipse rotation (major axis) and its thrust are perpendicular. Figure 5.4 shows a wheel-and-axle assembly with the ellipses and thrust axes superimposed.

Using AxonHelper to Determine Ellipse Exposure

As was previously mentioned, Adobe Illustrator does not provide a direct method for creating ellipses by exposure. As you saw, circles can be scaled independently in the horizontal (major axis) and vertical (minor axis) directions. AxonHelper (Figure 5.5) provides ellipse data for creating ellipses from 1- to 89-degree exposures. This is done in terms of minor–axis to–major axis percentage.

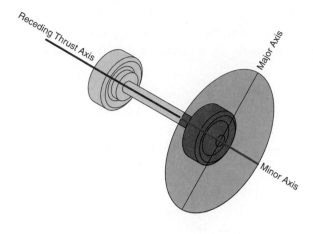

▼ Thrust Axis

You may wonder where the term *thrust axis* comes from. In aircraft illustration the axis through an engine is in the direction of its thrust. Because engines are usually comprised of many ellipses on the same axis, and the minor axes of those ellipses remain aligned to the thrust, we refer to the perpendicular to a circle as its *thrust axis*.

Figure 5.4   Minor axis remains concurrent (aligned) with the perpendicular thrust axis.

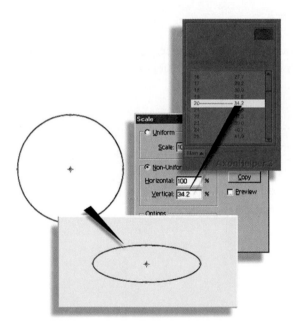

To create an ellipse of known exposure:

▼ Select the ellipse function in AxonHelper.

▼ Scroll down in AxonHelper until the desired ellipse is displayed. In this case a 20-degree ellipse is desired.

▼ Create a unitary circle in Illustrator and scale the circle 100 percent horizontally.

▼ Scale the vertical the percentage displayed in AxonHelper. The result is a horizontal ellipse of the desired exposure. This horizontal ellipse can subsequently be rotated into final position.

Figure 5.5   Use AxonHelper to determine ellipse axis percentages.

Determine the Exposure of an Existing Ellipse

You may need to determine the exposure of an existing ellipse in Adobe Illustrator. Knowing that ellipse exposure is expressed as a function of minor–to–major axis size, you can work backwards.

▼ Rotate the ellipse until the major axis is horizontal. You may want to create a guideline through the ellipse's center and rotate about the center until the rotate tool snaps to the guide (Figure 5.6). Note that the ellipse's bounding box remains oriented to the major and minor axes.

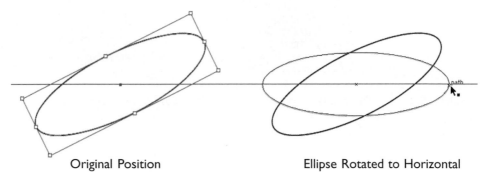

Original Position   Ellipse Rotated to Horizontal

Figure 5.6   Rotate ellipse until its major axis is horizontal.

▼ From the **Info Window** write down the ellipse height and width (Figure 5.7). In this case, the width is 4.062 in and the height is 1.354 in.

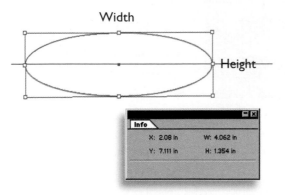

Figure 5.7   Determine height and width from Illustrator's **Info Window**.

▼ Divide the height by the width (1.354/4.062 = 0.333). The result is the minor axis expressed as a percentage of the major axis.

▼ Look up this percentage in AxonHelper. Use the nearest whole ellipse value (Figure 5.8).

Though the closest ellipse is 19 degrees, you would generally select 20 degrees because it better fits the standard views.

▼ Standard Ellipses

Although you can easily construct 19-degree ellipses, standard views use ellipses in 5-degree increments. This means that the ellipse isn't absolutely accurate, but it makes illustrations much easier to construct and match.

Figure 5.8   Look up closest ellipse value in AxonHelper.

Ellipse Axes and Perpendicular Thrust

You know that the minor axis of an ellipse is always concurrent with (parallel to or aligned with) the receding thrust of the circle. An axis is specified by the angle it makes with the 0-degree horizontal. As an example, Figure 5.9 shows an axis at an angle of 20 degrees. This establishes a 20-40-40 dimetric view. (You can check this with AxonHelper by selecting **Dimetric|Scale|20-40-40** and noting that both axis angles are 20 degrees.)

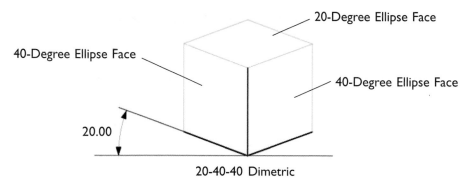

Figure 5.9   An axis set at 20 degrees to the horizontal.

We would like to correctly place a 40-degree ellipse in the right vertical plane of the cube. The left horizontal axis is perpendicular to the right face, so this is our thrust axis.

Follow these steps to correctly align an ellipse with its receding axis:

▼ Create a unitary ellipse of the correct exposure using minor axis scale data from AxonHelper.

▼ Position the center of the ellipse at the end of the axis as shown in Figure 5.10. You may want to turn on **View|Smart Guides** to facilitate this alignment.

▼ Determine the axis inclination. If you drew the axis, you should know this. If not, draw a line at the same inclination (snap!) and note the angle in the **Info** window.

▼ Select the ellipse.

▼ Double click the rotate tool. Because positive rotation is counter clockwise (right-hand rule), rotate the ellipse 90 degrees minus the axis inclination angle (Figure 5.11).

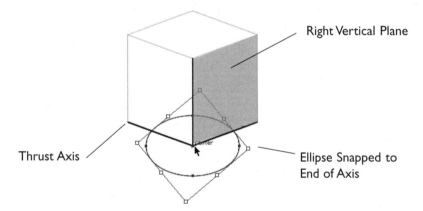

Right Vertical Plane

Thrust Axis

Ellipse Snapped to
End of Axis

Figure 5.10   Starting position of an ellipse and its perpendicular thrust axis.

▼ Tech Tip

Each of the three axonometric axes is perpendicular to one of the planes of the axonometric cube. The vertical axis is perpendicular to the horizontal plane and serves as the thrust axis for horizontal ellipses. The left horizontal axis is perpendicular to the right vertical plane and serves as the thrust axis for right vertical ellipses. The right horizontal axis is perpendicular to the left vertical plane and serves as the thrust axis for left vertical ellipses.

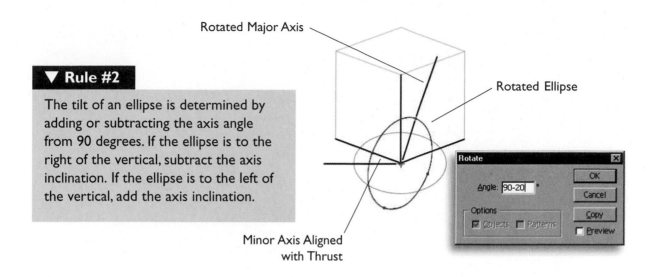

▼ **Rule #2**

The tilt of an ellipse is determined by adding or subtracting the axis angle from 90 degrees. If the ellipse is to the right of the vertical, subtract the axis inclination. If the ellipse is to the left of the vertical, add the axis inclination.

Figure 5.11   Ellipse is rotated until its major axis is perpendicular to the thrust line.

Measuring with an Ellipse in Adobe Illustrator

Everyone knows that one characteristic of a circle is that every point on the circle's circumference is the same distance from the circle's center. With this in mind, we can say that *every point on the ellipse's circumference is the same distance from the ellipse's center*.

Correctly aligned circles can be used to measure distances along lines in axonometric views because the axonometric diameters, though they appear of different lengths, are really the diameters of a circle and are the same length. This technique is called *measuring with an ellipse*.

Figure 5.12 shows an irregular cylindrical object (a spindle) that will be accurately constructed in an isometric view using ellipses to measure distances along the receding axes. The shape of the spindle has been sampled with a number of cross sections and each of these sections is circular. Table 5.1 records the pertinent information—distances along the axis from point 1, and diameters at each section or point. The diameters of each circular section are measured across the spindle's centerline to where the section crosses the outside profiles. Point 1 is the left end of the spindle. Point 4 is the right end. Two additional points are marked—points 2 and 4. These locate the far outside and inside of the profile curve. These final two points are important because it is at these points that the outside curve changes direction.

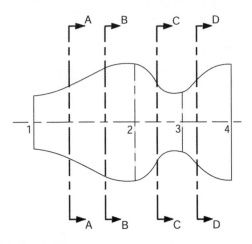

Figure 5.12   A cylindrical spindle with sectional cutting planes in place.

▼ Table of Diameters and Distances								
	I	A	B	2	C	3	D	4
Distance from Point I	0.00	0.375	0.750	1.06	1.30	1.55	1.700	2.06
Diameter of Section	0.55	0.750	1.125	1.20	0.80	0.62	0.875	1.20

Table 5.1   Tabular data for the spindle in Figure 5.12.

Create the Underlying Geometry

Open the file *iso_cube.ai* from the CD-ROM.

▼ Create a new layer named construction. Using the isometric unit *projection cube*, copy axis lines. Arrange these at a convenient position on the Adobe Illustrator sheet (Figure 5.13).

▼ From Table 5.1, you can see that the overall length of the spindle is 2.06. Uniformly scale the right and left horizontal axes 206 percent. The length of your object, from points 1 to 4, is the length of the right horizontal axis (Figure 5.14).

Verify Ellipse as a Measuring Device

If an ellipse can be used as a measuring device a 2.06 diameter ellipse should fit this line exactly.

▼ Copy a horizontal ellipse from the top of your projection cube.

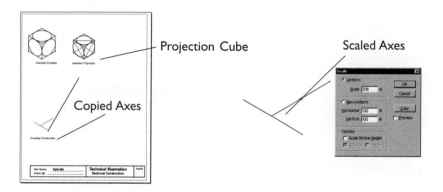

| **Figure 5.13** | Copy lines from the isometric cube to start your construction. | **Figure 5.14** | Scale the receding axis to the overall length of the object. |

▼ Uniformly scale this ellipse 206 percent and align the ellipse's center on the axis so that its circumference passes through point 1. You can see from Figure 5.15 that the ellipse also passes through the rear of the axis.

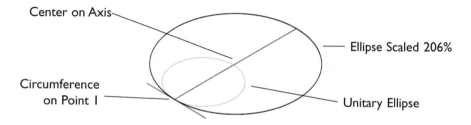

Figure 5.15   An ellipse is used to measure by scaling its diameter.

Create Measuring Ellipses for the Other Sections

To complete the measurements needed to illustrate the spindle we will refer to Table 5.1. Each circular section diameter has a distance associated with it.

▼ Since the unitary horizontal ellipse is on the clipboard, paste and uniformly scale using the number from Table 5.1 corresponding to the distance section A is from point 1. In this case, the value is 0.375, or a scale percentage of 37.5 percent.

▼ Align this ellipse along the receding axis so that its circumference passes through point 1 (Figure 5.16). *Where this ellipse's circumference passes across the axis is the center of the 0.75 diameter circular section.* This is point A. Continue this technique for each of the sections until all center points are found (Figure 5.17). Isometric axis lines have been added for clarity so you might better visualize where each circular section will be located.

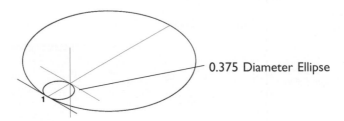

Figure 5.16   Scale the unitary ellipse to the distance of the first section.

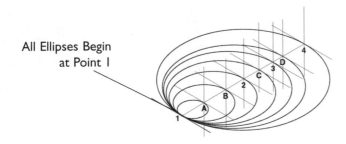

Figure 5.17   All distances are found by scaling the unitary ellipse.

Scale and Place Circular Sections

Now that we know the location of the various sections along the central thrust line, we can place the ellipses representing the sections.

▼ Create a new layer in Illustrator named sections. Copy the unitary ellipse from the left vertical plane of the isometric projection cube and paste it over your construction.

▼ From Table 5.1, you see that the first diameter at point 1 is 0.55. Uniformly scale this left vertical ellipse to 55 percent and grab its center to relocate the center to point 1. Repeat this same procedure for each of the circular sections until your construction looks like Figure 5.18.

Figure 5.18   Unitary ellipse in vertical plane is scaled at each section location.

Complete the Outside Profile

The spindle's shape will be defined by a curved tangent element connecting the cross sections in order.

▼   Hide the construction layer. Select a contrasting color (such as magenta) and substantial stroke.

▼   With the pen tool, create the top profile of the spindle. Begin and end the profile with a single click. Click-drag anchor points on each of the intermediate sections. Don't be overly concerned with accuracy when first placing the profile (Figure 5.19). Anchor points can be repositioned and Bezier handles adjusted until the profile fits the sections.

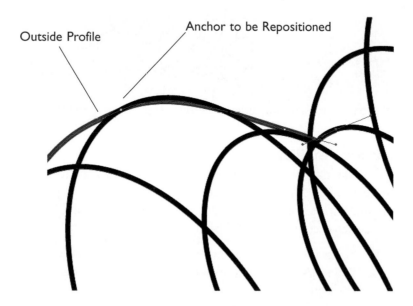

Figure 5.19   The first attempt at fitting a curve doesn't have to be perfect. Edit the curve later.

Our spindle requires two curves, one disappearing behind the first in the narrowest part. This is highlighted in the gray circle in Figure 5.20. To be really accurate, we would need more sections (data samples) between sections C and D. The finished solution in Figure 5.21 shows at this two-curve solution.

▼ When the top profile is completed and edited into final shape, **Reflect** and **Copy** this curve about a 30-degree axis. This corresponds to the receding axis of the spindle in isometric view. This operation is identical (but two-dimensional) to the **Lathe** operation in 3ds max. If you have been accurate in your construction, the reflected curve should match the sections on the bottom side of the spindle (Figure 5.20).

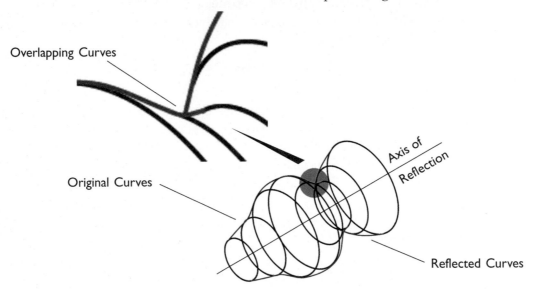

Overlapping Curves

Original Curves

Axis of Reflection

Reflected Curves

Figure 5.20   Finished profile is actually two curves. The top curves are reflected about a 30-degree axis.

Edit for Visibility

To make the spindle appear solid, section lines on the rear must be cut and deleted and all measuring ellipses and section axes hidden. If you have been accurate, the profile's anchor points should be at the very outside of each section. This is where each section ellipse will be cut.

▼ Select the ellipse at point A. Choose **Bring to Front** from the **Object|Arange** menu. Adobe Illustrator is sensitive to front-to-back position when cutting objects.

▼ Cut the ellipse at the top and bottom where the ellipse and outside profile lines are tangent.

▼ Delete the hidden portion. Continue this operation for each of the sections. Figure 5.21 shows the final spindle.

Figure 5.21   Final edited spindle.

The Ellipse and Spheres

A sphere will always appear circular in an axonometric view. However, a sphere must be scaled such that its "great circle" is tangent to the ellipses formed by vertical and horizontal sections through the sphere. Figure 5.22 shows a cube enclosing a sphere. Note that both vertical and horizontal sections touch the cube in the center of its faces. The sphere also touches the cube at these points and is centered within the cube, tangent to the ellipses forming the sections.

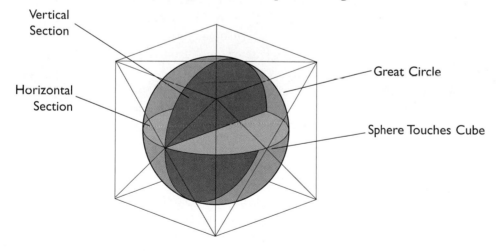

Figure 5.22   A sphere is tangent to its horizontal and vertical great circles.

Why is this of interest to illustrators? It's because every point on the surface of a sphere lies the same distance from the sphere's center; this allows the sphere to be used to measure distances that are not parallel to the axonometric axes.

▼ Illustration in Industry

K. Daniel Clark Communications
San Francisco, California

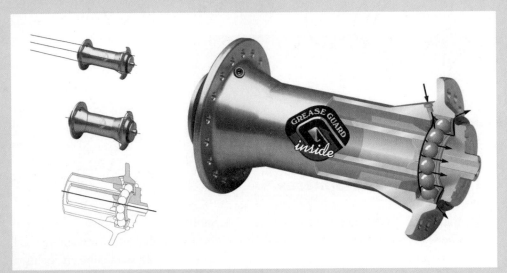

Images Copyright 2002 by K Daniel Clark Communications.

The Wilderness Trail Grease Guard system allows users to squirt grease into a port on the hub, which pushes dirt away from the bearings. To illustrate this feature, the photo tracing technique was employed. After securing a high-quality photograph of the hub from the client, Dan decided to use it as the base of the illustration and add Adobe Illustrator art on top of it to describe the system.

Perspective was established based on the photograph to ensure that the illustration would match the image. To do this, a series of long guides were used to locate the vanishing point (top step). Dan then had to establish curves in the Illustrator artwork that could be used as guides to build the wireframe (middle step). Each of these shapes were sheared into place using the **Rotate**, **Scale**, and **Skew** tools.

Using the perspective lines and cylindrical guides, Dan built the guts of the hub based on the client-supplied model (bottom step). This process started with an imaginary slice from the top edge to the center. Once this cross section was complete, it was duplicated and skewed into place for the lower slice. The bearing race (in yellow), seals (in blue), and other objects were manually extruded into solids using the cylindrical guides.

Software

Adobe Illustrator

Hardware

Macintosh G4
Digital Scanner

Single and Compound Angles Using Spheres and Ellipses

It is apparent that all lines originating from the center of a sphere and terminating on the sphere's surface will be the same length. With this knowledge, *nonaxonometric lines*, that is, lines not parallel to the axonometric axes, can be accurately constructed. But to do so, you need a set of axonometric protractors. A circular protractor has been provided for you on the CD-ROM in *tools/protractor.ai*. Single and compound angle construction can also be accomplished by circle projection. See Chapter 7 for an explanation of this technique.

Axonometric Protractors

Starting with a circular protractor, you can, with AxonHelper, create any elliptical protractor you desire. You simply scale the protractor identically to the ellipse in the face you want to measure. Numbers have been omitted from the protractor so that measuring can be started at any point. Figure 5.23 shows a correctly scaled 40-degree protractor superimposed on the right vertical great circle of the sphere in Figure 5.22. For convenience, the circular protractor was first rotated 20 degrees (the axis angle) so that major divisions would align with the axonometric axes.

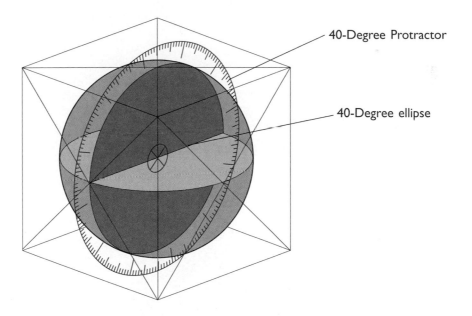

Figure 5.23   A circular protractor is scaled using AxonHelper ellipse values.

Single Angles

When a line is angled to one of the axonometric axes, we can refer to it as being a single angle line. To locate it correctly, we need to know its proper direction and length, and spherical construction accomplishes both of these requirements.

Figure 5.24 shows how a line of 100 units can be constructed parallel to the right vertical plane at 70 degrees above the horizontal. Note that the sphere has a 100 unit radius, so any line originating at the sphere's center and terminating on the surface of the sphere will be 100 units long.

Compound Angles

When a line is angled to two of the axonometric axes, we can refer to it as being a sompound angle line. To locate it correctly, we need to know its proper direction and length using two spherical constructions.

Figure 5.25 shows how a second rotation of the line in Figure 5.24 may be done. Assume that you want the final line 70 degrees above the horizontal and 60 degrees toward the front.

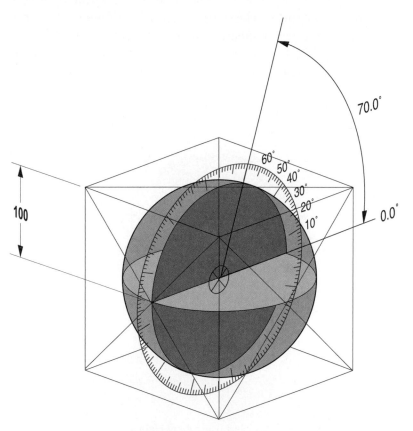

Figure 5.24   A line 70 degrees above the horizontal can be found using an elliptical protractor.

The circumference of the white ellipse in Figure 5.25 defines all possible 100-unit lines 70 degrees above the horizontal. Its radius passes through the point where the 70-degree line touches the sphere. A 20-degree protractor (the exposure of the top face) has been positioned concentric with this horizontal slice, and 60 degrees marked off. The heavy line is 100 units long, inclined 70 degrees above the horizontal, and rotated 60 degrees toward the front.

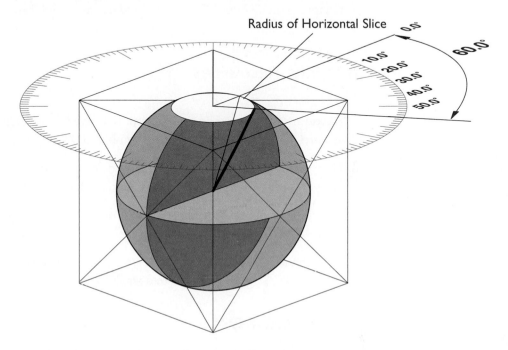

Figure 5.25   A compound angle requires the use of a second elliptical protractor.

Review

Because technical subjects contain circular and cylindrical features, technical illustrations contain numerous ellipses. These ellipses must be the correct exposure, tilt, and size. One guiding feature is that a thrust vector, perpendicular to the circular feature, will always appear parallel to the ellipse's minor axis. Because of the parallel nature of concentric ellipses, they can be used as measuring devices. Every point on the circumference of an ellipse is equidistant from the center. Additionally, knowledge that every point on the surface of a sphere is equidistant from the sphere's center, allows spheres to be used to measure angles and distances.

Text Resources

Uddin, Mohammed Saleh. *Axonometric and Oblique Drawing : A 3-D Construction, Rendering, and Design Guide*. McGraw-Hill. New York. 1997.

Bertoline, Gary R. and Wiebe, Eric N. "Axonometric and Oblique Drawings". Chapter 9 in *Technical Graphics Communication*. McGraw-Hill. New York. 2003.

Internet Resources

Axonometric drawing and projection. Retrieved from: *http://www.ul.ie/ ~rynnet/keanea/HomePage.html*. September 2002.

Drawing circles in Photoshop. Retrieved from: *http://www.bizcreator.com/ designtips/photoshop/circlesandsquares.htm*. November 2002.

Eclectic glossary of historical drawing tools. Retrieved from: *http:// www.daube.ch/docu/glossary/drawingtools.html*. November 2002.

Right hand rule. Retrieved from: *http://www.arch.usyd.edu.au/~paul/courses/dc-c/intro_acad/3Dintro.html*. September 2002.

On the CD-ROM

For those of you who would like to try out the procedures described in this chapter, several resources have been provided on the CD-ROM. Look in *exercises/ch5* for the following files:

ellipses.ai.   This file contains examples of ellipses and axes for alignment.

spindle.ai   Use this file to follow along with the instruction on using an ellipse to measure.

Look in *tools/* for the following file:

protractor.ai   Use this file to create axonometric protractors of any exposure.

Perspective Drawing

Axonometric Projection

Technical Animation

Technical Illustration

Rendering

3D Modeling

Technical Drawing

Chapter 6

Axonometric Scale Construction

In This Chapter

Creating 2D axonometric views from scale views is a fundamental technical illustration technique. It requires an ability to "read" orthographic views (like those presented in Figure 6.1). To be perfectly honest, this is an extremely difficult task because you need to assemble a 3D mental picture from flat 2D view information. It takes practice and experience to do this efficiently.

This technique, thankfully, is only one of several techniques for arriving at accurate axonometric pictorials. Still, you must be prepared to employ this technique when you only have numerical information. The alternative to this direct scale construction is to reconstruct the views at scale so that they can be projected or sheared or to model the data in three dimensions so that an axonometric viewing position can be assumed. If you do not want to do either of these, then direct scale construction is the answer.

Chapter Objectives

In this chapter you will understand:

▼ How to extract necessary size and placement information from dimensioned orthographic views

▼ Appropriate choices of axonometric views based on object dimensions and features

▼ The procedure for correctly associating orthographic dimensional information with axonometric axes

▼ The construction of an accurate unitary axonometric cube based on AxonHelper information

▼ Axonometric construction including the boxing-out and centerline methods

▼ Analysis of spatial positioning to edit the construction for visibility

Views Not Drawn to Scale

Figure 6.1 shows three views of a crank weldment. There is a general note on the drawing stating that the views are not to scale, so you can't assume that the views themselves can be measured to accurately create an axonometric drawing. You might quickly create a scale and compare dimensions to see if the views are, in fact, drawn to scale. For example, there is a dimension in the top view that is 1 radius. You could use this unitary dimension to develop a scale and quickly determine the scalability of the views. In the case of Figure 6.1, the views are simply representational, and cannot be directly scaled.

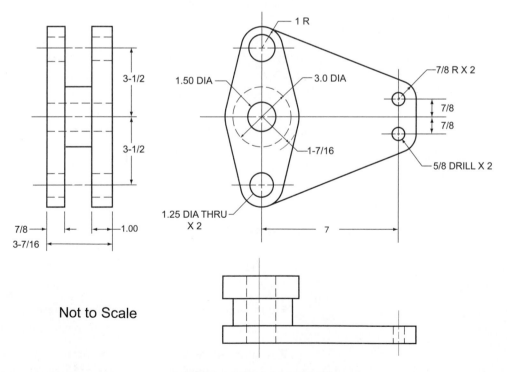

Figure 6.1 Three views of a crank weldment. Orthographic views that are marked not to scale cannot be directly measured (R = radius, DIA = diameter, DRILL = diameter).

View Selection

The geometry of the crank weldment in Figure 6.1 dictates which axonometric view might be most appropriate. Because the major features of the weldment (its mounting holes) are in horizontal planes and little identifying shape is evident in vertical planes, an axonometric orientation revealing more of the top and less of the sides seems appropriate. Because the front and sides have equally unimportant features, we will use a *dimetric view*.

The standard dimetric views available in AxonHelper are shown in Figure 6.2. Only two views, 55-25-25 and 70-10-10, satisfy our need for significantly greater exposure of the horizontal planes. The 70-10-10 view severely foreshortens the sides (resulting in a near top view), making it less desirable than the 55-25-25 view.

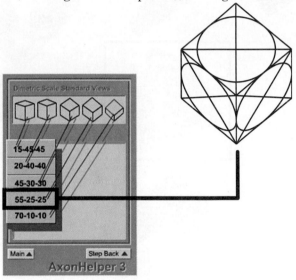

Figure 6.2   Standard dimetric views from AxonHelper.

Orient Object Geometry to the Axes

There are a number of possible orientations within the 55-25-25 dimetric axis system, as shown in Figure 6.3. The only difference in these views is in how the object's geometry is aligned to the axes. Most of these views can be dismissed because too many features on the object are obscured. The view surrounded by the dark frame, however, satisfies the following constraints:

▼   The object is shown in its traditional spatial orientation. The top orthographic view remains the top of the object.

▼   The fewest number of features (holes, slots, radii) are hidden.

▼ The object is viewed from small to large, and from front to back, which presents least risk of important features being covered.

Figure 6.3   Variations on a 55-25-25 dimetric orientation. The most advantageous view is outlined.

Create a Correctly Scaled Unitary Cube

The first step in creating an axonometric view by scale construction is to create a correctly scaled unitary cube in the desired orientation. Using AxonHelper, determine the correct axis scale and inclination (Figure 6.4). The results are X and Y 0.910 scale at a 39.32-degree inclination and Z 0.610 scale at a 90-degree inclination. We know that this is a *projection cube* because none of the axis scales have been normalized to 1.00. Begin your cube construction by turning your rulers on and dragging vertical and horizontal guides to intersect at a convenient location on the Adobe Illustrator page.

A 1.00 x 1.00 x 1.00 cube in axonometric position with 1.00 diameter circles on each of the faces.

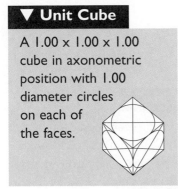

X Axis: 0.910 scale, 39.32176°
Y Axis: 0.910 scale, 39.32°
Z Axis: 0.610 scale, 90.00°

Figure 6.4   Scale information from AxonHelper for a 55-25-25 dimetric view.

X-Y-Z Axes Scale and Inclination

Continue the construction of the unitary cube by creating correctly scaled and rotated axes using AxonHelper information.

▼ Create a horizontal line representative of the left horizontal axis that is 0.910 in long (Figure 6.5). Don't try to make the line this exact length to start out with. Open the **Transform Window** and enter for width 0.910 and 0.0 for height.

▼ Rotate this line –39.32 degrees in a clockwise direction as shown in Figure 6.6. Remember the right-hand rule.

▼ **Tech Note**

The right-hand rule is used to determine positive and negative rotations about an axis. In Illustrator, because it's a 2D drawing program, the axis of rotation comes straight out at you. If you cup your fingers, they point in the positive rotation direction.

▼ Move this line by its end to snap to the intersection of the two ruler guides (**View|Smart Guides|Snap to Point**). Because the desired view is dimetric, the Y axis can be formed by reflecting and copying this first axis about the vertical and positioning its end at the intersection of the two ruler guides (Figure 6.7).

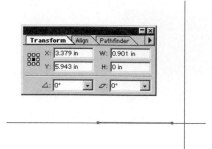

Figure 6.5   X axis correctly scaled to 0.910.

▼ Rotation

Illustrator will rotate objects about their geometric centers by default. To move the axis of rotation to the end of the line, simply click there first before dragging. Unfortunately, the axis is returned to the object's center when values are entered from the keyboard.

Figure 6.6   X axis rotated clockwise 39.32 degrees.

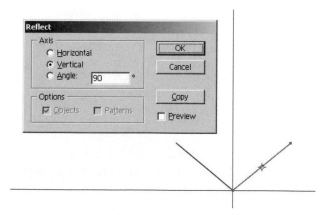

Figure 6.7   X axis reflected about the vertical (90 degrees) to form the Y axis.

▼ Create the vertical axis in the same manner as you did the first horizontal axis. Create a vertical line sized at 0.610 and move it to the intersection of the ruler guides (Figure 6.8). These three axes, when duplicated and moved to snap to the ends of the originals, complete a correctly scaled 55-25-25 unitary cube (Figure 6.9).

▼ Note

You may want to choose **View|Guides|Hide Guides** to assure that you are snapping to the ends of the axes.

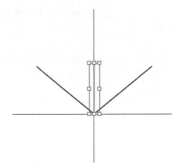

Figure 6.8   Z axis created at 0.610 and moved to the ruler guide intersection.

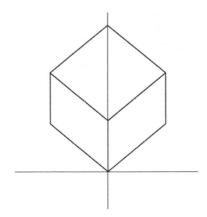

Figure 6.9   Unitary cube completed by duplicating and moving axes.

Construct Correct Ellipses

Because the cube we just constructed is a 1-in axonometric projection cube (all three axes are scaled or foreshortened), 1-in major diameter ellipses of 55- and 25-degree exposures will fit perfectly on the faces. For this we use AxonHelper again. Begin all ellipse constructions with a 1-in-diameter circle and scale the height axis.

▼ By using the ellipse function in AxonHelper and scrolling to the desired exposures we determine the following minor axis ellipse percentages:

▼ 55- and 25-Degree Ellipse Exposures

degree exposure	percent of major axis (from AxonHelper)
25	41.9
55	81.9

▼ Figure 6.10 shows a 1-in-diameter circle whose height is set at 0.819, the minor axis percentage for a 55-degree ellipse.

▼ Turn your guides back on. Create a horizontal guide across the middle of the top plane. Move this ellipse by its center to the intersection of the top face's guides. As you can see from Figure 6.11, the ellipse fits perfectly on the top plane.

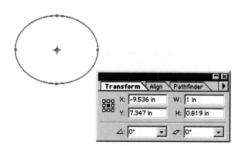

Figure 6.10   Circle height scaled to 81.9 percent produces a 55-degree ellipse.

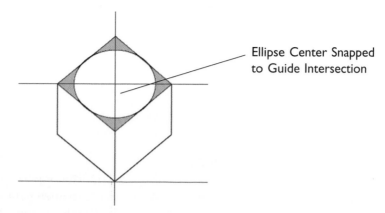

Ellipse Center Snapped to Guide Intersection

Figure 6.11   Ellipse moved to top plane of 55-25-25 dimetric cube.

▼ **Tip**

Because we are using every 5-degree ellipse exposure for our standard views (55 as opposed to, say, 55.367), you may have to adjust the ellipses slightly for a perfect fit.

▼ Construct the ellipses in right and left vertical planes in the same manner, using 0.419 as the minor axis scale for 25-degree ellipses.

▼ Position each ellipse at the correct tilt. Obey the rule that the minor axis of an ellipse is parallel to its receding thrust axis. The horizontal axis is inclined 90 + 39.32 degrees.

▼ Use the **Transform Window** to adjust the minor axis and the inclination of a 25-degree ellipse in the same operation (Figure 6.12).

▼ Reflect the first ellipse and position it correctly on the other face (Figure 6.13). Save this file as *55-25-25 master.ai* so that you'll be able to use it at any time.

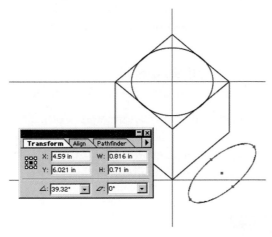

Figure 6.12   Adjust height and inclination in the same transform operation.

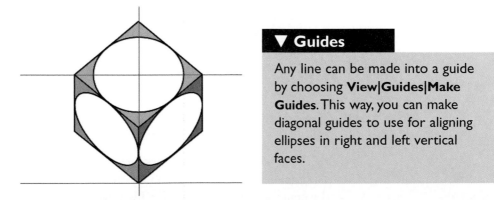

▼ **Guides**

Any line can be made into a guide by choosing **View|Guides|Make Guides**. This way, you can make diagonal guides to use for aligning ellipses in right and left vertical faces.

Figure 6.13   Correctly positionedellipses in vertical planes.

▼ **Note**

Once the ellipse is inclined, the width and height values change to reflect the reduction of the bounding box that surrounds the ellipse. The major and minor axes remain 1.00 and 0.419, respectively.

Boxing Out the Crank Weldment

The crank weldment from Figure 6.1 is constructed using two techniques: the *boxing-out method* for its overall size, and the *centerline method* for the position of individual features within the box. By combining these two methods we can quickly determine the location of the crank weldment's major features.

Traditionally, an illustrator might make specialized paper or plastic scales to use to measure distances along the axes. But because computer graphics makes it so easy to enlarge or reduce by a numerical factor, we will use the unitary cube and scale it as needed. For example, the unit cube scaled to 37.5 percent yields a line that has a length of 0.375. The same cube scaled 375 percent yields a line 3.75 in.

Boxing Out the Overall Shape

Refer to Figure 6.1 as we construct the crank weldment. Follow these steps:

▼ Determine the overall height, width, and depth. The dimensions are mostly in fractions, but it may be convenient to convert to decimals. The width, for example, can be expressed as $7.00 + 1^7/_8 + ^7/_8 = 9.75$. The crank weldment then has these overall dimensions:

▼ Boxing-Out	
Height	3.4375
Width	9.75
Depth	9.00

▼ On a copy of the 55-25-25 master scale the right axis to 900 percent, the left axis to 975 percent, and the vertical axis to 343.75percent.

▼ Move these axes as was done in Figure 6.7 to form the front corner of a cube.

▼ By duplicating and moving these scaled axes, finish a 55-25-25 axonometric volume as you did the unitary cube (Figure 6.14). You know the crank weldment is contained within this shape.

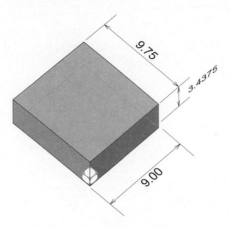

Figure 6.14 A volume defining height, width, and depth is constructed by scaling axes of the unitary cube.

Centerline Construction

From an inspection of Figure 6.1 we see that several features are aligned on centers.

▼ By studying the crank weldment views, centerlines that control the placement of holes, diameters, and radii can be determined and their distances from the edges of the volume measured off (the edges are actually control datum planes). These are shown in red in Figure 6.15.

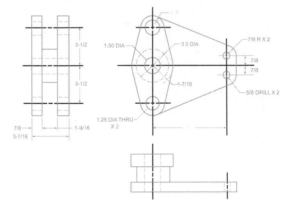

Figure 6.15 Centerlines are identified on the orthographic views. Refer to Figure 6.1 for dimension values.

▼ Using the unitary cube, duplicate and scale copies of its axes and use these to position the red center lines in an axonometric view. For example, scale a right axis 450 percent to determine the middle of the 9.00 depth. Scale a left axis 87.5 percent ($^7/_8$ or 0.875) to determine the centerline of the two $^5/_8$ (0.625) holes (Figure 6.16).

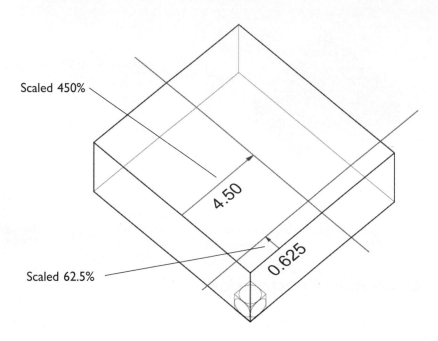

Scaled 450%

4.50

0.625

Scaled 62.5%

Figure 6.16   Centerlines are positioned on the top datum plane by using scaled axes to measure.

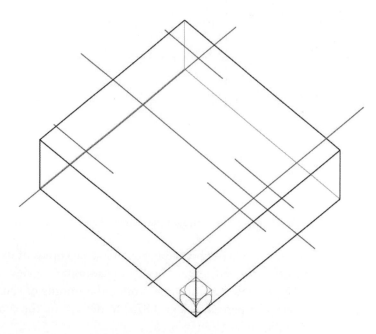

Figure 6.17   Complete set of centerlines are positioned on the top plane.

▼ Continue placing centerlines on the top surface by scaling unitary axis lines (Figure 6.17). Don't bother duplicating these centerlines at different heights because after the geometry on the uppermost surface is constructed, it can easily be duplicated and moved downward to the correct position.

Scale the Ellipses

Because ellipses on your unitary cube are 1.00, they can also be scaled by a factor equal to the diameter of the desired circular feature. Partial arcs are specified by their radii, so you'll have to double radius values. You can then trim the ellipse to the desired arc.

▼ Start with the centerline intersection at the 1.500 diameter hole. Duplicate a 55-degree ellipse from the unitary cube and move it by its center to the intersection of the centerlines.

▼ Scale this ellipse 150 percent (Figure 6.18). It is now a 1.500 diameter ellipse.

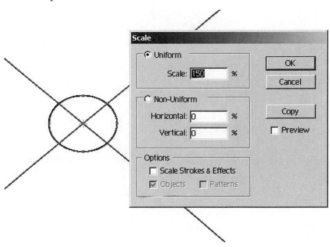

Figure 6.18   Ellipse from unitary cube is positioned and scaled to diameter shown on dimensioned orthographic view.

▼ Working in Units

Unitary cubes should be constructed in the units appropriate for the way objects are dimensioned. For example, if you work in the metric system, your unit cube would be 1 mm on each side; if you work in feet and decimal inches, you unit cube would be 1.00 feet on each side.

▼ Continue duplicating, moving, and scaling ellipses until the construction looks like Figure 6.19.

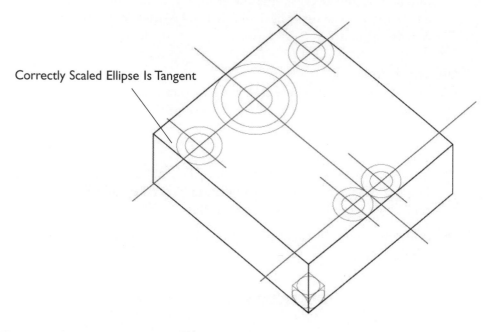

Correctly Scaled Ellipse Is Tangent

Figure 6.19   Completed elliptical construction.

▼ Tech Tip

It is evident now why we boxed out the overall shape. You can see in Figure 6.19 that several ellipses, when properly scaled, touch the outside of the box.

Complete Tangent Geometry

▼ With the circular geometry in place on the top plane, create tangent lines with the **Pen Tool** as shown in Figure 6.20. Don't trim the ellipses yet because it's difficult to determine ahead of time what will be visible and hidden as you extrude the geometry downward.

Extrude Geometry to Correct Heights

▼ When you analyze the orthographic views, you see that the geometry we just constructed (at least some of it) is repeated 1.00 below the top datum plane and 0.875 above the bottom. Figure 6.21 displays the geometry duplicated and moved to scaled height lines.

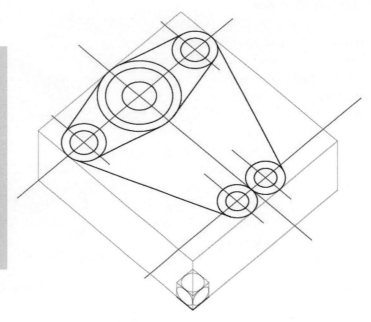

▼ **Tech Tip**

Adobe Illustrator doesn't find tangents as does a CAD program. It does, however, determine when an anchor is on a path or at the intersection of two elements. It may be helpful once one tangent is located to place a guide through that point. That way a symmetrical tangent can be found more easily.

Figure 6.20   Tangent line geometry is added using the **Pen Tool**.

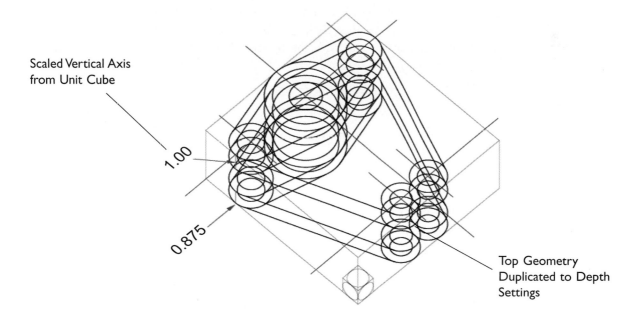

Scaled Vertical Axis from Unit Cube

1.00

0.875

Top Geometry Duplicated to Depth Settings

Figure 6.21   Geometry is extruded to correct depths.

▼ Illustrator Tip

When assembling an object in Adobe Illustrator, pay close attention to areas where planes or lines meet. Remember that when paths are stroked, the stroke ends sharply at the exact end of the path—at the path's anchor (**Butt Cap**). At corners, a sharp intersection is formed (**Miter Join**). Even when perfectly aligned, lines may not appear to smoothly blend together. You will usually want to choose the **Round Cap** and **Join** options from the **Stroke Palette**. This extends the line past the anchor by a distance of one-half the stroke width.

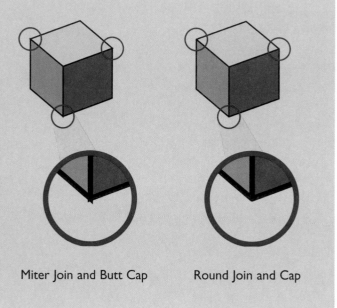

Miter Join and Butt Cap   Round Join and Cap

Edit for Visibility

By referring to the orthographic views, start at the top of the object and delete all geometry that is not seen. You may want to move the unitary cube and boxing-out volume to a hidden layer. Just click and delete while viewing the object without any zoom; cut any lines or arcs with the **Scissors Tool**, leaving the geometry slightly long (Figure 6.22).

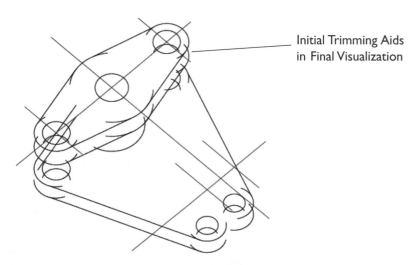

Initial Trimming Aids in Final Visualization

Figure 6.22   Edit roughly at extended zoom for efficiency.

Complete a final edit by zooming in on intersections, carefully trimming, and adding any missing geometry such as limiting elements of cylindrical surfaces. After moving centerlines to the hidden layer, you should have the final 55-25-25 dimetric projection shown in Figure 6.23.

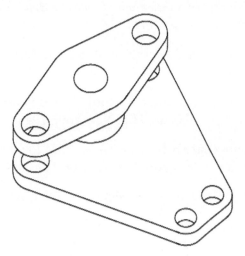

Figure 6.23   Finished crank weldment axonometric pictorial.

Review

If you have information about size and feature location, but don't have scaled orthographic views, the axonometric construction method provides a way of creating accurate pictorials. In this method, distances are measured in the directions of the axonometric axes.

The foundation of any accurate scale axonometric construction is an accurate unitary cube with its attending ellipses. By duplicating, moving, and scaling elements from the unitary cube (axes and ellipses), accurate distances and sizes can be determined and accurate geometric features can be represented.

It is helpful to construct overall volumes that enclose the part being illustrated. This provides a way of visually checking whether certain measurements are valid and allows distances to be projected to the edges of the form, further shortening construction time. It is also helpful to recognize the centerline geometry with which features are aligned. By combining boxing-out and centerline construction, an accurate axonometric view is attainable.

An accurate construction can be duplicated, moved, and quickly edited to display visibility. Finally, the construction can be edited and added together to arrive at a final accurate axonometric representation.

Text References

Duff, Jon M. *Industrial Technical Illustration*. Brooks/Cole Engineering. Monterey, California. 1982.

Duff, Jon M. *Technical Illustration with Computer Applications*. Prentice Hall. Englewood Cliffs, New Jersey. 1994.

Gibby, Joseph C. *Technical Illustration*. American Technical Society. Chicago. 1962.

Hoelscher, Randolph P., Springer, Clifford H., and Pohle, Richard F. *Industrial Production Illustration*. McGraw-Hill. New York. 1946.

Nicyper, Raymond. *Graphics Underlay Guides with Reference Data for Scale Drawing.* Graphicraft. Westport, Connecticut. 1973.

Thomas, T.A. *Technical Illustration*. McGraw-Hill. New York. 1978.

Treacy, John. *Production Illustration*. John Wiley and Sons. New York. 1945.

Internet References

Automated Isometric Scaling in CorelDRAW. Retrieved from: *http://www.isocalc.com/isocalc3/index.htm*. November 2002.

Marquis, Philippe. "Graph Paper Printer." This trial version utility prints axonometric grid paper from AxonHelper scale function information. Retrieved from: *http://www.jumbo.com/pod/1999/04/043099.html*. November 2002.

Very informative instructional Web site covering all aspects of axonometric drawing and projection. Retrieved from: *http://www.ul.ie/~rynnet/keanea/HomePage.html*. September 2002.

On the CD-ROM

For those of you who would like to try out the procedures described in this chapter, several resources have been provided on the CD-ROM. Look in *exercises/ch11* for the following files:

crank.pdf   This file contains orthographic views of the crank weldment used throughout this chapter.

55-25-25.ai   This file contains a completed axonometric unit cube. However, you are encouraged to create your own.

Perspective Drawing

Axonometric Projection

Technical Animation

Technical Illustration

Rendering

3D Modeling

Chapter 7

Axonometric Projection

In This Chapter

Axonometric views are created when orthogonal views (top, front, side) are projected onto a plane not parallel to the X, Y, or Z axes. The resulting image is a foreshortened paraline pictorial projection called an *axonometric projection*. Axonometric projection is similar to the projection of principal orthographic views in that view geometry must be at the same scale and properly aligned. Projection is a fundamental method by which axonometric views are created.

Chapter Objectives

In this chapter you will understand how to:

▼ Use axonometric projection to visualize potential views

▼ Identify when an axonometric projection is appropriate

▼ Create an accurate axonometric cube and projection diagram

▼ Use the diagram to set existing scale views and project common features

▼ Locate centers and tangents of circular features

▼ Complete visibility using common extruded projections and editing

The Theory of Axonometric Projection

Chapter 3 reviewed the fundamental relationships in an orthogonal projection system—where adjacent views were mutually perpendicular. This spatial orientation created views that were 90 degrees apart—as in top, front, and side views. It also produced normal (perpendicular) or point (parallel) views of a stable world axis system.

When the plane of projection is not perpendicular to its neighbors (Figure 7.1), it is said to be *axonometric*, and a view taken perpendicular to that plane is called an *axonometric projection view*.

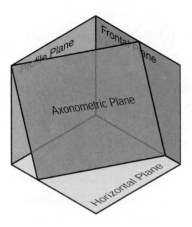

Figure 7.1 Relation of axonometric plane of projection to the principal planes.

As you might guess, there are an infinite number of axonometric views. But in practice, a number of set or *standard views* are used most often because they can be easily matched at a later time. Reserve nonstandard axonometric views for subjects whose geometric features don't match well with the exposure of standard faces.

Some concessions are made in these standard views. In order to arrive at views that have three easily produced ellipses (say, 20-20-40 and not 20-20-41.329) axis scales and inclinations have been adjusted slightly.

When the axonometric plane of projection is equally inclined to the X, Y, and Z axes, the resulting view is referred to as an *isometric projection*. When the axonometric plane of projection is inclined equally to two of the axes, the resulting view is referred to as a *dimetric projection*. When the axonometric plane of projection is unequally inclined to all three axes, the resulting view is referred to as a *trimetric projection*. All axonometric projections produce pictorials where the orthogonal dimensions of height, width, and depth are foreshortened (smaller than true size).

Axonometric projection is reserved for cases where orthographic views already exist in a form that can be easily used in a particular computer graphics program. It is not efficient to draw scale views from either unscaled drawings, or from views at different scales, and then perform projection. However, for developing an understanding of axonometric projection and how it foreshortens axes and features, every technical illustrator should at least understand its methods. When same-scaled orthographic views are available and in a format easily used by Adobe Illustrator, axonometric projection is very efficient.

The Axonometric Projection Diagram

Just as a missing orthographic view can be projected from two adjacent orthographic views in an *orthographic diagram*, an axonometric view can be projected from adjacent orthographic views in an *axonometric diagram*.

Figure 7.2 compares an orthogonal diagram with an axonometric diagram. In Figure 7.2a, the front view is in the plane of the paper and the top and side views are revolved. The top view is revolved about its intersection with the front view (X axis), perpendicular, until it is coplanar. In Figure 7.2b, the axonometric plane (1-2-3) is in the plane of the paper and the top, front, and side views are revolved until a coplanar condition results. The top view is revolved about its intersection with the axonometric plane (1-2) until it is coplanar. The other planes are revolved in a similar fashion.

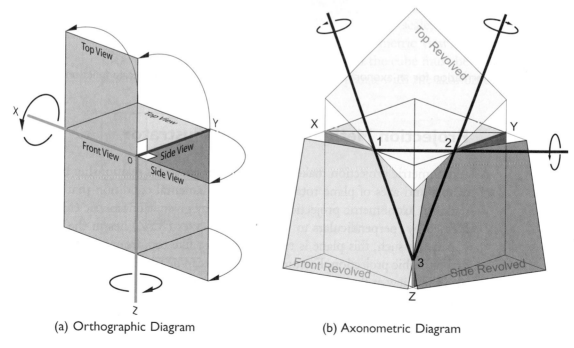

(a) Orthographic Diagram   (b) Axonometric Diagram

Figure 7.2   Orthographic and axonometric views are both created from projection diagrams.

An easy way to locate the midpoint of an angled line is to recognize that the line can be considered the diagonal of a rectangle. The rectangle's corners can be snapped to the ends of the line, and its center becomes the midpoint of the angled line (Figure 7.10).

▼ Choose the **Rectangle Tool** and turn off the fill.

▼ Turn on **Smart Guides** (**CTL-U**) and let the cursor snap to the ends of line 1-2 as you draw the rectangle.

▼ Move the cursor to the midpoint of the line. The center label should pop up.

▼ Create a marker (I use a little perpendicular line) and grab its approximate middle. Move the line to the centerpoint, and when the center label appears, release the line.

▼ Because our example is a dimetric view, **Reflect-Copt** the marker about the vertical and with the **Shift Key** presses, move it to intersect line 1-3.

▼ The midpoint of line 1-2 can be found using the same rectangle method.

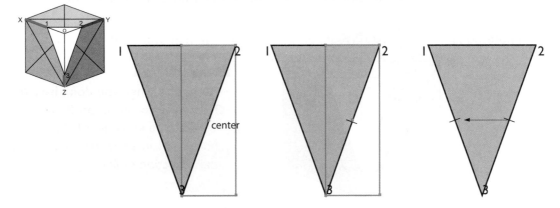

Figure 7.10   Locate the midpoints of the edges of the axonometric planes.

Continue by rotating the ZY plane into the axonometric plane as shown in Figure 7.11.

▼ Create a circle whose center is at the midpoint of line 2-3 and whose circumference passes through points 2 and 3.

▼ Tech Tip

Enlarge or reduce the circle by 0.1 percent and use **Ctrl-D** to repeat the transformation as many times as needed to get the circle's path to pass through the corner of the cube.

▼ Revolve point O along line O-1 until it intersects the circle. Do this by creating a line from this intersection that when extended passes through point 2.

▼ Extend this line until it intersects point Y as projected parallel to O-1. This is at the same angle as the X axis.

▼ Extend the revolved front edge of the side plane through point 3 until it intersects an X axis projection from corner Z.

▼ Complete this plane by copying the front edge to the back and the top edge to the bottom and snapping to the end points.

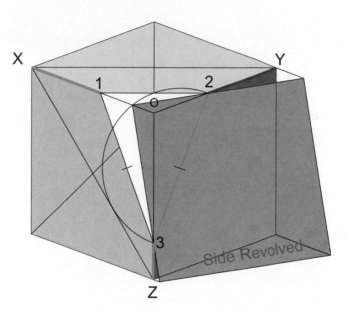

Figure 7.11   The Y-Z plane is revolved into the axonometric plane.

Repeat the preceding steps to revolve the XZ and XY planes as shown in Figures 7.12 and 7.13.

Set the Orthographic Views

An inspection of the revolved planes will show how the orthographic views of Figure 7.5 will be positioned. Line O-3 of the ZY plane pivots at point 3 as it is revolved into the plane of the paper. This line represents the front of the side view. You can then assign meaningful descriptors to the other sides of the view such as *bottom, top,* and *rear*. Align the bottom of the side view with the bottom of the revolved plane. This aligns the front in alignment with the front, the top with the top, and the rear with the rear.

Do this for the front and top views also. The actual position of the view on the drawing surface is *not* critical, but its alignment with the view *is* critical.

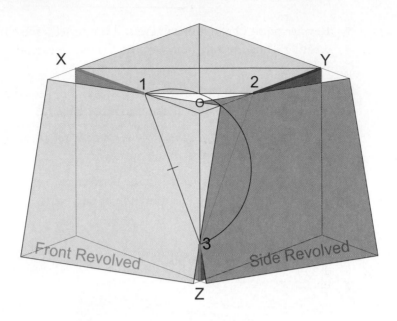

Figure 7.12   XZ plane revolved.

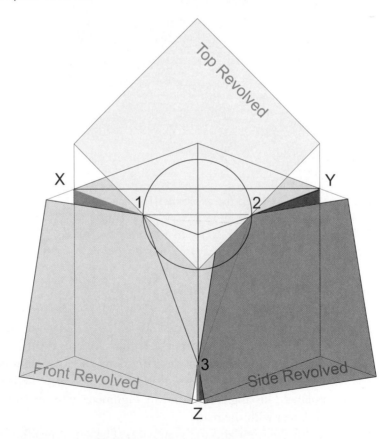

Figure 7.13   XY plane revolved. All three principal planes now lie in the axonometric plane.

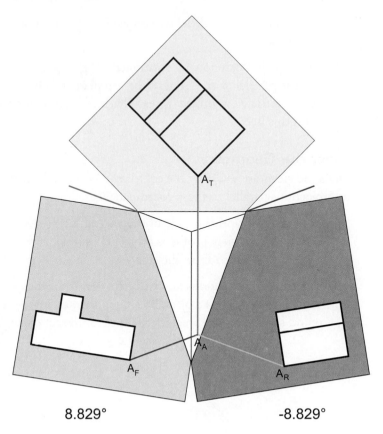

$8.829°$ $-8.829°$

Figure 7.14   Views are aligned with the edges of the rotated planes.

Figure 7.14 shows the three views in correct alignment with the revolved planes. In normal circumstances, all height, width, and depth information will be contained in any two of the three principal views. The third view may be used to verify projections or when non-normal information is displayed (as in angular features.

Set the Views

Begin by placing one of the views, say, the front view.

▼ Align the bottom of the front view to the bottom of the revolved plane. This has been determined to be 8.829 degrees. Once you know this for a given axonometric view, you don't have to actually create the diagram.

▼ Find a common starting point such as the bottom, front, right corner. This has been identified as point A in Figure 7.14.

▼ Project this corner into the axonometric view parallel to the receding (red) axis.

▼ Align the side view to the revolved side plane and position it so that its projection (blue) intersects the first projection conveniently.

▼ The intersection of these two projections locates point A in the axono-metric view.

▼ The top view can now be positioned. Align the top view such that its front is parallel to the front of the top plane revolved. From point A in the axonometric view, project (green) back to the top view and position point A on this line.

Project the Geometry

Once the views are set, copy the coordinate axes and hide the diagram. Make all your projections parallel to these coordinate axes.

▼ Project a descriptive profile of the part (Figure 7.15) if one exists. In this case, the front view profile is projected as it is located on the front of the object as projected from the side view.

▼ Turn on **Smart Guides** and with the **Pen Tool**, click a continuous polyline at each corner of the object. You should get an intersection label at each point, signifying you are, indeed, at the intersection of two projectors.

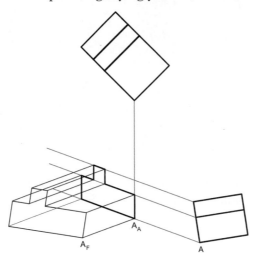

Figure 7.15   Begin construction with a dominant feature, in this case, the object's profile.

▼ Continue the projection of point B from the front view until it intersects the depth as projected from the side view. You can see in Figure 7.16 that this point is verified with a projection from the top view.

▼ Select the axonometric profile. Hold down the **ALT** key and duplicate the profile. Grab point A and move the profile to the intersection of the projectors at point B (Figure 7.17).

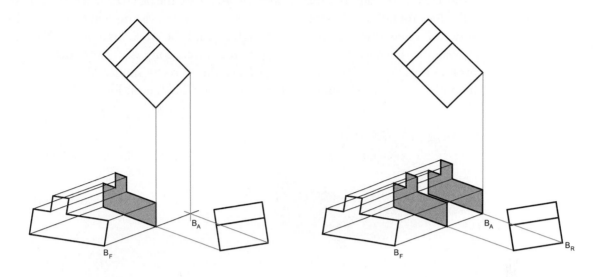

Figure 7.16 Locate the depth of the object.

Figure 7.17 Front profile is duplicated and moved into position.

▼ Tech Tip

Always complete the final axonometric illustration with shapes rather than individual lines. This allows the axonometric planes to be rendered, giving the object even more solidity. If the illustration is comprised of only lines, the planes can't be rendered.

▼ Complete the top and end planes (Figure 7.18). Using the **Pen Tool**, create planes that span the open areas of the object.

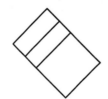

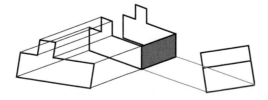

Figure 7.18 Complete the planes.

Illustrator will want to combine planes as you place the pen over existing anchors. If the **Pen Tool** displays a small circle when you use an existing point to create the corner of a new plane, move the pen slightly until the circle disappears. You can always move anchors so that they snap to the correct corners. Finally, move planes front and back (**Object|Arrange**) to achieve the correct visibility. When finished, the projected 20-40-40 dimetric view should look like that in Figure 7.19.

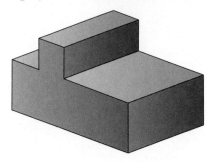

Figure 7.19   The complete object with planes adjusted front-to-back for visibility.

Figure 7.20 shows a more complicated part, this time including cylindrical features. The views for this object can be found in *exercises/ch7* on the CD-ROM. Follow the steps in Figures 7.14 through 7.18 in order to complete an axonometric projection.

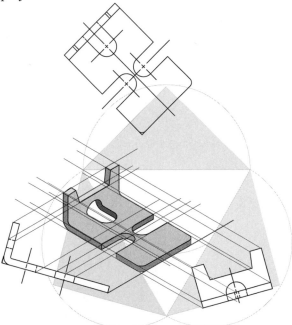

Figure 7.20   An object with cylindrical features constructed using an isometric projection diagram.

Solve Angles in an Axonometric Diagram Using Circle Projection

We can apply our knowledge of the axonometric diagram to locating angular features. Remember, angular features can't be directly measured or constructed in an axonometric diagram when the angular measurements are not parallel to any of the axes.

Circle projection can be used with any axonometric technique: scale construction, projection, or shearing. But because it is an extension of the axonometric diagram, it is presented here.

Figure 7.21 shows the views (not to scale) of an object that requires angular construction. Two angles and two important dimensions are marked. The rest of the construction is assumed. You could, of course, construct scale views and project the angular features as you did in Figure 7.14. But there is a more direct method: *circle projection*.

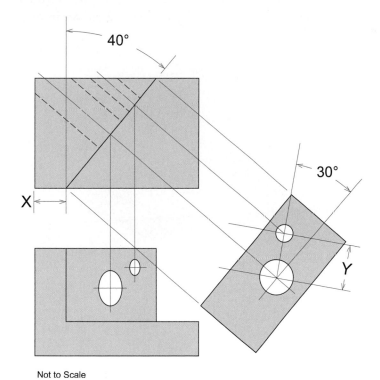

Not to Scale

Figure 7.21   An object with angular features appropriate for circle projection.

▼ Create the basic form (without angular features) in an axonometric view, this time using direct scale construction (Figure 7.22). In this case, we are using a 15-55-30 trimetric view.

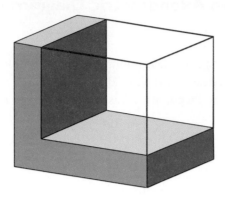

Figure 7.22   Basic object is constructed using axonometric scales and parallelism.

▼ Fit an ellipse that is the same exposure as the top plane (15 degrees from AxonHelper, any convenient size) to the top of the enclosing form (Figure 7.23). Place its center at the start of the 40-degree angle. The baseline for this required angular measurement is shown in red.

Any Convenient Size   Baseline for Angle

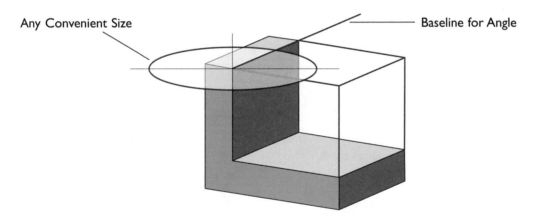

Figure 7.23   Axonometric ellipse on top of object.

▼ Project a true circle from the major axis of this ellipse (Figure 7.24). This is like the top of the axonometric diagram in Figure 7.13.

▼ Circle Projection

Circle projection is like creating a temporary orthographic (circular) view of the ellipse for the purpose of measuring true angles. Because we know that an ellipse rotates about its major axis, and the major axis remains equal to the circle's diameter, any angle measured on the circle and projected back to the ellipse is the axonometric projection of the angle.

Note: You must establish a starting point on the circle by projecting a point on the ellipse to the circle in order to begin the angular measurement.

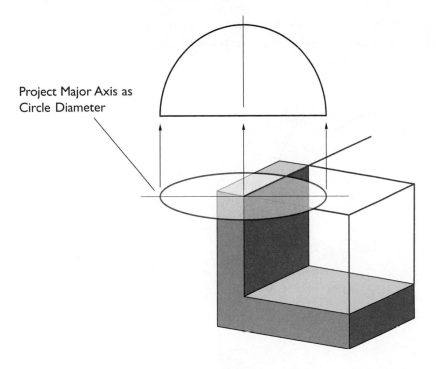

Project Major Axis as Circle Diameter

Figure 7.24   True circle projected from ellipse.

▼   Project a starting point from the centerline of the ellipse to the circle and create a radius at this point. The angle will be measured from this radius (Figure 7.25).

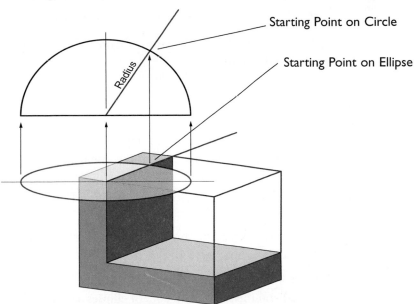

Starting Point on Circle

Starting Point on Ellipse

Radius

Figure 7.25   Starting point projected from the baseline.

▼ Rotate a radius minus 40 degrees from this starting point and project it back to the ellipse (Figure 7.26). This is the 40-degree measurement in the axonometric view.

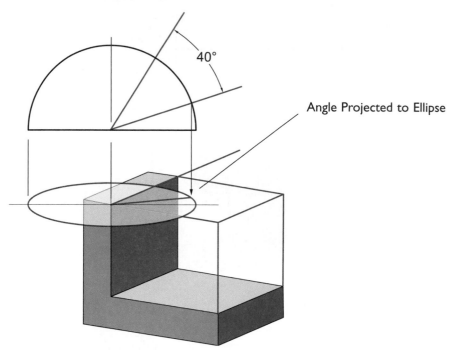

Figure 7.26   Radius rotated 40 degrees and projected back to the ellipse.

▼ Extend the axonometric radius to the edge of the box (Figure 7.27).

▼ Complete the construction (Figure 7.28) by duplicating and moving parallel elements.

▼ Tech Tip

You may want to make use of Illustrator's guides at this point. Rather than drawing lines and then creating a plane over the lines, drag ruler guides out with **Smart Guides** turned on. Then, when you use the **Pen Tool** to make the final shapes, you can snap to the intersections of the guides. When finished, choose **View|Guides|Clear Guides** to get rid of them.

In order to solve the angular problem on the 40-degree face, you need to determine the ellipse exposure on that face. Here's an easy way to do that:

▼ You know the axonometric is a 15-55-30. This means that the right face is 30 and the left face is 55. All ellipses as they rotate about the vertical axis follow concentric circular paths. The desired ellipse exposure is between 30 and 55 degrees.

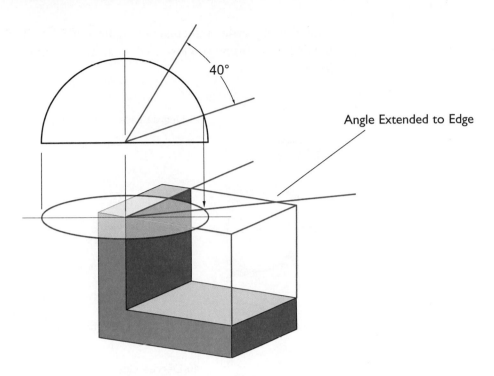

Figure 7.27   Radius at 40 degrees is extended to the edge of the box.

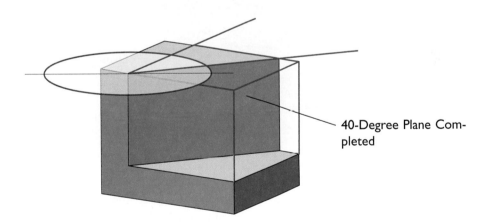

Figure 7.28   Angular feature is completed.

▼ You also know that each of these ellipses is contained in a unitary enclosing form. Figure 7.29 shows these two ellipses and the 15-degree arcs that control the height and width of the forms.

▼ The ellipse on the 40-degree face is between these two. Use the arcs and the 40-degree centerline to find the desired enclosing form (Figure 7.29). Place a trial ellipse in the middle of the form, rotate, and nonuniformly scale until tangent (Figure 7.30). You don't need to know the degree exposure; you just need an accurate ellipse that fits the unitary form.

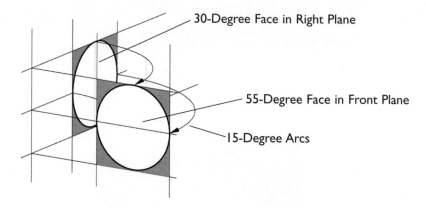

30-Degree Face in Right Plane

55-Degree Face in Front Plane

15-Degree Arcs

Figure 7.29   Concentric nature of ellipses as they rotate about a common axis.

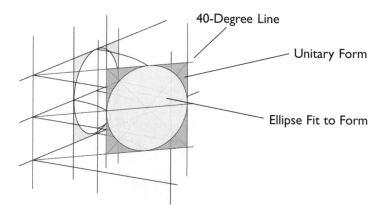

40-Degree Line

Unitary Form

Ellipse Fit to Form

Figure 7.30   Unitary form and ellipse at 40-degree position.

▼ Locate the center of the 40-degree angular plane by constructing diagonals. Place a copy of the ellipse from Figure 7.30 in the center and enlarge appropriately (Figure 7.31). Project the major axis of this ellipse perpendicular to form a circle projection.

▼ Determine a starting point on the axonometric ellipse and project that point to the circle. Angular measurements begin from this point. Construct the two diameters and distances in full scale in the circle projection (Figure 7.32).

▼ Project the true angles, true distances, and true diameters back on the axonometric ellipses (Figure 7.33). Copy and scale the 40-degree ellipses to fit the projected diameters.

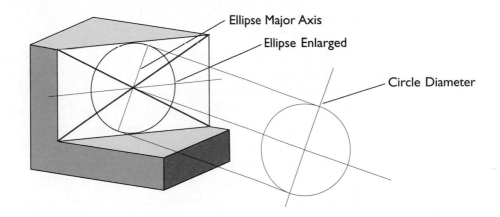

Figure 7.31   Ellipse from Figure 7.30 transferred to the 40-degree face.

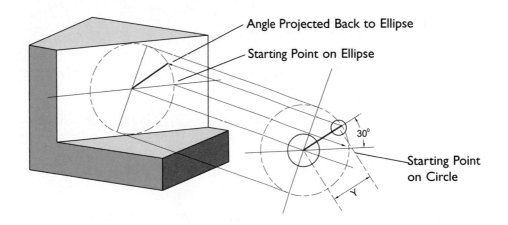

Figure 7.32   Starting point found and orthographic information constructed.

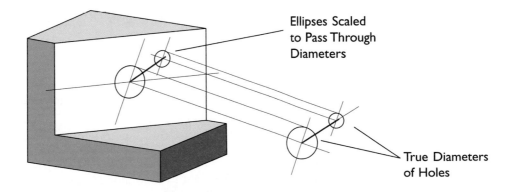

Figure 7.33   Orthographic information projected back onto the 40-degree plane.

Review

Axonometric projection using a PostScript drawing tool is appropriate when scale orthographic views are readily available. Axonometric projection creates correctly scaled pictorial representations; these projections can be integrated among 2D constructions and 3D modeling with accurate results. In 2D construction, axonometric projections involve developing an axonometric diagram based on an accurate axonometric cube. This diagram is used to set scale views and project common features. Projections begin as extruded shapes from principal views that when combined form accurate pictorials.

An application of the axonometric diagram is circle projection. Circle projection uses the major axis of an axonometric ellipse to project a circle that can be used for orthographic construction. True measurements, angles, and ellipse diameters can be projected back to the axonometric ellipse with correct foreshortening.

Text Resources

Duff, Jon M. *Industrial Technical Illustration*. Brooks/Cole Engineering. Monterey, California. 1982.

Duff, Jon M. *Technical Illustration with Computer Applications*. Prentice Hall. Englewood Cliffs, New Jersey. 1994.

Gibby, Joseph C. *Technical Illustration*. American Technical Society. Chicago. 1962.

Hoelscher, Randolph P, Springer, Clifford H., and Pohle, Richard F. *Industrial Production Illustration*. McGraw-Hill. New York. 1946.

Thomas, T.A. *Technical Illustration*. McGraw-Hill. New York. 1978.

Treacy, John. *Production Illustration*. John Wiley and Sons. New York. 1945.

On the CD-ROM

Several resources can be found in *exercises/ch7* for additional practice:

projects	Use the files in this directory for additional projection assignments.
t_block.ai	Use this file to follow along with the instruction covered in Figures 7.6-7.19.
trans_plate.ai	Open this file for scaled views of the object featured in Figure 7.20.

Chapter 8

Axonometric Shearing

In This Chapter

Chapters 6 and 7 covered what have been, until the advent of universal computer graphics, the two main methods for generating axonometric views—scale construction and projection. In this chapter you will be introduced to *axonometric shearing*, a method that automates the transformation of orthographic views into correctly foreshortened axonometric projections. Shearing takes orthographic views and by a combination of scaling and rotation, produces flat axonometric projections that can be copied and edited into standard axonometric views.

This method requires that you have same-scale views. If you don't, it may be more efficient to use scale construction techniques or to model the geometry in three dimensions. Shearing produces such immediate results that it may be used to sketch out the basic layout of an axonometric view. Scale construction can then be used to finish the illustration.

Chapter Objectives

In this chapter you will understand how to:

▼ Use axonometric shearing to shorten construction time when scaled orthographic views are available

▼ Construct scaled views in Illustrator from unscaled dimensioned orthographic views

▼ Use AxonHelper shearing information to create correctly sheared views

▼   Copy and extrude geometry along axonometric axes

▼   Edit the construction for visibility

▼   Use the sheared construction as the basis for line and tone illustration

The Theory of Axonometric Shearing

If you have orthographic views, at the same scale and in a form your program can open or place, axonometric shearing is the immediate choice. Even when orthographic views are not available (as with paper engineering or architectural drawings), it may be more efficient to draw just enough of the layout to provide correctly projected centerlines, diameters, and basic dimensions, and then shear this into axonometric position. You may choose to draw freehand directly over this *blockout*, knowing that angles and distances are correct.

Because projection causes features parallel to the front, top, and side planes to be both skewed (leaned over) and scaled, using the skew or shear tools available in most vector illustration programs does not produce correctly scaled projections. In fact, only isometric drawings (sheared but not scaled) can be easily produced using this standard tool, and indeed this simplistic approach is documented in many books covering Illustrator, FreeHand, and CorelDraw. AxonHelper provides information that lets you create hundreds of correctly scaled dimetric and trimetric views through a process of rotation and scaling. We call this combination of rotation and scaling *axonometric shearing*.

Shearing is based on how a circle is correctly scaled and rotated so that it is exposed and tilted to correctly fit in its axonometric face. Isometric projection best illustrates this, but the application is the same in dimetric and trimetric projection.

Figure 8.1 displays the top, front, and right side views of a cube having a circle centered on each face. To keep track of the views the view name is also included on each face. The relevant information for isometric shearing is that:

▼   Each face is scaled identically.

▼   The minor axis of an isometric ellipse is 57.7 percent of its major axis (from AxonHelper).

▼   The receding axes are set at 30 and 150 degrees to the horizontal (from AxonHelper).

▼   The major axes of ellipses must be perpendicular to these receding axes. This means that one ellipse is rotated 60 degrees (90 minus the axis angle) and one is rotated at negative 60 degrees (90 minus the axis angle but on the other side of vertical). Axis angles are also determined from AxonHelper.

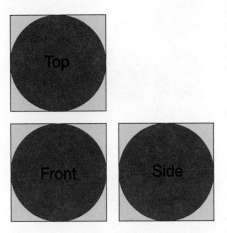

Figure 8.1   Orthographic views ready for shearing.

Use AxonHelper (Figure 8.2) to determine the correct scaling and rotation data to produce the desired view. Be advised that the rotation of the top view (either positive or negative) determines the positive or negative rotation of the other two views.

▼   Make a copy of each view before shearing.

▼   Begin with the top view. Rotate and then scale the top view until it is sheared into the correct projection. In this case the top view is rotated negative 45 degrees and scaled vertically 57.73 percent.

▼   Rotate, scale, and then rotate the front view until it is sheared into correct projection. In this case the view is rotated 45 degrees, scaled vertically 57.73 percent and then rotated negative 60 degrees.

▼   Rotate, scale, and then rotate the side view until it is sheared into correct projection. In this case the view is rotated negative 45 degrees, scaled vertically 57.73 percent, and then rotated 60 degrees.

▼   Align the planes by their common front vertex, and you have a correctly scaled isometric projection.

▼ Note

In isometric shearing the horizontal plane is rotated and then scaled. The vertical planes are rotated, scaled, and then rotated. These operations must be done in the assigned order.

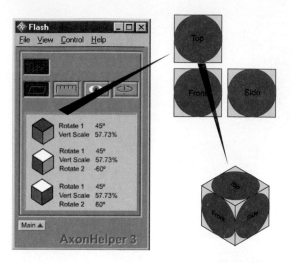

Figure 8.2   The shearing operation applied to an isometric view.

You can see from Figure 8.2 that the views were correctly sheared into their axonometric (in this case, isometric) faces. Rotating the top view negative 45 degrees necessitated rotating the right side view negative 60 degrees. Any of the standard dimetric or trimetric views contained in AxonHelper can be produced by this method.

▼ **Note**

You can see from Figure 8.2 that not only was the geometry correctly sheared into axonometric position, but that the text was also. Many times an illustration's success hinges on the accuracy of text in axonometric projection. With shearing, you don't have to rely on text in the plane of the paper; you can put text right on the object!

Geometry That Can Be Sheared

Many times you may find that correct geometric information is available, not in the form of *scaled views*, but in sketches, drawings, or tables. To make use of shearing, you must have enough correctly scaled geometry to make shearing efficient.

With experience, you will be able to anticipate which features will be hidden in a particular standard view. These features can simply be omitted from the scale views used for shearing. However, if you find out later that you are missing something, you'll have to go back to the original orthographic view, add the detail, and shear again. For this reason, always make a copy of the view before shearing.

Figure 8.3 shows the top and front views of an actuator handle with associated dimensions. The drawing is not marked drawn to scale nor is there a scale noted.

You can quickly determine if the drawing is, in fact, drawn to scale by following these steps:

▼ Locate a dimension that is convenient and possibly an integer value, such as the overall handle depth of 3.

▼ Develop a small scale that subdivides this into three equal divisions and one subsequently into quarters and eighths.

▼ Check a sample of dimensions with this scale to determine if the views are drawn to scale.

If the views are in fact drawn to scale, the drawing can be scanned and used as a template to quickly guide the drawing of a sufficient amount of the information to make shearing efficient. This is covered in detail in Chapter 11.

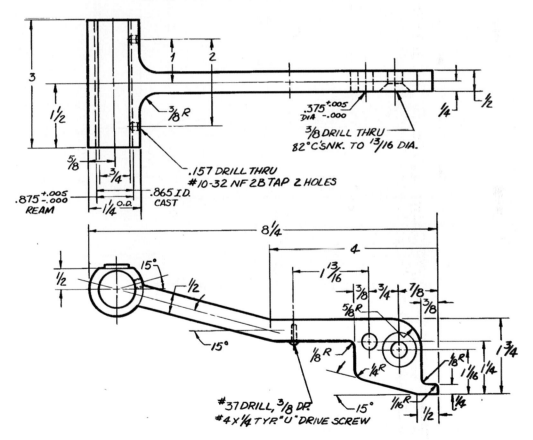

Figure 8.3   Dimensioned orthographic views not drawn to scale.

In the case of Figure 8.3, the views were not drawn to scale. The dimensional information is used to create scaled views for shearing. Figure 8.4 shows the result of creating scaled views in Illustrator. Note that the views are slightly abstracted—not every detail has been drawn. For example, most of the hidden lines have been omitted. The hidden detail of the countersink is constructed to give the necessary depth information. You may wish to review Figures 3.14 through 3.19 where the orthographic views of the actuator handle were drawn in Adobe Illustrator.

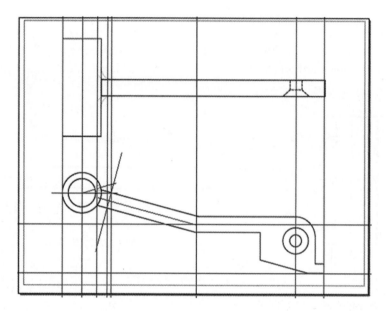

Figure 8.4   Scale views in Illustrator ready for shearing.

Pick a Standard View for Shearing

Sheared standard views are identical to projected standard views. This means that some of an illustration can be produced by one method and some by another, depending on what type of data you have to start with. To demonstrate the flexibility of the shearing method, a 50-15-35 trimetric view is chosen with an orientation that puts the length of the handle in the 35-degree ellipse face (Figure 8.5). The data for shearing is unique to this view and is different for each of the three faces because of its trimetric orientation.

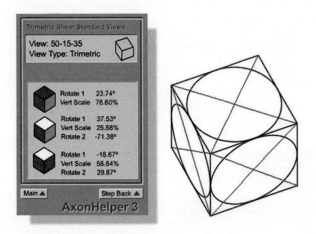

Figure 8.5   50-15-35 trimetric shearing data produces the standard view shown to the right of AxonHelper.

▼ Illustrator Tip

Illustrator's **Actions Pallette** is the place to automate the shearing method. Choose **New Set** from the **Actions Menu.** You can see from the example on the right that three actions with descriptive names have been created. For each action choose **Start Recording** and perform the transformations from AxonHelper. Choose **Stop Recording** when you are finished with the face.

You'll want to save the action in a location on your hard disk that's easy to find. Create a folder named "Actions" in *c:/programs files/adobe/ illustrator.* That way the action will be available whenever you are in illustrator.

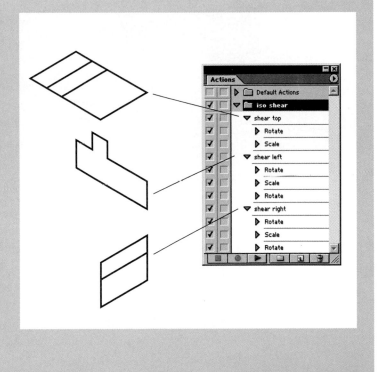

Shear the Views

The front view requires rotation-scaling-rotation as shown in Figure 8.6. Perform these operations in the same order as presented in AxonHelper.

▼ Rotate the front view negative 18.67 degrees, scale the vertical 58.84 percent, and then rotate 29.87 degrees. The completed sheared view is shown
superimposed over the trimetric cube so that you can see that it is in the correct orientation.

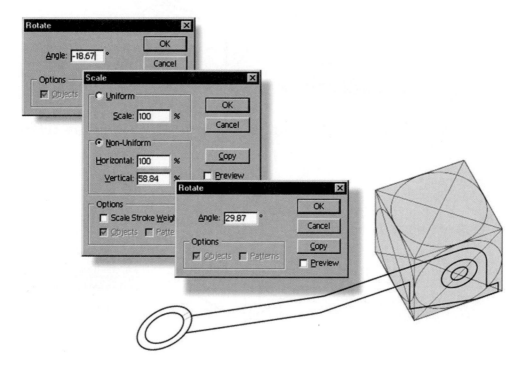

Figure 8.6   The front view is sheared into correct orientation by rotation-scaling-rotation using information from AxonHelper.

▼ The top view requires rotation and scaling as shown in Figure 8.7. Rotate at an angle of 23.74 degrees and then scale the vertical to 76.6 percent. The view, when superimposed over the trimetric cube, again demonstrates the accuracy of the shearing method.

Extrude the Geometry

Getting the front and top views sheared into correct position is just the start of completing an axonometric view by this method. In fact, shearing is the easiest part. Once the view is sheared, you must use your visualization skills to correctly copy and position the views to actually build the geometry. This procedure is quite similar to the CAD method of *extrusion,* and you may find that CAD skills spill over to illustration skills as you manually move profiles along axes.

Figure 8.8 displays the sheared front profile extruded along the depth of the top view. In other words, the top view determines where copies of the front view are placed along the depth axis.

In many cases this can be done either by eye or by using the object snaps available in Illustrator. For more accuracy, set Illustrator's constraint to the angle of the receding left axis. This value is found in AxonHelper's *scale function* for the 50-15-35 trimetric view. By setting the constraint to this value, front profiles can be **Copy-Moved** along the depth axis by holding down the **Alt-Shift** keys. Figure 8.9 shows where this number is found in AxonHelper and entered in Illustrator's constraint preference.

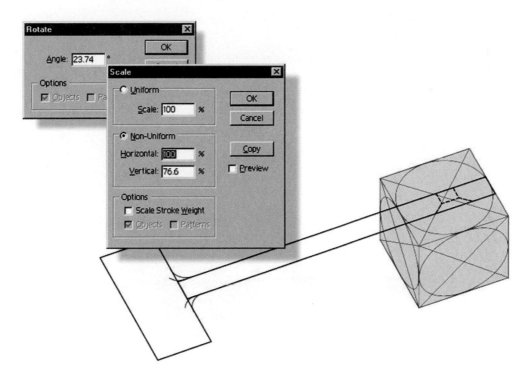

Figure 8.7   Top view is sheared into correct orientation.

▼ Note

The constraint angle impacts operations in Illustrator even when the **Shift Key** isn't held down. Unfortunately, there is no visual indicator on the Illustrator interface that alerts you to the fact that the constraint is toggled on, or, what the active constraint is. If rotations or scaling appear to be performing oddly, choose **Edit|Preferences|General** and make sure no constraint angle is set.

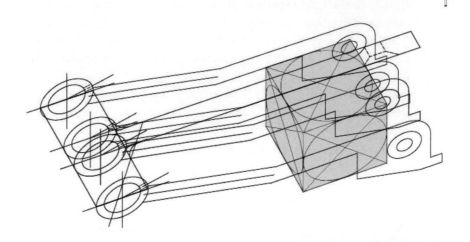

Figure 8.8   Front view shear is duplicated along receding axis.

Because AxonHelper reports acute axis angles, you can enter a simple equation (180–60.16) for Illustrator to set the constraint at positive 119.84 degrees.

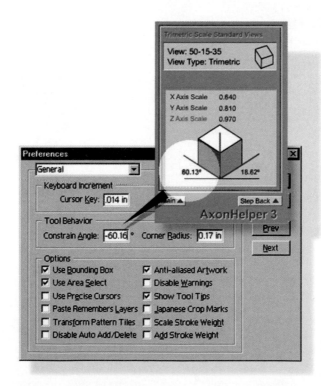

Figure 8.9   Constrain the extrusion of the front profile geometry by using Illustrator's constraint preference.

Edit the Geometry

Once height and depth geometry has been extruded to correct depth positions (Figure 8.8), edit the geometry for visibility. It is also helpful to have copies of the unedited profiles available on hidden layers so if you make a mistake that is too laborious to **Undo**, you always have original sheared profiles to fall back on.

▼ Tech Tip

Copies are a beautiful thing! Always copy critical geometry to another layer before you start to edit. Hide that storage layer, knowing that if you do something terrible, you can retrieve the sheared views.

Figure 8.10 displays the result of correctly editing the sheared geometry for visibility. This line work can be rendered using line weights (Chapter 10) or by using PostScript gradient fills (Chapter 13). Just to give you an idea of how axonometric shearing is used in a finished illustration, Figure 8.11 shows the result of applying gradient fills to render the geometric shape of the object.

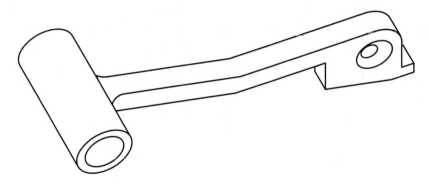

Figure 8.10   Sheared profiles are edited for visibility after extrusion.

A preliminary note on the eventual task of rendering. PostScript drawing tools perform two operations: they stroke lines and fill areas. You can create effective line illustrations (like the actuator handle in Figure 8.10) by making sure that the lines all meet at their anchors and are assigned an appropriate weight for the final display size. But to create photorealistic illustrations like that in Figure 8.11, these lines need to be joined into contiguous shapes that can be filled with various PostScript routines. So as you shear and trim, its best to plan ahead, knowing that the lines you create now will become the shapes you will fill later.

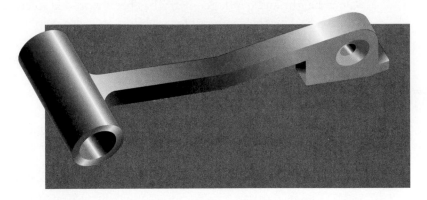

Figure 8.11   PostScript gradient fills are applied to areas to better represent the form of the object.

Make Planes Renderable

After editing the sheared profile, you usually end up with various pieces and parts of lines, curves, and planes. In order for solid and gradient fills to be applied in ways that result in realistic renderings the plane elements must be joined into single contiguous outlines. The realism achieved in Figure 8.11 is the result of carefully joining these discontinuous elements into continuous outlines.

Observe the cylindrical end of the handle in Figure 8.12. What appears as a single outline in Figure 8.12a is really a collection of pieces that though visually aligned, are not joined (Figure 8.12b).

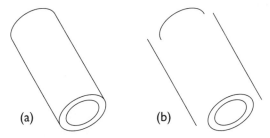

(a)   (b)

Figure 8.12   After editing, sheared outlines are a group of discontinuous lines.

▼   Begin by working on a copy of the geometry.

▼   Cut the front ellipse at the tangent limiting elements (Figure 8.13). Use **Smart Guides** to find the intersection point and the **Scissors Tool**.

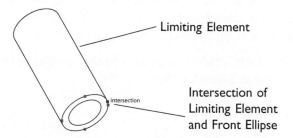

Limiting Element

Intersection of
Limiting Element
and Front Ellipse

intersection

Figure 8.13   Front ellipse is cut to provide the curve for the shape.

▼   Copy the rear curve, tangents, and top of the front ellipse (Figure 8.14).
Although the anchors *appear* to be aligned, use the **Open Selection Tool**
to move one anchor off position and then snap it back into place.

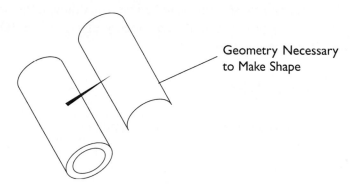

Geometry Necessary
to Make Shape

Figure 8.14   Copy the geometry necessary to create the shape.

▼   With the anchors perfectly aligned, join the curves. Use the **Open Selection Tool** to draw a marquee selection window around two coincidental
anchors. Make sure you only select one set of anchors (Figure 8.15).

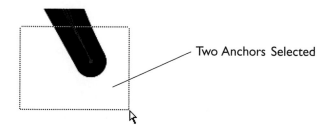

Two Anchors Selected

Figure 8.15   Select two coincidental anchors with the **Open Selection Tool**.

▼   Choose **Object|Path|Join** and select either the **Corner** or **Smooth Point**
option depending on the shape of the curve. In our case, we have all
corners.

▼ Work around the shape, selecting anchors with the **Open Selection Tool**, until all coincidental anchors are joined. This contiguous path can be filled with a gradient fill to produce the results in Figure 8.11.

▼ The filled path is then moved back to the original geometry by an anchor (use the **Filled Selection Tool**).

Review

Axonometric shearing automates the process of axonometric projection when correctly scaled orthographic views are available. Because the shearing procedure is so effective, it may be profitable to spend time to construct 2D orthographic views (or partial views) from not-to-scale information rather than use other methods—projection, scale construction, or 3D modeling. The AxonHelper utility provides useful information for shearing standard views. Although the orthographic views appear correctly projected onto axonometric planes, these 2D projections must be extruded to create pictorial depth.

On the CD-ROM

For those of you who would like to try out the procedures described in this chapter several resources have been provided on the CD-ROM. Look in *exercises/ch8* for the following files:

cube.ai	Use this file to practice isometric, dimetric, and trimetric shearing.
handle.ai	This file contains scale views of the actuator handle you can use while following the instructions in this chapter.
shear1.ai through shear5.ai	Use these files for additional shearing assignments.

Chapter 9

Perspective Techniques

In This Chapter

This book would not be complete without a discussion of perspective, even though many technical illustrators may go their entire careers without ever making a perspective drawing. Chapter 11, makes use of many of the topics in this chapter. Because cameras (other than extremely long focal lengths) introduce perspective into a photograph, almost all photo tracings will exhibit the convergence of perspective.

Additionally, as illustrations made from 3D computer models become more prevalent, the difficulty of creating accurate perspectives is minimized. Architects often relied on axonometric (or worse, oblique) projection for large buildings because accurate perspectives were too difficult to produce. They now choose axonometric or perspective views of their modeled work depending on their needs, not based on the difficulty of the projection method.

PostScript applications such as Macromedia Freehand or CorelDraw may have automated perspective tools; 3D modelers make perspective a simple viewing choice. Adobe Illustrator, unfortunately, is devoid of automated perspective tools: perspective views must be constructed.

In any of these cases, a solid understanding of perspective theory and practice will make you a more productive illustrator. But absent automated tools that easily produce perspective views from orthogonal views or numerical data, accurate perspective views can be achieved only by perspective projections and constructions. But as you will see, perspective views aren't all that different from orthographic or axonometric views.

Chapter Objectives

In this chapter you will understand how to:

▼ Choose between axonometric and perspective views for individual illustrations

▼ Become familiar with perspective terminology

▼ Create a flexible perspective template in Adobe Illustrator

▼ Accurately construct 2D perspectives from scale views using the visual ray method

▼ Create accurate perspectives from numerical data using measuring points

▼ Create and use accurate perspective grids

▼ Set up perspective cameras in a modeler such as 3ds max

Examples of Perspective Solution

It may be helpful to see two examples of when perspective is warranted. Figure 9.1 shows a preliminary sketch of a scrubber system to be installed on a power plant. Because it is an architectural subject, our eyes would naturally expect perspective convergence. Without perspective, the illustration would appear to be a small model.

Another reason to use perspective is when the various detail of a subject can't be shown by a single axonometric view. The hydraulic roof support in Figure 9.2 presents a unique illustration problem: how to show the features under the canopy and on top of the support feet at the same time. Any axonometric view would limit the viewing direction to either from above or below, not both simultaneously. You can see in Figure 9.2 that by placing the upper portion of the support above the horizon line, that both the underside of the canopy and the top of the feet can be illustrated.

Perspective and Axonometric Views

A perspective view assumes a known, discrete viewing position, while an axonometric view assumes a viewing position set at infinity. This produces a most fundamental difference between the two types of drawing: in an axonometric view, parallel lines (edges, intersections, and limits) remain parallel and at the same scale; in a perspective view, these parallel features visually converge at a *vanishing point*, a point at infinity, and cannot be directly scaled.

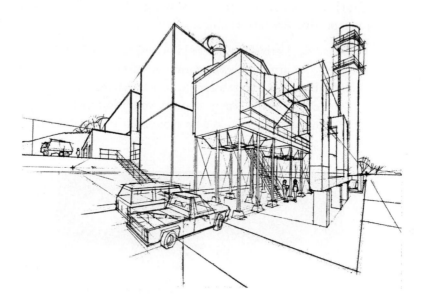

Figure 9.1   Perspective is effective in illustrating architectural subjects.

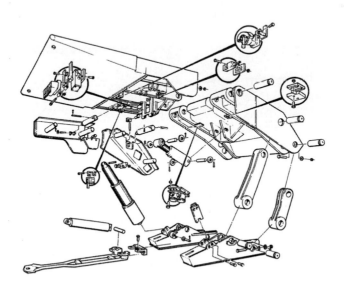

Figure 9.2   Perspective allows multiple viewing directions.

What this means in technical illustration goes beyond the obvious. In fact, experienced illustrators can visually approximate accurate perspective views without laborious constructions. Knowing the differences in axonometric and perspective views, a technical illustrator should keep in mind that:

▼ Perspective views should be reserved for objects that you can't hold in your arms. This is a common mistake made by inexperienced illustrators. Figure 9.3 shows perspective and axonometric views of a small electrical component. The perspective rendition makes the component look like a large storage tank. The axonometric view gives the impression of a small part, viewed up close.

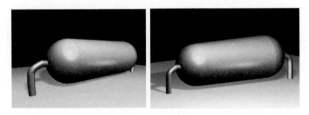

Perspective View   Axonometric View

Figure 9.3   The addition of convergence to small objects gives a false impression of great size.

▼ Likewise, large objects, when drawn in axonometric view, may look like small toys (Figure 9.4). Adding perspective to large objects often produces a better sense of scale.

Axonometric View   Perspective View

Figure 9.4   The absence of convergence makes large objects appear smaller.

▼ Many dimensions can be directly compared in an axonometric view; parallel perspective measurements cannot, in most cases, be directly compared (Figure 9.5). In an isometric view, distances parallel to any axis can be compared; in a dimetric view distances parallel to the two equal axes can be compared; in a trimetric view, distances parallel to the same axis can be compared. In a perspective view, distances can't be directly compared at all (other than in a few special situations).

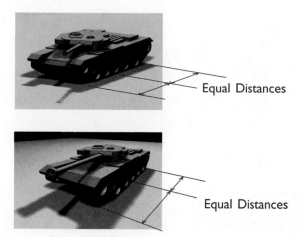

Equal Distances

Equal Distances

Figure 9.5   Many dimensions in an axonometric view are comparable; parallel dimensions in a perspective view generally are not.

▼ Identical ellipses in parallel axonometric planes remain the same exposure, tilt, and scale wherever they are drawn (see Chapter 5). Circular features in parallel perspective planes will vary in size, shape, and tilt (Figure 9.6).

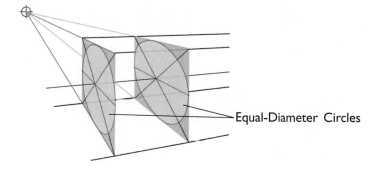

Equal-Diameter Circles

Figure 9.6   Circular features in parallel perspective planes vary in shape, size, and tilt.

▼ Identical ellipses in parallel perspective planes aren't true ellipses at all—although we will still call them ellipses and try to use ellipses whenever possible. The shape enclosing an axonometric ellipse is symmetrical, while in a perspective view (Figure 9.7) the enclosing shape is not symmetrical. For ease of construction, a true or slightly edited ellipse *may* approximate a perspective circle. But as you can see in Figure 9.7, an axonometric ellipse is a poor approximation of a perspective circle.

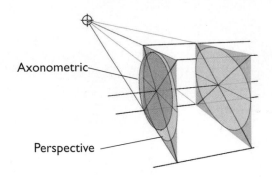

Axonometric

Perspective

Figure 9.7   A true ellipse may be used to roughly approximate a perspective circle.

▼   In a perspective view, distortion increases as geometry digresses from a point (the *center of vision* or CV) directly in front of the observer (Figure 9.8). This is analogous to decreasing the focal length of a camera or widening the field of view. A camera with a short focal length (say, 20 mm) and a wide field of view (say, 80 degrees) increases distortion as objects move out from the center of vision. In an axonometric view, there is no comparable distortion.

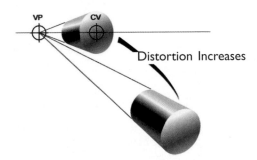

Distortion Increases

Figure 9.8   Distortion increases from a point directly in front of the observer.

Understanding Perspective

Figure 9.9 establishes the relationships between the elements of a perspective drawing. Study the terms below in their positions shown in Figure 9.9.

▼   **Center of Vision (CV)**. In the plan view, the intersection of the central or shortest possible visual ray with the picture plane; also, the vanishing point for horizontal elements perpendicular to the picture plane. In the perspective view, the center of vision represents the position of the observer's eyes on the horizon line.

- ▼ **Cone of Vision**. The conical volume (filled with visual rays) describing the field of view and acceptable distortion. Traditionally, a cone of 30 degrees (horizontal and vertical) results in acceptable distortion.

- ▼ **Ground Line (GL)**. The position of the ground relative to the observer's eyes. If an average observer is standing on the ground, the ground line will be approximately 5 $\frac{1}{2}$ feet below the horizon line. When the ground line is *below* the horizon line you get a view from above. When the ground line is *above* the horizon line you get a view from below.

- ▼ **Horizon Line (HL)**. A line described by two or more vanishing points of lines lying in the same or parallel planes. In Figure 9.9, there is a right vanishing point (RVP) and a horizon line for horizontal elements.

▼ Note

There are many horizon lines in a perspective drawing, only one of which is the *horizontal horizon line*. The parallel inclined eaves of a house would have their own horizon line—in this case an *inclined horizon line*. But for ease of discussion, the horizontal horizon line is referred to simply as the *horizon line*.

- ▼ **Picture Plane (PP)**. The plane on which the perspective is projected. In the top view, the picture plane appears as an edge. Visual rays are projected from the station point to each feature in the plan view. The intersection of these visual rays and the picture plane forms the projection of width and depth in the perspective view.

- ▼ **Plan View**. The top view of the object (width and depth) set at a desired angle to the edge view of the picture plane. This angle is a function of how much, or little, of the object's various faces are desired.

- ▼ **Profile View**. The side view of the object set in the desired relationship to the ground line. The profile view establishes height in the perspective view.

- ▼ **Station Point (SP)**. In the top or plan view, the point at which all projections (visual rays) converge. In the perspective view, the center of vision.

- ▼ **Vanishing Lines**. True heights projected to their common vanishing point. The intersection of these true-height vanishing lines and the projection of visual ray–picture plane intersections, form the image in the perspective view.

- ▼ **Vanishing Point (VP)**. The apparent point of convergence of parallel elements.

▼ **Vertical Measuring Line**. A line along which true heights can be projected from the profile view. In Figure 9.9, this height line is where the object touches the picture plane in the plan view.

▼ **Visual Rays**. Projectors (eyeball sights) that converge at the station point, forming the perspective drawing by their intersections with the picture plane.

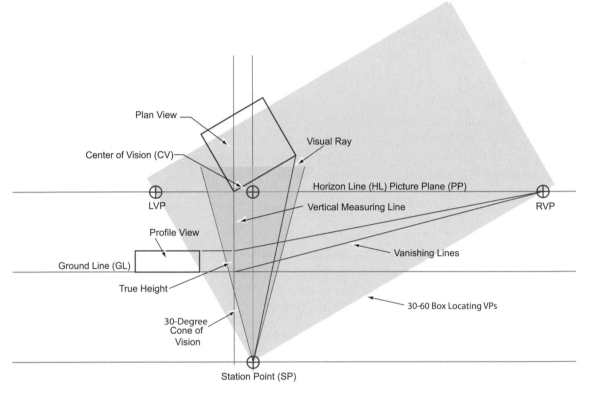

Figure 9.9   Elements of a perspective drawing.

▼ Note

Geometry can be parallel and perpendicular to the picture plane (one-point), parallel and angular to the picture plane (two-point), or oblique to the picture plane (three-point). However, the number of vanishing points used is strictly a matter of convenience. You'll find no mention of one-point, two-point, or three-point perspective in this chapter. The reason for this is that the number of vanishing points used to create a perspective is totally irrelevant (as long as you have one) and a reference to one-point, two-point, or three-point perspective has been, and continues to be, incorrect and unnecessary.

Comparing Axonometric and Perspective Views

An axonometric view may be considered an orthographic projection where the views have been tipped and rotated or where a nonprincipal viewing direction is assumed. A perspective view then becomes simply an orthographic projection with the viewer moved from infinity to a known position.

Figure 9.10 shows this orthographic-perspective relationship. In Figure 9.10a, points in a traditional top and profile view are projected from infinity—parallel to each other—to assemble an orthogonal front view. The projection (where the projectors cross the picture plane) is shown in red on the edge of the top and side views. These red edges are actually the frontal plane in the top and side views.

In in Figure 9.10b, the projectors from the top and side views are no longer parallel but converge at the *station point*. This station point is the same distance from the frontal plane in both the top and side views. The projection of these intersections of visual rays with the picture plane (both in the top and profile views, shown as red edges) forms the perspective. So you can see that perspective projection is very similar to orthographic projection. When the object is aligned parallel to the PP, we call this a *parallel perspective*.

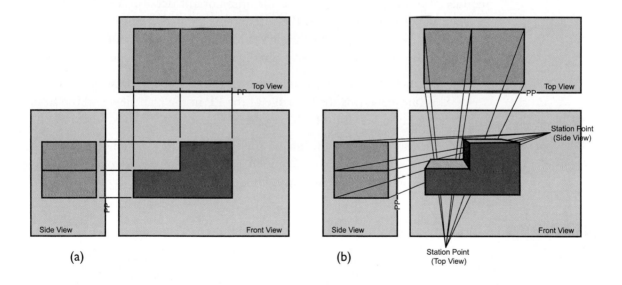

(a)　　　　　　　(b)

Figure 9.10　　(a) An orthographic view is created by projecting adjacent views from infinity. (b) Projecting adjacent views from a known position forms a perspective view.

Setting Up a Perspective View

Just as there are an infinite number of axonometric views, there are an infinite number of perspective views. Still, there are several standard perspectives that can be used over and over.

▼ Note

Deciding on the *best* perspective view in 2D illustration is much more critical than in 3D illustration. The last thing you want is a perspective where an important feature is obscured (see Figure 9.50) because you picked an inappropriate angle or elevation. You may want to block out the perspective view with basic forms to see if you arrive at an advantageous view. In 3D illustration, you just have to move the camera until any visual problems are minimized (Figure 9.53).

The Standard 30-60 Perspective Template

A perspective with the plan view set at 30 and 60 degrees to the picture plane affords a convenient perspective view because one face is greatly exposed to the viewer, while another is not. This lets you place important features on the 30-degree face and use the 60-degree face for unimportant details. When the vertical edges are parallel to the PP, and the sides are at an angle, we refer to this as an *angular perspective*. Figures 9.11 through 9.15 show how this standard 30-60 perspective is constructed. We will use a simple box to illustrate the perspective setup.

▼ Construct a horizontal ruler guide near the top of your drawing sheet. This line represents the top edge of the PP in the plan view and the front (perspective) view of the HL (Figure 9.11).

▼ Place the top view of the box on the PP and rotate it 30 degrees (Figure 9.12). Place a vertical ruler guide through the front corner of the box. This will become a vertical measuring line for heights.

▼ Note

The object *doesn't have to* touch the picture plane. This is done simply for convenience. The object could have just as easily been set behind, through, or in front of the PP. Any feature touching the picture plane, however, will be true height and easily measured; points in front of the PP will be larger than true height; points behind the PP will be smaller than true height.

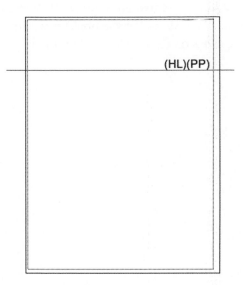

Figure 9.11   Begin your perspective view with a line representing the PP and HL.

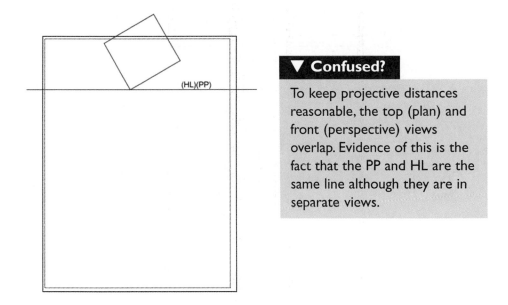

▼ **Confused?**

To keep projective distances reasonable, the top (plan) and front (perspective) views overlap. Evidence of this is the fact that the PP and HL are the same line although they are in separate views.

Figure 9.12   The top view is rotated 30 degrees and placed on the PP.

▼   In the approximate center of the object, construct a vertical guide (Figure 9.13). This establishes the CV on the HL and puts the object in the center of the field of vision. This minimizes possible distortion.

▼   Construct a horizontal line, rotate it 75 degrees, and **Mirror-Copy** it about the vertical axis. Join these two lines into a 30-degree angle and position the vertex on the CV line so that the object is within the 30-degree cone of vision (Figure 9.14). This establishes the SP in the top or plan view.

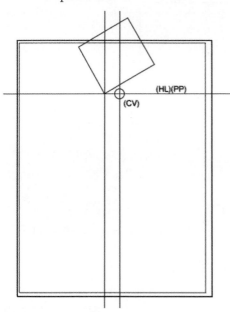

Figure 9.13   Place the CV in the approximate center of the object.

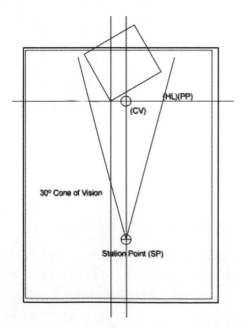

Figure 9.14   A 30-degree cone defines the position of the SP in the top or plan view.

▼ Tip

To keep large objects that are not balanced on the CV within the 30-degree cone, the SP must be moved a greater distance from the PP.

▼ From the SP, rotate a large rectangle to 30 degrees so that it intersects the HL (Figure 9.15). This locates the right vanishing point (RVP) and the left vanishing point (LVP) on the HL.

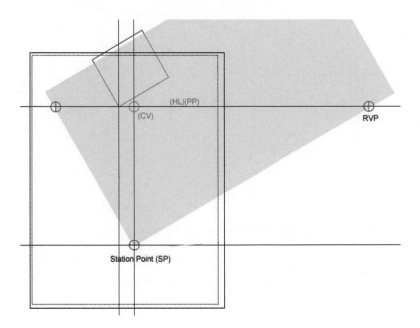

Figure 9.15   Locate the RVP and LVP at 30 and 60 degrees from the SP.

Other perspective views may call for the object to be rotated at angles other than 30 degrees. By simply rotating the box to the same angle as the plan view, this perspective layout can be modified to produce an infinite number of perspectives. Generally, an angle less than 30 degrees places the VP at a position so distant that convergence is hardly noticeable and detail in the narrow face is indistinguishable. You might as well choose an axonometric view.

Use the Template for a Test Perspective

Any perspective setup should be saved as a perspective template. In fact, you may want to load *30-60_template.ai* from the CD-ROM and use it to complete a test perspective. Begin your test perspective by establishing height information from the profile view.

▼ Test a possible elevation position by placing GL below the HL. Place the profile view on the GL(Figure 9.16). Remember, the lower the GL, the more of the top will be seen.

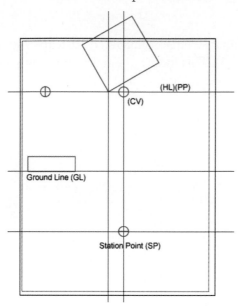

Figure 9.16   Place the GL and profile view below the HL to see the top of the object.

Note the vertical ruler guide from the front corner of the box (Figure 9.17). This is your *vertical measuring line* and marks the place on the PP where the perspective begins.

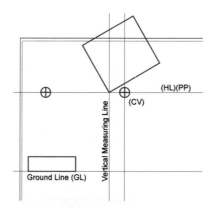

Figure 9.17   A vertical measuring line starts the perspective.

▼ Project heights from the profile view to the vertical measuring line (Figure 9.18). This establishes the true height of the box on the PP because as you can see in the plan view, that's where the box actually touches the PP.

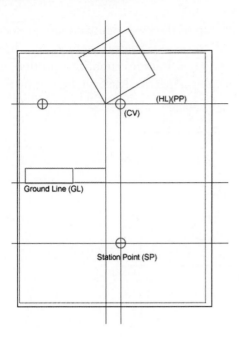

Figure 9.18   Heights projected from the profile view to the vertical measuring line.

▼ From this true height, construct the right and left sides of the box (vanishing lines) to the respective vanishing points (Figure 9.19).

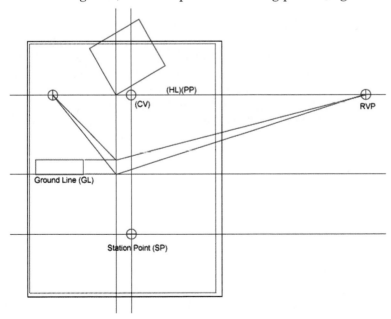

Figure 9.19   Sides of box vanish to their respective VPs.

▼ Locate the position of the box's right and left corners by projecting visual rays from the SP to those corners (Figure 9.20). The only point that you'll need is the visual ray–PP intersection. The portion of the visual ray extending over the perspective itself can be trimmed and removed.

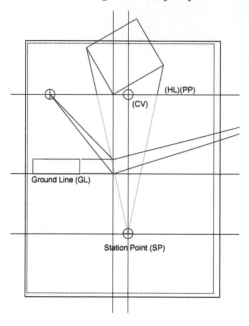

Figure 9.20   Visual rays establish width and depth on the PP.

▼ Drag ruler guides to these visual ray–PP intersections (Figure 9.21). These lines define the right and left corners of the box as projected on the PP and are marked with black circles.

▼ Complete the box with projections to the LVP and RVP (Figure 9.22).

▼ Edit the box for visibility (Figure 9.23). In this case, polygons have been snapped to the construction intersections and filled to better show the object. This is identical to how planes in axonometric drawings are completed.

▼ Hint

Because boxes (prisms) are so easy to draw in perspective, enclose nonrectilinear geometry in tangent boxes. Locate the nonlinear features by transferring a grid structure into perspective. Construct the boxes and then find the more difficult geometry within them. This is identical to the boxing-out method previously discussed in Chapters 4 through 8.

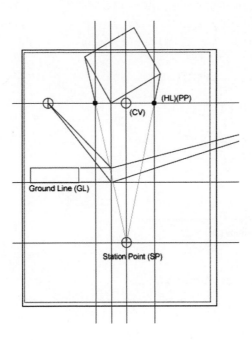

Figure 9.21   Ruler guides project corners into the perspective view.

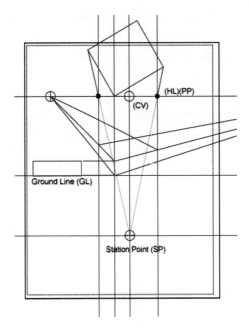

Figure 9.22   Complete the box.

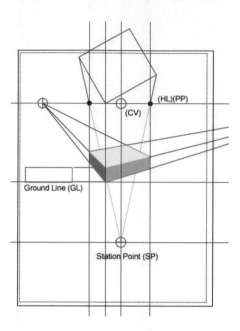

Figure 9.23   Edit the box.

Now that you have a test position for the box, you can see more or less of the top by moving the GL up or down. (The amount you can see of the sides was determined by your original decision of a 30-60 perspective and its position within the cone of vision.) If you locate the GL *above* the HL, you'll get a view from below; if you lower the GL, you'll see more of the top (Figure 9.24).

▼ Note

One way of determining if you have unacceptable *vertical distortion* is to check the angle of a perpendicular corner. If the angle is less than 90 degrees, the distortion is too great. See this for yourself by inspecting the corner of a sheet of paper as it rotates toward an edge view. The range the corner can take is 90 to 180 degrees. It can never be less than 90 degrees. Horizontal distortion is discussed in the section on *measuring points*.

Apply Your Perspective Knowledge

Rectilinear objects, like those shown in Figures 9.11 through 9.23 are relatively easy to draw. However, technical illustrations usually contain much more complex geometry. Figure 9.25 shows a cylindrical spindle in a perspective view. Upon inspection, you should be able to identify the important elements of this perspective view. We will use this spindle as the subject for the remaining perspective constructions. You can use the file *perspective.ai* on the CD-ROM to follow along with the construction.

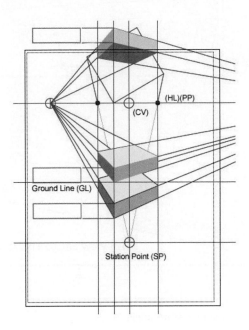

Figure 9.24   Alternate positions for the GL produce different perspectives.

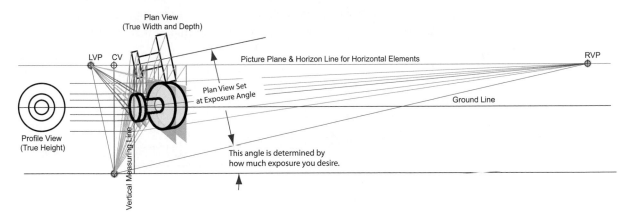

Figure 9.25   Completed spindle perspective.

▼ Modify the 30-60 template into a 15-75 template by setting the large rectangle at 15 degrees. This was done because the left-facing ellipses have no detail and require little exposure. Locate the RVP and LVP using the existing SP.

▼ Set the spindle at 15 degrees and place its left center on the PP (Figure 9.26). This provides a convenient reference point for all vertical measurements along the spindle's axis.

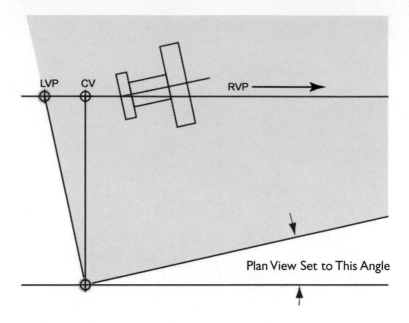

Figure 9.26   Spindle and vanishing-point box set at same angle.

▼ Slide the spindle along the PP until only the left portion is within the 30-degree cone of vision (Figure 9.27). This places the right portion of the spindle outside the cone of vision and makes it susceptible to distortion.

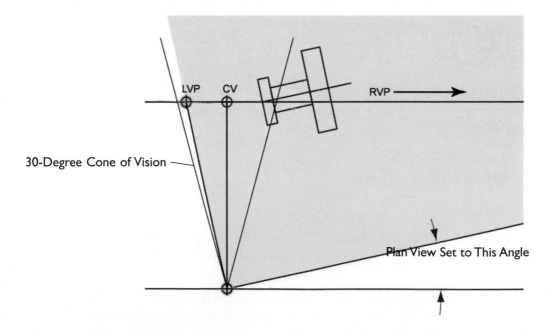

Figure 9.27   Only the left portion of the spindle is within the 30-degree cone of vision.

▼ Place the spindle's profile view on a GL below the HL (Figure 9.28).

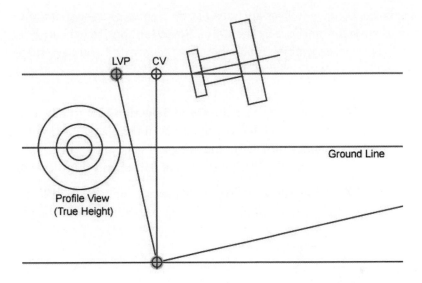

Figure 9.28   Position of GL controls the upward or downward direction of the view.

▼ On the bottom of *perspective.ai* you'll find a number of *magic vanishing points*. Move a magic vanishing point to the LVP, RVP, and SP (Figure 9.29). Use **Smart Guides** to position the magic vanishing point into exact position.

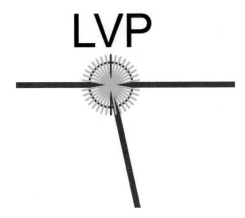

Figure 9.29   Magic vanishing point positioned at LVP.

▼ **Note**

Magic vanishing points contain 36 lines anchored at the center of a registration mark. These lines, when pulled by their outer anchors by the **Open Selection Tool**, remain anchored at the center. By using these magic vanishing points, you never have to worry if you have snapped to a VP or SP!

▼ From the SP, use the **Open Selection Tool** to pull a visual ray to each center point and depth corner in the plan view (Figure 9.30). With **Smart Guides** turned on, the lines will snap to the corners. For the points in front of the PP, use the **Filled Selection Tool** to extend the line to the PP.

▼ Trim each visual ray to its intersection with the PP (Figure 9.31). This is done to minimize the line work extending over the perspective view.

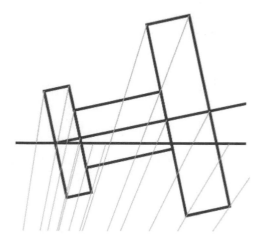
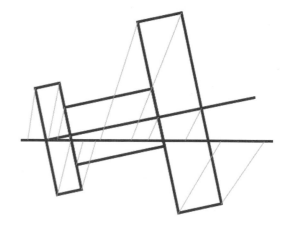

Figure 9.30 Visual rays pulled from the SP.   **Figure 9.31** Visual rays trimmed to the PP.

▼ Project heights from the profile view to the vertical measuring line. The easiest way to do this is to use **Ruler Guides**. Pull vanishing lines from the RVP to the true heights on the vertical measuring line (Figure 9.32). When you have all your vanishing lines, you can select the guides and delete them. Select the guides off the document (on the pasteboard) so that you don't run the risk of selecting and deleting your construction.

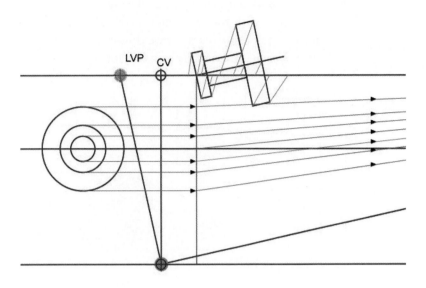

Figure 9.32   True heights from profile view are taken to the RVP.

▼ Create the perspective of the rectangular forms corresponding to each of the spindle's circular features by combining heights (profile view) and widths (visual ray–PP intersections). These rectangular forms enclose the circular features (Figure 9.33).

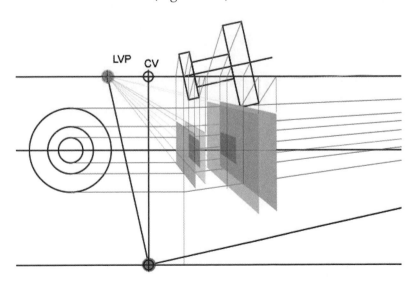

Figure 9.33   Draw the rectangles that enclose the spindle's circular features.

▼   By trial and error, rotate and scale an ellipse (pick the largest so that it's easier to work with) to fit its enclosing form. Start with the ellipse's minor axis aligned with the receding central axis (Figure 9.34). This shiuld be familiar because this is what we do for an axonometric view. Note that a true ellipse doesn't fit the enclosing form.

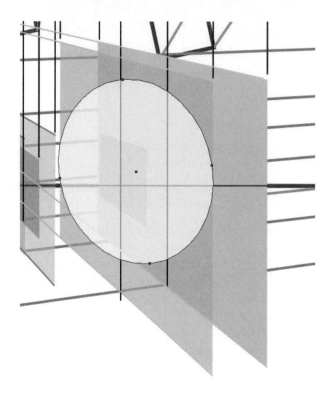

Figure 9.34   A regular ellipse inside a perspective square.

▼   Using the **Open Selection Tool**, select the ellipse and edit its handles for a better fit. Rotate the ellipse so that its major axis better follows the enclosing form's long diagonal. Figure 9.35 shows the result of editing the ellipse into a perspective circle tangent to the enclosing form. You can see that the minor axis is no longer on the thrust line. The minor axis bisects the true ellipse, but more of the enclosing form is below the receding axis because of its position below the horizon line. That's perspective!

▼ **Note**

Editing an ellipse into a perspective circle having no lumps or flat spots is not an easy task. Always work with a copy of the true ellipse and try to keep the shape somewhat symmetrical about its major axis or long diagonal.

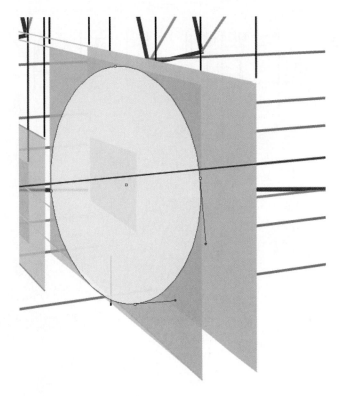

Figure 9.35   Edited ellipse that fits the enclosing form.

▼ **Copy-Paste** this first ellipse (Figure 9.36). Reposition and edit inside the other enclosing forms. Repeat for subsequent circular features.

▼ Standard Perspectives

Along with a 30-60 standard perspective, you might want to consider these other standard perspective views. Once you have determined the components as in Figure 9.25, save this as a template file. You will also want to create perspective grids at the same angles so you have an entire set of perspective tools.

45-45   Use this for objects that have important information in both the right and left vertical planes. By positioning the profile view you can achieve what might be considered an isometric perspective.

25-65   Use this to provide variety from the 30-60 perspective setup.

15-75   Use this for objects that have detail in one face and little in the other.

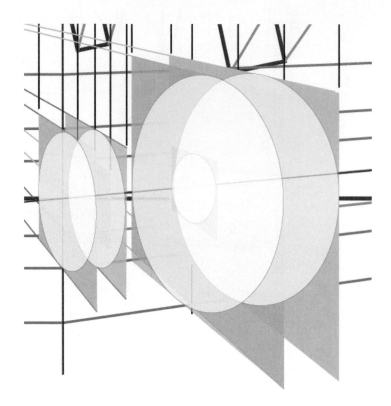

Figure 9.36   Copies of the first ellipse form the basis for subsequent ellipses.

▼ **Note**

You can see from Figure 9.36 that as ellipses move to the right and outside the cone of vision their exposure and tilt change dramatically. To minimize editing, it's more efficient to make the next ellipse out of the previous one.

▼ Pull limiting elements from the RVP and align them tangent to the top and bottom of the cylindrical features (Figure 9.37). You may have to go back and fine-tune one or more of the perspective circles to assure that the ellipses are tangent limiting elements.

▼ **Copy-Move** the geometry to a new layer. Edit line work for visibility (Figure 9.38).

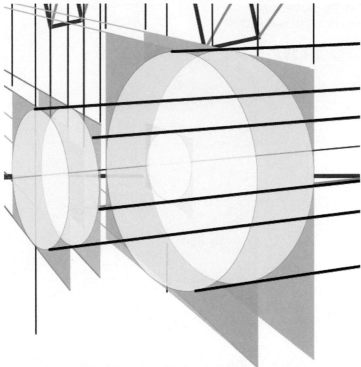

Figure 9.37   Limiting elements are pulled from the RVP.

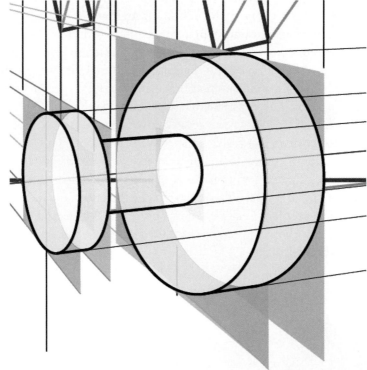

Figure 9.38   The edited spindle.

▼ Illustration in Industry

Greg Maxson Illustration
Urbana, Illinois

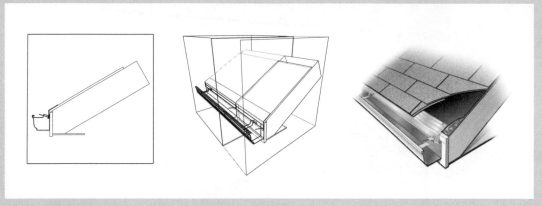

Software

Adobe Illustrator
KPT Vector Effects

Hardware

Macintosh G4
Epson Proofer

Practicing illustrators quickly learn that a number of software tools are available that will automate many of the construction operations necessary in creating effective technical illustrations. One interesting product is the KPT Vector Effects plug-in because it operates on vector geometry in Adobe Illustrator.

The first step in creating this illustration of a gutter installation for *The Family Handyman* magazine was to construct accurate geometry. A section or profile was generated first, based on a rough sketch and the various material dimensions supplied by the client. This line art was then modified using the KPT Vector Effects plug-in via Adobe Illustrator, whereby a suitable viewpoint could be established.

Using KPT Vector Effects, the gutter profile was extruded while the X-Y-Z axis values were adjusted to arrive at a desirable perspective orientation. This orientation was critical in order for the reader to be able to look down into the gutter to see the brackets, yet still recognize the soffit profile at the right of the illustration.

After the approval of one of several different layouts by the client, the final illustration was rendered and proofed. Realistic colors and textures were added using Adobe Illustrator and Photoshop to mimic the actual building materials that would be encountered by the homeowner.

Perspective Without Scale Views—Measuring Points

Both the box and spindle examples relied on existing scale views for projection. However, many technical illustrations are constructed from sketches, notes, or tabular information from vendors' catalogs; illustrations may contain *standard parts* for which no drawings exist. Additionally, many engineering drawings are not drawn to scale but with accurate dimensions. It may not be efficient to draw scale views when only several height, width, and depth measurements are needed to build the enclosing forms that control the perspective.

You know by now that you can't directly measure dimensions in perspective views like you can in axonometric views. You *can* measure in top and profile views because they are orthographic. How can we measure dimensions in a perspective view using the measurements obtained from the ortrhographic view?

The answer is to find two special vanishing points, called *measuring points*. Figure 9.39 shows how these measuring points are found.

▼ Create a perspective setup as usual.

▼ Revolve the RVP-to-SP and LVP-to-SP distances into the PP (Figure 9.39).

▼ The RVP finds the right measuring point (RMP) while the LVP finds the left measuring point (LMP). (You don't want to mix these up.) Because we use the same line for the PP and HL, these rotated points are on the HL as well. Place a magic vanishing point at each measuring point.

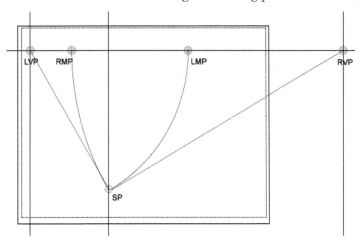

Figure 9.39   Measuring points are rotated into the PP.

▼ Figure 9.40 demonstrates how width (to RMP) and depth (to LMP) are found for a simple 2 x 1 box using measuring points.

▼ Measure the width (2 in) to the *right* of a starting point (vertical measuring line) on a horizontal measuring line. For convenience, consider this horizontal measuring line to be the GL.

▼ Drag a projector from the RMP to this true width measurement.

▼ Measure the depth (1 in) to the *left* of the starting point and drag a projector from the LMP to this true depth measurement. These measuring lines cut off correct perspective distances on vanishing lines from the starting point. True heights are measured as usual along the vertical measuring line using the same scale.

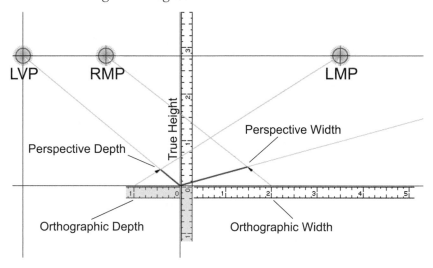

Figure 9.40   Construction using measuring points.

An Example Using Measuring Points

Assume you work for a spindle manufacturer who makes lots of spindles. They all have the same design but are of different sizes. There wouldn't be scale drawings of any of the individual spindles, rather there would be tabular data like that in Table 9.1.

▼ Numerical Spindle Data							
	Model	A	B	C	D	E	F
	AA						
	AB	1.25	1.25	0.50	1.00	0.25	2.00
	AC						
	AD						

Table 9.41   Tabular spindle information.

Figure 9.41a shows the construction of the spindle's *perspective plan view* from the data in Table 9.1. Use the file *measuring.ai* to practice this measuring point technique. Follow the steps previously outlined for setting up measuring points and finding width, depth, and height.

Figure 9.4b shows how this plan view has been expanded to create the enclosing forms necessary to find the circular features. True heights are measured on the vertical measuring line using the same scale as was used to construct the perspective plan. The result is the same as Figure 9.33 where the enclosing forms were found by the visual ray method. The difference is that with the measuring point method, we didn't need scale views, only correct dimensional data. Complete the spindle's circular features as was done in Figures 9.34 through 9.36.

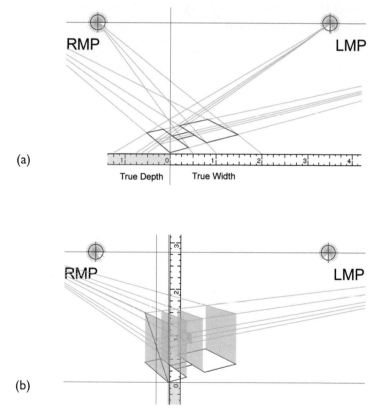

Figure 9.41   (a) Perspective plan construction using tabular data and measuring points. (b) True heights are measured on the vertical measuring line to complete the enclosing forms.

Create a Perspective Grid

It may be helpful to use a perspective grid when designing an illustration, especially when exacting engineering drawings or photographs (for photo tracing) are not available. Because planning a perspective is so important, you may want to create a perspective grid for each setup you make. That way you can print the grid and sketch a preliminary solution. A sample 30-60 perspective grid can be found on the CD-ROM in *exercises/ch14/30-60_grid.ai*. A perspective grid isn't difficult to create. Figures 9.42 through 9.46 walk you through the steps. You'll want to create a directory where you can store grids as you create them.

▼ Create a perspective setup as you did for the block or spindle. Rotate the rectangle that finds the vanishing points to the desired perspective orientation. Locate the measuring points as you did in Figure 9.39. You'll need magic vanishing points at the LVP, RMP, LMP, and RVP.

▼ Pick a grid scale that gives you an appropriate number of divisions (**Edit|Preferences|Guides & Grid**). If there are too few divisions, it will be difficult to use the grid; if there are too many, it will be difficult to see anything. In the example, a 1 in grid was used with 10 subdivisions.

▼ Position 0, 0 at the SP. Turn on the grid and **Snap to Grid**. Use the grid and magic vanishing points to create half-inch unit horizontal grid lines to the vanishing points (Figure 9.42). Make these lines light blue and 1-point stroke.

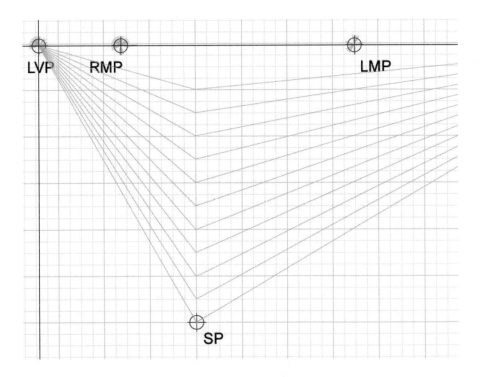

Figure 9.42   Orthographic grid positioned on ground line to space horizontal grid lines.

▼ Turn off **Snap to Grid** and turn on **Smart Guides**. To the right of the SP, pull projectors from the RMP to half-inch measurements (Figure 9.43a). Pull vertical ruler guides to their snap intersections with the bottom horizontal grid line (marked with red crosses) as shown in Figure 9.43b.

▼ Delete the measuring point lines and release the vertical guides (**View|Guides|Release Guides**). **Trim** the guides to the top and bottom of the grid (Figure 9.44), and use the **Eyedropper** to pick up the grid line style.

▼ Repeat the previous steps for vertical grid lines on the left side (Figure 9.45).

▼ Using magic vanishing points for the RVP and LVP, complete the top of the grid, and trim away all excess. You should have the final grid shown in Figure 9.46.

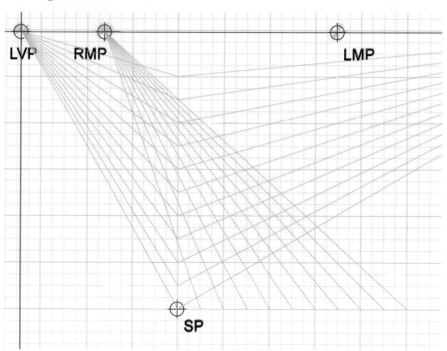

Figure 9.43a   Use ruler guides to form vertical grid divisions.

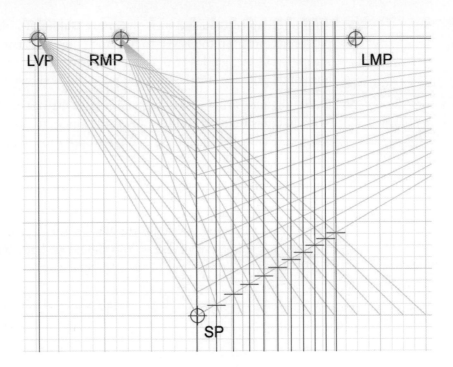

Figure 9.43b   Use ruler guides to form vertical grid divisions.

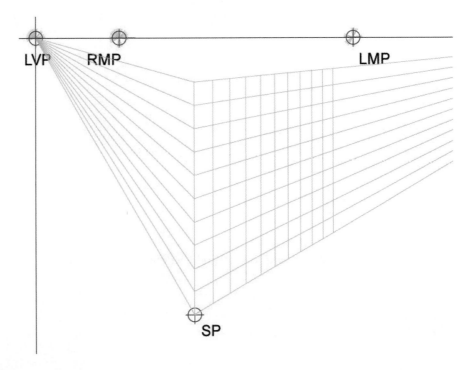

Figure 9.44   Released ruler guides are trimmed to the top and bottom of the grid.

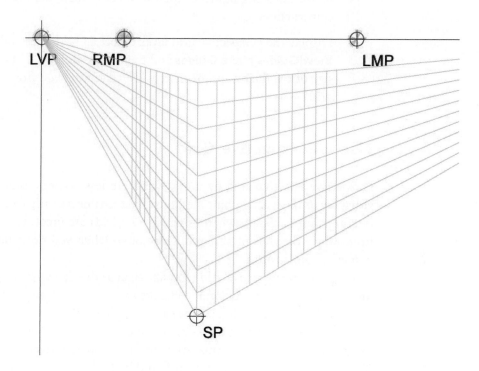

Figure 9.45   Vertical lines completed on both sides of the grid.

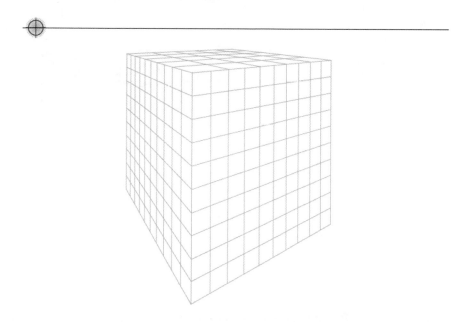

Figure 9.46   Completed perspective grid.

▼ Move vanishing points to their own layer where they can be used for construction.

▼ Display the perspective grid layer. Select all grid lines and choose **View|Guides|Make Guides** and then **Lock Guides**. This will reduce the size of the file by approximately 20% and keep you from selecting the grid.

Use a Perspective Grid

Perspective views created by using a grid are less accurate than those created by visual ray or measuring point methods. That isn't necessarily bad, but there is some interpolation necessary when using a grid. Grids are great when you are working from rough sketches or verbal descriptions—when you don't have scale views or numerical dimensions.

Start with a perspective grid template such as that in *exercises/ch14/30-60_grid.ai* on the CD-ROM. It is helpful to change the color of the magic vanishing points so that they are distinguishable from the template grid. In our example, they have been changed to red (Figure 9.47). You'll be doing your rough construction on the vanishing point layer and final work on the geometry layer. We will be using the numerical values for the spindle as found in Table 9.1.

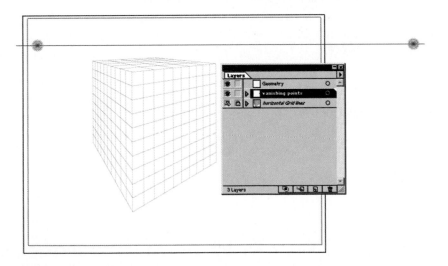

Figure 9.47   Grid file ready for construction.

▼ Turn **Smart Guides** on. Pull a smart vanishing line out from the LVP to serve as a centerline (Figure 9.48). Locate this in the center of the cube's left face.

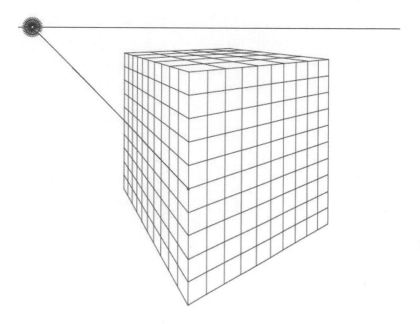

Figure 9.48   Centerline starts the construction.

▼ Determine a scale for your perspective. (Make a note that every division is 0.25, or 0.5 or 1.0, etc.) Remember, the larger the scale, the greater the chance for unacceptable distortion because more of the spindle will be out of the cone of vision. We will consider each grid division to be 0.5 in.

▼ Using the height and depth values in Table 9.1, construct a perspective profile view of the enclosing forms on the left side of the grid. You'll have to estimate values smaller than the grid divisions (Figure 9.49). Trim vanishing lines and release ruler guides (for verticals) to finish the form.

▼ Tech Tip

Having the perspective grid reduces the number of visual rays you have to project from the station point to points in the plan view. Once you locate a point on the grid, parallel points can be found by using vanishing lines to "chase" the point around the grid. In fact, once you gain experience using grids you'll find that by locating only a few critical measurements with visual rays the majority of construction will happen right on the grid.

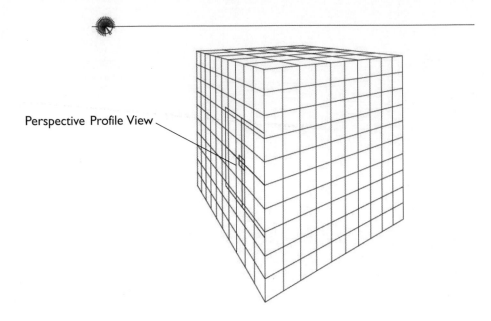

Figure 9.49   Enclosing forms constructed on left side of grid.

▼ Working on the right side of the grid, complete a perspective front view with the width and height values from Table 9.1 (Figure 9.50).

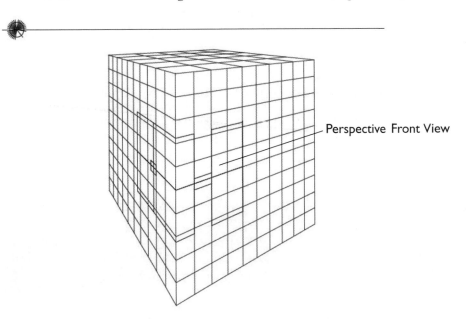

Figure 9.50   Perspective of width and height.

Figure 9.57   Camera view in 3D program in default perspective projection.

▼ Illustrator Tip

Transparency may be problematic when you are trying to print an Adobe Illustrator file saved in EPS format using a PostScript printer. You can avoid the **Transparency** setting by using the **Multiply** mode. In this life vest example, the transparency of the plastic bag remains at 100 percent. By selecting **Multiply**, an even mix of the top and bottom colors is achieved. You can see this where the plastic bag covers the vest. Finish the transparent effect with surface reflections that stretch across several surfaces, again using **Multiply** or possibly the **Overlay** mode.

▼ Complete the enclosing forms by dragging vanishing lines and ruler guides to appropriate intersections. These forms are shown in blue in Figure 9.51. Note that the middle cylinder of the spindle is hidden behind the second form and will not be seen from this perspective view. This is due to our initial choice of object angle relative to the picture plane.

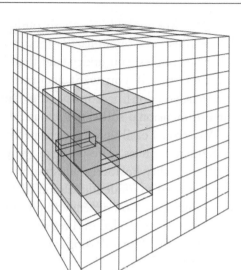

Figure 9.51   Completed enclosing forms.

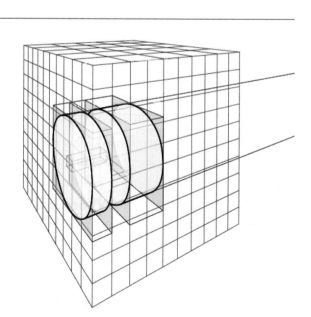

Figure 9.52   Complete spindle from perspective grid.

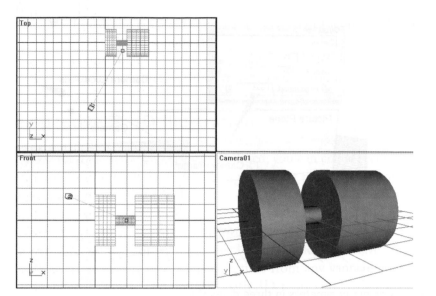

Figure 9.55   Translated camera produces a more representative view.

The Three-Dimensional Camera

Cameras in 3D programs can also be switched between perspective and orthographic views. The two views of a water purification system in Figures 9.56 and
9.57 were actually taken with the same camera. In Figure 9.56 the view has been
orthogonalized; projectors have been made perpendicular to the plane of projection.
In Figure 9.57 the projectors all meet at the camera's focal plane, the same as the
station point in perspective drawing.

The inadvisability of using an axonometric view for large objects can be seen
in these two examples. Although Figure 9.56 accurately displays dimensional
relationships, it fails to convey the sense of scale that Figure 9.57 is able to
communicate.

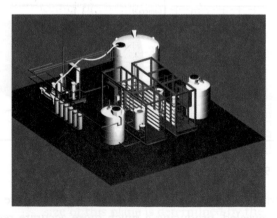

Figure 9.56   Camera view in 3D program set to orthogonal projection.

Review

If you have closely followed the material presented in this chapter, you have, by
now, formed certain opinions about perspective illustration. You understand that
perspective should be reserved for large objects which when viewed normally exhibit convergence of parallel elements. To introduce perspective into the representation of small objects may give a false impression of scale.

You also understand that perspective can be achieved by four methods: (1) visual
ray projection from scale orthographic views, (2) measuring point construction from
accurate dimensional data, (3) approximate construction over a perspective grid
from dimensional data, verbal description, or mental image, and (4) by viewing 3D
computer models from a camera position.

If you create a number of standard perspective setups and their matching perspective grids in order to have a number of possible perspective orientations, the
grids can be used for planning your perspective constructions. This is especially
inportant in the location of the ground line because the ground line controls how
much, or little, of the top or bottom will be seen.

Perspective projection and construction require significant knowledge of perspective and may scare an illustrator from using manual methods. Perspectives from
3D modeling programs are essentially automatic, but require accurate models. Once
accurate models are at hand, any and all perspective views are possible by simply
moving the camera or changing its parameters.

Although perspectives made by viewing computer models with a camera are simple
to make, modeling programs are not designed to create publication-quality resolution-independent PostScript output. Given sufficient computer resources, extremely
high quality (1440 dpi and above) raster illustrations *can* be made. However, objects
in PostScript illustrations can be combined and edited more easily, making raster
illustrations less flexible.

Text Resources

Burden, Ernest E. *Perspective Grid Sourcebook: Computer Generated Tracing
Guides for Architectural and Interior Design Drawings* . John Wiley & Sons.
New York. 1991.

Edwards, Betty. *The New Drawing on the Right Side of the Brain.* J.P. Tarcher
(Putnam). New York. 1999.

Linton, Harold, and Sutton, Scott. *Sketching the Concept: Perspective Illustration
for Architects, Designers and Artists* . Design Press (McGraw-Hill). New York.
1993.

Mitooka, Eiji . *Illustration and Perspective in Pantone Colors: Perfection Presentation: Illustration and Perspective (Illustration and Perspective, No 3).* Books Nippan. Carson, California. 1988.

Montague, John. *Perspective Drawing: a visual approach.* John Wiley & Sons. New York. 1998.

O'Connor, A. , et al. *Perspective Drawing and Applications* (2nd Edition). Prentice Hall Publishers. Englewood Cliffs, New Jersey. 1997.

Internet Resources

A Primer on Perspective. Retrieved from: *http://forum.swarthmore.edu/sum95/math_and/perspective/perspect.html*. November 2002.

Drawing in Perspective. Retrieved from: *http://www.unict.it/dipchi/05Didattica/Corsionline/Coloranti/03_ColorCostit/paint/draw.html*. November 2002.

Measuring Point Perspective. Retrieved from: *http://apostle.stpauls.nsw.edu.au/~IndArts/mppone.html*. September 2002.

On-line Inside Illustrator journal where you can use the word perspective to search for relevant articles. Retrieved from: *http://www.elementkjournals.com/iai/* . November 2002.

Perspective Drawing. Retrieved from: *http://webdeveloper.com/design/design_drawing_perspective.html*. September 2002.

Perspective drawing tutorial by John Robertson. Retrieved from: *http://www.conceptdesignforum.com/board showthread.php?s=b75f4433 d77d038ede3372ea03cc1fb3&threadid=860*. November 2002.

Perspective Illustration. Retrieved from: *http://www.perspectivearts.com/*. September 2002.

On the CD-ROM

For those of you who would like to try out the procedures described in this chapter, several resources have been provided on the CD-ROM. Look in *exercises/ch14* for the following files:

30-60_template   Use this file to practice perspective constructions and alternate ground line positions.

perspective.ai   This file contains scale views ready for perspective construction.

measuring.ai   This file is used to construct the perspective of the spindle from the data in Table 9.1.

30-60_grid.ai   A completed perspective grid for you to use.

Part Three

Technical Illustration Rendering

▼ **Line Rendering**

▼ **Photo Tracing**

▼ **Emphasis with Color**

▼ **Color Rendering**

▼ **PostScript Materials**

▼ **Text and Technical Illustrations**

Chapter 10

Line Rendering

In This Chapter

Up to this point, you have studied various methods for constructing accurate geometry. Because technical illustration is used in engineering, construction, architecture, medicine, and science, it is important that the underlying geometry be dimensionally and visually accurate. That's why you studied these topics first. Every effective technical illustration is based on accurate geometry. This accurate geometry is first rendered for presentation as lines that show the intersection of surfaces, the edges of surfaces, and the limits of surfaces. We call this technique *line rendering*.

As was described in Chapter 2, printed technical documents are often reproduced in black and white because of cost considerations when the information in the document has a short life cycle. It's cheaper and quicker to use black-and-white line illustrations. But as more and more technical documents find their way on-line, the reproduction constraints are eliminated, making color as inexpensive to distribute (though not to create) as black and white. Still, one can make a persuasive argument that a black-and-white line illustration may communicate technical information as well as, or better than, color.

Chapter Objectives

In this chapter you will understand how to:

▼   Analyze object geometry for the line work necessary to show shape and describe function

▼ Determine line work strategies based on orientation and position of the sun

▼ Apply the concept of *minimalism* to line rendering

▼ Understand the limitations of line rendering with various reproduction processes

▼ Develop a library of standard line weight styles

▼ Apply standard line styles to basic object geometry

▼ Understand the trade-off between one, two, and five-line-weight rendering strategies

▼ Be able to combine block shading with line rendering to create effective black-and-white illustrations

▼ Add spatial gaps and hot spots to further enhance the realism of line rendering

▼ Create smooth transitions using brushes and duplicate strokes

Line Rendering Theory

Objects themselves don't have lines. We use lines to represent various features of geometry that without shading would not be visible. For example, study the jet engine filter assembly in Figure 10.1. The photograph in Figure 10.1a appears fully shaded and realistic. The features of edges, intersections, and limits are discernible because of changes in tonal value. *No actual lines are present or needed.* But take away the tonality and, obviously, you would have nothing. Lines, like those in Figure 10.1b, are used to represent what otherwise would be changes in tonal values, and as you can see, the communication of technical information actually can improve with the abstraction of information.

Figure 10.1b is called a *line rendering* because lines are used to define the object's geometry. Line rendering is used for several reasons:

▼ Line rendering is the natural result of accurate construction. When you finish accurate construction, an effective line rendering is about 75 percent complete.

▼ Line rendering is the foundation of more sophisticated rendering techniques. Realistic shading, blending, and gradients are based on an accurate line rendering.

▼ Line rendering consumes the fewest resources in terms of time, talent, and computer processing.

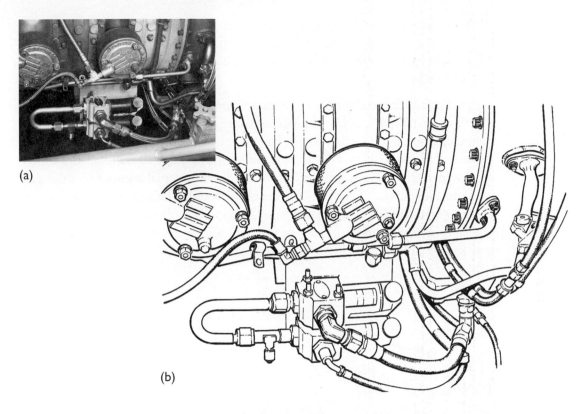

(a)

(b)

Figure 10.1   Lines are used to represent object geometry in the absence of tonal values.

▼ Line rendering can (and should) be reproduced in print without halftones, reducing the cost associated with creating, stripping, and holding the quality of halftone art. Additional savings occur when manual stripping of art and halftones—the taping of pieces of negatives together—is replaced by digital prepress.

▼ Line rendering is supported by the digital Graphic Interchange Format (GIF) for distribution over the Web. GIF utilizes lossless LZW compression, so line-rendered illustrations (which are mostly white pixels) are efficiently compressed. When they are decompressed, no data is lost, and the artifacts associated with Joint Photographic Experts Group (JPG) format decompression are absent.

▼ Tech Note

The GIF format has been augmented by the *Portable Network Graphic* (PNG) format because GIF uses a proprietary compression algorithm for which royalties are due with use. Most illustrators shouldn't have to be concerned with this, however, unless they are involved in massive distribution of GIF images.

▼   Finally, line rendering removes extraneous variables such as color, materials, texture, and environment from an illustration. This makes the illustration easier to read and, in many cases, more effective than photorealistic images.

An effective line rendering adheres to a *minimalist* approach to line work. Consider Figure 10.2. The object is rendered using selective line work. As you might imagine, a machine like this would have much more complexity than is actually shown on the illustration. For example, hoses, fittings, and all fasteners not distinguishable at the drawing scale are omitted. Likewise, actual components not germane to the purpose of the drawing are also left out. The base art can be used to show enlarged details, as is the case here. The ability to distinguish which lines are crucial to an illustration is developed through experience. A general rule of thumb is to start simple and add line work as necessary.

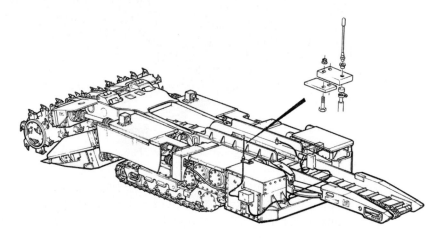

Figure 10.2   Effective line rendering is characterized by selective line work.

Line Rendering and Reproduction

Line-rendered technical illustrations—those reproduced by traditional print methods as with books, manuals, and catalogs—are less susceptible to reproduction problems than are illustrations reproduced on CD-ROM as multimedia training or delivered over the Web and displayed on a monitor. In either case problems arise when choosing a line thickness that when reduced is smaller than the display resolution. With traditional silver-based film, this resolution is approximately 4000 ppi or 0.018 point, thinner than any line you would ever need. That's good.

▼ Tech Note

The resolution of a display monitor is expressed in *dots per inch* (dpi). Printer resolution is expressed in *printer dots per inch* (ppi).

Digital copying can be as coarse as 300 ppi (0.24 point) or as high as 1000 ppi (0.07 point). Digital offset is usually 2540 ppi (0.02 point). At the low (and dangerous) end, digital displays are 72 to 90 dpi (1.0 to 0.8 point). That's bad.

Problems arise when an illustration is reduced from its original size before printing or display. Because most of you have experience with 300-ppi laser printers, let's use that as an example. You can make the same comparisons for any output device.

Determine the Thinnest Line Weight

We know that the thinnest line a 300 ppi laser printer can produce is 72/300 = 0.24, or about a quarter of a point. This is a line that is one printer dot wide. Horizontal and vertical lines align perfectly with the printer's matrix of dots. Diagonal lines (other than 45 degrees) do not. Three things may happen to a line thinner than this, depending on the decisions made by the printer's rendering engine.

▼ The line, if *not* perfectly horizontal or vertical, will break up. Solid lines will appear broken as the rendering engine tries to match the line geometry with available printer dots.

▼ Note

The phenomenon of raster stepping to match curves or angles is referred to as *alias*. When a program, such as PhotoShop adds transition pixels to smooth out these steps, it is called *anti-alias*. Anti-alias is a function of raster graphics. PostScript programs such as Illustrator don't anti-alias because they are resolution independent.

▼ The line, if horizontal or vertical, may completely disappear. Because no printer dots match the coordinates of the line, no line is created.

▼ The line, if horizontal or vertical, appears much thicker than desired. If the line happens to match a row or column of printer dots, the smallest dot possible (larger than you wanted) is used.

Determine the Impact of Reduction

In our 300-ppi example, assume that you used a 1-point line and the illustration is reduced 75 percent for printing. The 1-point line becomes a 0.25-point line, right at the smallest size that's printable at 300 ppi (72/300 = 0.24). However, had you chosen a 0.5-point line, it would be reduced to 0.125-point, thinner than can be printed on a 300-ppi printer. You would expect some of the lines to break up when they failed to match the printer's matrix of dots.

Experienced illustrators are aware of these limitations and develop close relationships with printers and manufacturers so that the least amount of rework is necessary. Unfortunately, an illustration originally intended for high-resolution reproduction (2540 ppi) may find its way onto the Web for 72-dpi display. Figure 10.3 demonstrates how severe this line breakup can be when displayed on a 72-dpi monitor.

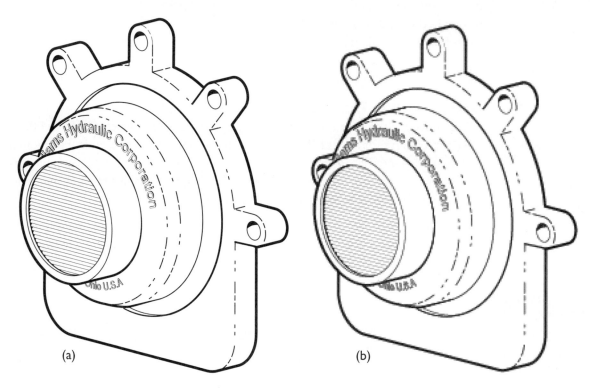

(a)   (b)

Figure 10.3   Line detail is lost when display resolution is coarser than the illustration resolution.

Line Rendering Fundamentals

Line rendering is based on giving an object's features different weights so that an appearance of solidity can be formed without tone. In order to demonstrate the effectiveness of increasingly sophisticated line techniques, the same object is used for the following examples.

Single-Line-Weight Rendering

The simplest strategy is to first carefully construct the geometry and move and copy that geometry to a new layer (so you have access to the original construction if need be). Edit the duplicate geometry for visibility and detail, and make sure that the anchors (end points) are in fact coincidental. Assign a line weight to all the lines appropriate for the scale of the object and how it will be reproduced. Figure 10.4 shows a single-line-weight rendering.

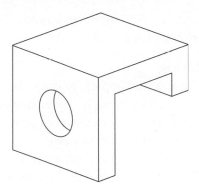

Figure 10.4   A rendered illustration with a single line weight.

Although this method is fast and effective, it has a serious limitation: it doesn't really look like an illustration. In fact, it's really just a pictorial drawing, ready to be turned into a technical illustration.

Two-Line-Weight Rendering

By using a single line weight, like in Figure 10.4, the visual distinction of the object from the background is lost. A line on the perimeter that separates object the from the background is just as visually dominant as a thin interior intersection line. Figure 10.5 shows the result of assigning a heavier line to the outside of the objcct. Now, the object stands out from the background and projects a sense of solidity. As you can see from the example, this simple change adds significant clarity to the illustration and is appropriate for a large percentage of line illustrations. If the **End Cap** and **Corner Stroke Styles** in Adobe Illustrator are set to *rounded*, and all anchors are coincidental, the outline will be smooth and continuous.

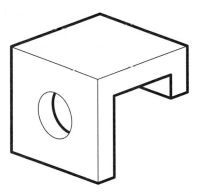

Figure 10.5   By adding a bold outline, the object is more easily identified from the background.

It is important to achieve sufficient a difference between the outline and other object lines. A rule of thumb is to make the outline three 3thicker than the inside lines. However, extremely bold outside lines connote weight and mass and are appropriate only for large, heavy objects (trains, planes, construction equipment, buildings, and so on). As you might then imagine, light, transparent, or soft objects require lighter perimeter lines.

Five-Line-Weight Rendering

Large, significant technical illustrations, especially assemblies of many parts, may require more than two line weights to accurately depict spatial relationships. But because applying five separate line weights increases the cost of a technical illustration dramatically, it should be reserved for special cases.

When using five-line-weight rendering, you must make a careful analysis of the function of each line on the object and an initial decision as to the position of the sun. The sun's position determines which exterior lines are the heaviest as well as the position of any highlights. Table 10.1 displays a chart explaining these relationships, from the thinnest to heaviest lines.

▼ Line Weight Strategy

Value	Title	Function
.25	Interior Blend	Any line within the object that represents an intersection greater than 90 degrees. For curved intersections and tangents, break this line.
1.0	Interior Sharp	Any line within the object that represents a perpendicular intersection.
2.5	Overhang	Any line within the object that represents the limit or extent of a feature.
3.0	Outside	Any line on the perimeter of the sides or top.
5.0	Backside	Any line on the perimeter, of the back, bottom of the object, away from the sun.

Table 10.1   Rationale for a line rendering method based on five steps.

The values in Table 10.1 are relative and can be applied to line measurement in points (1 point = 1/72 in), millimeters, or decimal inches. Were you to consider the numbers to represent actual points, they would be appropriate for an illustration reproduced full size in 8.5 x 11.0 format. Use the interior sharp line (1.0) as the unitary basis for all adjustments and factor the other values to adjust line work for larger or smaller final sizes. To better understand how these multiple line weights are assigned, let us first use a test object that clearly reveals the function of each line. Figures 10.6 through 10.12 show the application of five line weights to a simple object.

▼ First determine the thinnest line you can tolerate at the target output resolution. If you don't know the actual resolution, assume 72 dpi for display, 300 ppi for work produced on laser printers, and 1240 ppi for offset publications.

▼ Figure 10.6 shows a single-line-weight rendering, the basis for our example, and the position of the sun. The position of the sun determines on which side the heaviest lines will be assigned. Refer to Table 10.1.

▼ Figure 10.7 shows the outside style applied to the perimeter of the object. Notice how the hole now appears to exit the back of the object because it shares this outside line weight.

▼ Figure 10.8 completes the perimeter with a backside line weight (note the position of sun). Now you can see the difference of the heavy lines on the bottom and backside and how it adds weight.

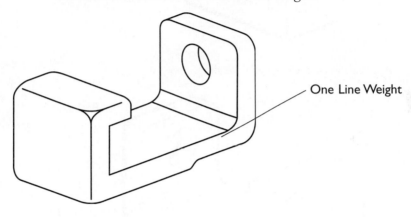

Figure 10.6   Single-line-weight illustration.

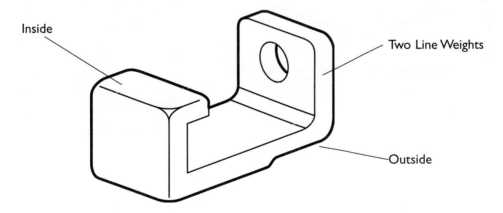

Figure 10.7   Application of outside line weight.

▼ Figure 10.9 creates all perpendicular intersections (interior sharp).

▼ Figure 10.10 uses the thinnest line (interior blend) to represent intersections that are curved or not at 90 degrees.

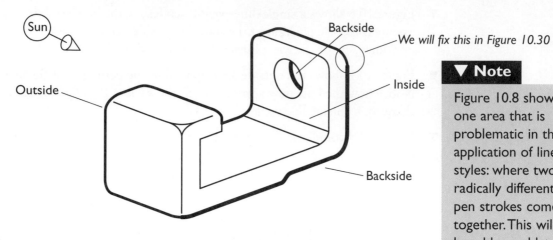

Figure 10.8   Application of backside line weight.

▼ **Note**

Figure 10.8 shows one area that is problematic in this application of line styles: where two radically different pen strokes come together. This will be addressed later in the sections on brushes and duplicate strokes.

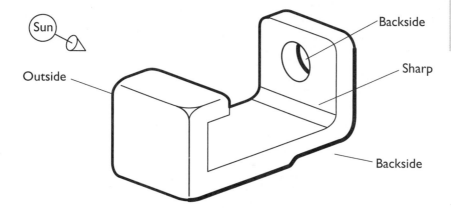

Figure 10.9   Application of interior sharp line weight style.

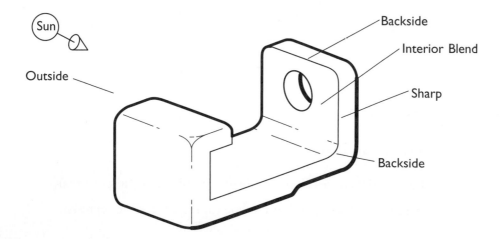

Figure 10.10   Application of interior blend line weight style.

▼ The final line (overhang) and done purposefully out of order shows where one part of the object overlaps another part of the object *within its perimeter* (Figure 10.11).

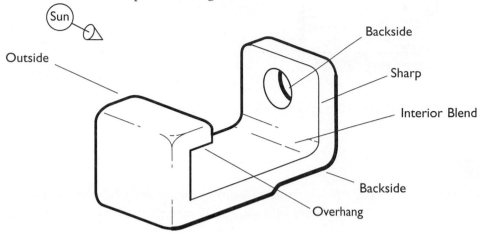

Figure 10.11   Application of overhang line weight style.

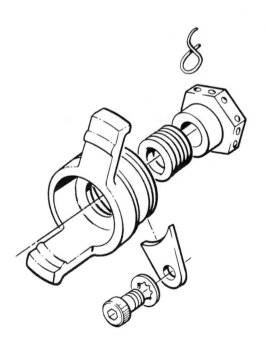

Create a Library of Standard Line Styles

Several tools in Adobe Illustrator make the application of standard line weights easy and consistent. Aside from manually selecting each line and assigning line weights via the **Stroke Palette**, you can also build a set of standard line weights by using the **Styles Palette**. This palette is shown in Figure 10.13. Note that the styles follow the line weight strategy described in Table 10.1.

▼ Create a new document in Illustrator, and show the **Styles Palette** via the **Window** pull-down menu. Delete unwanted styles.

Figure 10.12   Identify the five types of line weights in this hose clamp illustration.

Figure 10.13   The **Styles Palette** in Illustrator.

▼ Draw a series of boxes within your new Illustrator document, each stroked with a line weight from Table 10.1. You can draw as many different boxes as line weights you intend to use in your illustration.
Five line weights have been used in Figure 10.14. Select each square with your cursor and drag it into the **Styles Palette**.

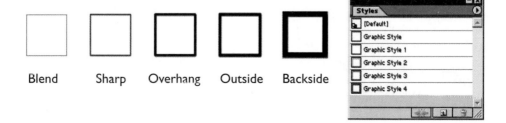

Blend   Sharp   Overhang   Outside   Backside

Figure 10.14   Line weights dropped into the **Styles Window**.

▼ Once in the **Styles Palette**, each style can be named as you choose (Figure 10.15). The new styles can also be displayed in various ways, in swatch or list view. The styles in Figure 10.15 are displayed in list view so that the style names can be recognized at a glance. Save this file so you can access it later.

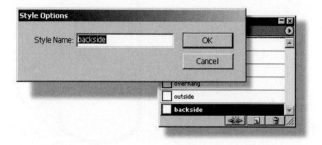

Figure 10.15   Line rendering styles named in the **Style Options** dialog box.

Using Line Styles

Successful technical illustrators are able to see basic shapes in complicated objects and then apply line rendering to those basic shapes so that when taken as a whole, the object is effectively rendered.

Render a Prism

Figure 10.16a shows how line rendering is applied to a prism. Note the position of the sun and how that impacts placement of the backside line weight. In Figure 10.16b, a more complicated prism is rendered. Make note of overhang and interior sharp lines.

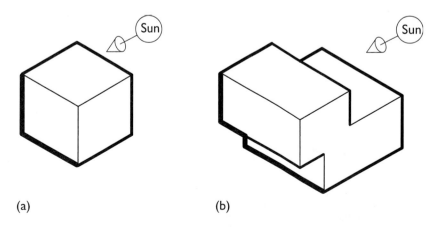

(a)   (b)

Figure 10.16   Line rendering applied to a prism.

Render a Cylinder

The cylinder in Figure 10.17a shows effective basic line rendering. When combined with other cylindrical shapes Figure 10.17b, the full range of line styles are implemented.

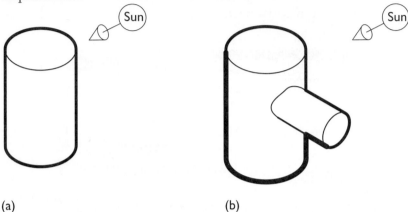

(a)   (b)

Figure 10.17   Line rendering applied to a cylinder.

Render a Cone

The cone in Figure 10.18 is rendered using *outside* line weights. When truncated and combined with cylinders in (b), the full range of line styles are used.

Render a Sphere

A drawong of a sphere may need a little help for its geometry to be perceived as spherical. Figure 10.19a shows a sphere rendered with *element lines* on the surface. These lines help define the sphere's shape and because they are simply *on* the surface and not at the actual edges, intersections, or limits, they are represented by broken interior blend lines. In Figure 10.19b these lines help define where one spherical surface intersects another.

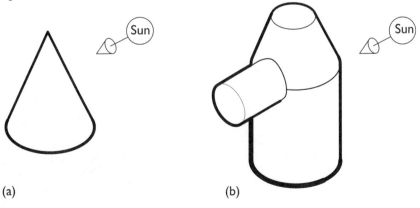

(a)   (b)

Figure 10.18   Line rendering applied to a cone.

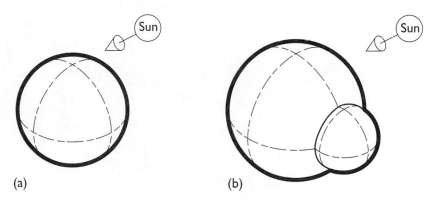

Figure 10.19   Line rendering applied to a sphere.

Render a Torus

The torus in Figure 10.20a also receives help from surface elements. In Figure 10.20b we have combined a torus and a sphere, making use of surface elements and line weights.

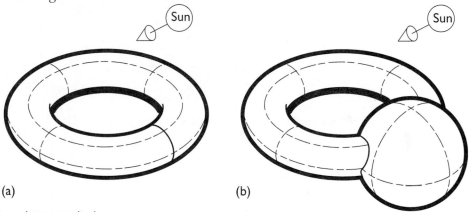

Figure 10.20   Line rendering applied to a torus.

Apply Basic Shapes to More Involved Geometry

The trick, of course, is to apply an understanding of basic shape rendering to more complicated geometry. In fact, by breaking down an illustration job into its basic components you'll be able to quickly determine what techniques might best be applied and in what order. Study Figure 10.21 and identify the basic shapes that make up the subject.

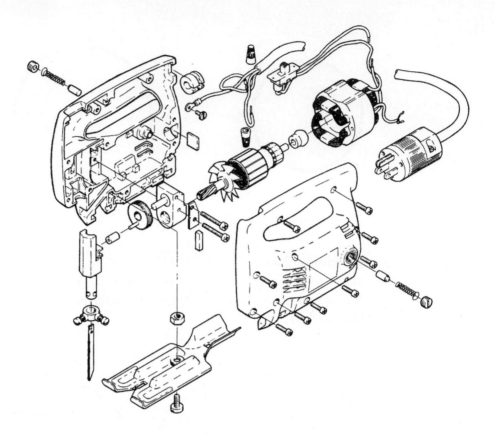

Figure 10.21   Basic shapes make up more complicated geometry.

Block Shading

Block shading isn't line rendering per se, but because the judicious application of block shading can add additional realism to line renderings, it is covered here. It is also, like line rendering, a non-gradient technique.

Block shading actually adds more emphasis to the boldest lines on your illustration. Take for example the objects in Figure 10.22. Four examples of block shading are utilized. A word of caution: a little block shading goes a long way. Excessive block shading creates the illusion that the object itself is black and may actually decrease the informational quality of the illustration. You don't have to block shade everything, only critical places where lines might be misunderstood. For example, an ellipse on a surface can hardly be perceived as a hole. But the addition of block shading removes any confusion. The same applies for overlapping wires.

Hole Shading

The portion of a hole closest to the sun receives a small crescent of shadow, giving the holes a sense of depth (Figure 10.22a). Note that the shadow emanates from the hole's coordinate axis and is slightly curved.

Sphere Shading

A sphere also receives a crescent of block shading on the lower portion away from the sun (Figure 10.22a).

Spring Shading

The realism of a spring is increased with the careful application of shadows where a turn of the spring casts its shadow on the next turn (Figure 10.22b).

Hose Shading

Like the turns of the spring, a small shadow can add significant dimensionality to hoses or tubes that overlap (Figure 10.22c).

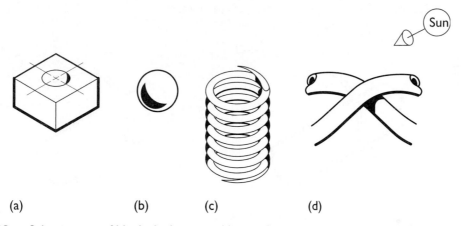

(a)   (b)   (c)   (d)

Figure 10.22   Selective use of block shading can add to realism.

Gaps and Hot Spots

A final touch can be used to add snap and sparkle to your line illustration. Observe the differences between the two chamfered cylinders in Figure 10.23. By the position of the sun, a reflection or *hot spot* would be formed at the top of both ellipses. By varying the size and position of this hot spot, you can portray different materials under different lighting conditions. Don't go overboard with hot spots. If there are too many, the geometry tends to fall apart.

When one line on an illustration is in front of another (closer to the viewer), a break in the line to the rear gives a better impression of spatial depth. Figure 10.24 shows how breaking lines before they intersect gives a visual cue that one part (a wire) is in front of the other (a cylinder). These breaks should be small. If they are too large, they give the impression of two separate objects, rather than one continuous object.

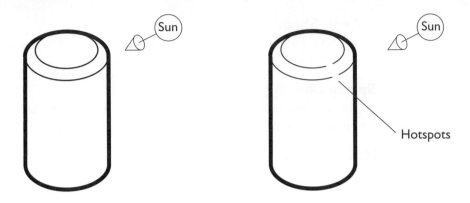

Figure 10.23   Hot spots can add a greater feeling of cylindrical shape.

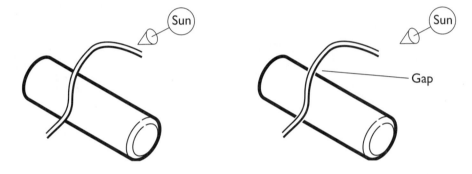

Figure 10.24   A break in a line represents spatial depth.

▼ Tech Tip

You may not want to actually cut holes in line work because this changes how Adobe Illustrator will eventually fill the shape. Instead, create a small white patch (a circle or box) that can be duplicated and positioned as needed. If you do punch holes in lines to represent hot spots or gaps, make a copy of the unedited illustration on a new layer. That way, you have a copy of the complete geometry should you want to apply gradients at a later date.

Step-by-Step Line Rendering

On the accompanying CD-ROM, you'll find the file *doorcloser.eps* for you to use to complete a line rendering. The next section will get you started rendering this illustration. With the *doorcloser.eps* file open, and your new line weight styles displayed in the **Styles Palette** (Figure 10.25), you can now begin applying the line weight rendering described earlier in the chapter.

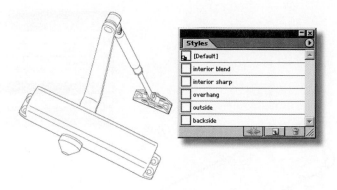

Figure 10.25   The *doorcloser.eps* Illustrator file open with your new styles displayed in list view.

▼   First select the interior lines of the object with your cursor. With these selected, click on the blend line style in the **Styles Palette**. Figure 10.26 illustrates this step.

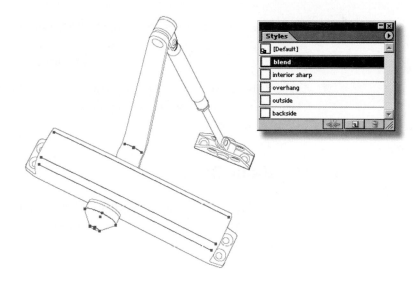

Figure 10.26   Specific line weight applied to interior blend lines of the door closer.

▼   Additional interior blend lines can be selected and the style applied at any time.

▼   Select a few of the outside lines of the object—those that separate the object from the background. Again, with these selected, click on the outside style in the **Styles Palette**. Figure 10.27 illustrates this step and the resulting change to the illustration.

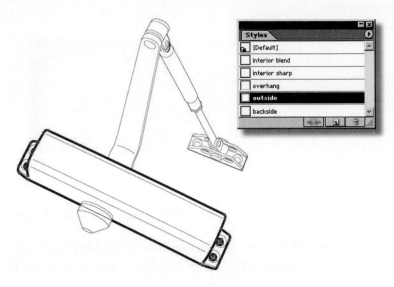

Figure 10.27 Specific line weight applied to outside lines of the door closer.

▼ Continue in this manner, utilizing the line styles you have built. Pay particular attention to edges that overhang within the boundaries of the geometry. Use backside emphasis to add additional solidity.

▼ Inspect the illustration for places where selective block shading might add clarity. Finally, add line breaks to pull some lines to the front and send others to the rear.

A Note about Brushes

A brush is an artistic technique created by the **Brush Tool** or applied to a path created by the **Pen Tool**. Because the brush can have a variable shape it is possible to use a brush to arrive at results similar to that of traditional pen-and-ink illustration. Traditional line illustrations are done with technical pens (Rapidograf, Mars, Castel, and so on), and because the technique is analog, an infinite number of transitions between various line weights are possible. Indeed, in some cases illustrators may achieve a high degree of success using brushes. If you always work in isometric views you may find a set of isometric brushes to be effective. The problem is that brushes lack flexibility. Each brush must be designed for a specific kind of transition. You could create a whole set of brushes for, say, isometric views. You would need maybe a dozen to account for all the transitions in the three isometric planes. Nonisometric transitions would require new brushes, as would every dimetric and trimetric view. Unless you always produce the same view, using brushes simply isn't cost-effective.

Take the hole in Figure 10.28 for example. It is rendered with brushes such that a smooth transition is achieved between the thicker overhang-outside front edge of the hole and the thinner interior sharp of the rear half ellipse.

▼ **Illustrator Tip**

The angle of a brush is critical because it is the angle that controls the thick-to-thin orientation. This angle is relative to the drawing's XY coordinates and stays fixed when you rotate geometry. That's why you can't simply rotate our example from Figure 10.28 into the right and left vertical planes. You must cut the ellipse in two at the ends of its major axis. At this point the brush must be the same width where the thinner rear ellipse contacts the thicker front ellipse.

▼ Create a **Calligraphic Brush** (Figure 10.29) with **Roundness**, **Angle**, and **Diameter** appropriate for this line.

▼ Create another brush that starts with this end thickness, becomes thinner, and ends with the same thickness. Do this by setting the **Angle** to 0 degrees.

▼ Create a backside brush to represent where the hole exits the object.

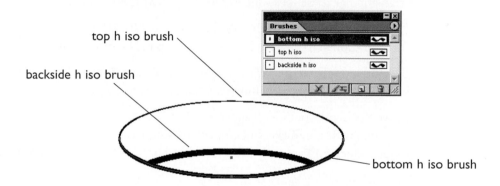

Figure 10.28   Hole rendered with variable-thickness calligraphic brushes.

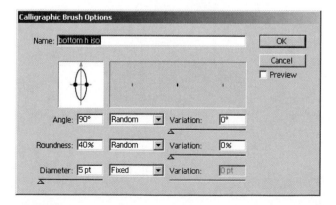

Figure 10.29   Brush parameters for the bottom horizontal ellipse.

You now have the three brushes that represent an elliptical hole in the horizontal plane. The results are effective. However, two more sets of brushes must be created for the right and left vertical planes *and* brushes must be created for all non-cylindrical features such as prisms, toruses, cones, spheres, and irregular shapes. So you can see, the more variable your illustration tasks (I didn't even mention perspective), the more work is required to create the correct brushes.

Variable Lines through Duplication

Many times smooth transitions aren't worth the effort because an illustration is reproduced at a scale where transitions can't be discerned or because the importance isn't that critical. But if smooth transitions are justified and you don't want to create calligraphic brushes, there is a simple alternative.

Figure 10.11 is reproduced in Figure 10.30 with two areas identified where one line thickness and another make sudden changes Figure 10.30a. We can arrive at a smooth transition at these locations Figure 10.30b by the following steps.

▼ Duplicate the thinner of the two lines (Figure 10.31). Cut this line where the transition starts. Move it directly on top of the original line.

▼ Select the duplicate line with the **Open Selection Tool**. Select the handle on the transition end and move the handle away (Figure 10.31a).

▼ Move this handle until it offsets from the original line and make a smooth transition with the thicker line (Figure 10.31). Because you are using the **Open Selection Tool**, only the selected end of the line moves Figure 10.31b.

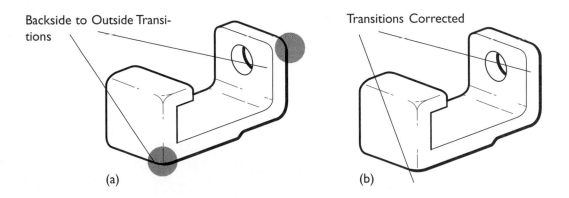

Figure 10.30   (a) Locations of line style transition and (b) transition corrected by duplication.

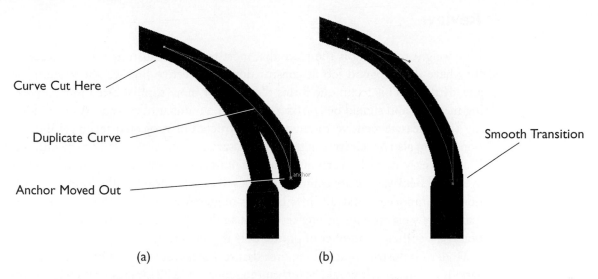

(a)　　　　　　　　　　**(b)**

Figure 10.31　　Duplicate line end repositioned to make first transition.

▼ Duplicate the original line again and move it directly on top of the original line. Move its anchor to the other side of the original line (Figure 10.32a), completing the smooth transition (Figure 10.32b).

By triple-stroking, the outside line now makes a smooth transition with the *backside* line. Reexamine Figure 10.3a. You should be able to identify several locations where this duplication technique has been utilized.

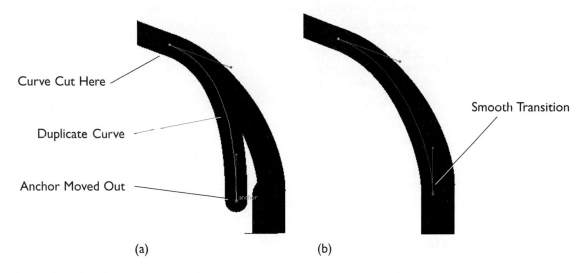

(a)　　　　　　　　　　**(b)**

Figure 10.32　　Second duplicate line end repositioned to finalize transition.

Review

Line weight rendering is the place that most technical illustrators start because if they have done a good job at construction, an effective line weight rendering is already there. This technique forms the basis for more sophisticated tonal rendering methods and should be considered first for a number of reasons. A line rendering is the least-expensive method for constructed or projected technical illustrations. It can also be selective in that unnecessary or confusing detail can be omitted. Plus, it's easy to make versions of a line illustration, reducing future illustration overhead. Because you are counting on lines to communicate all the form and function, it's important that all lines be fully rendered at final resolution. You must match line weights with printer resolution so that the thinnest line on the illustration has a sufficient number of printer dots for rendering.

Modern industrial practice requires that technical illustrations be repurposed at various resolutions. It would be difficult to repurpose a 72-dpi raster Web image for quality print publication. Likewise, there are problems in taking a PostScript illustration with fine line work needing 1240 ppi or more, and rasterizing it at 72 dpi for the Web.

Still, even with these limitations line weight rendering remains an indispensible rendering technique for the complete technical illustrator.

Text Resources

Davis, Jack and Dayton, Linnea. *Adobe Photoshop 7 One-Click Wow!*. Peachpit Press. Berkeley. 2002.

Kelby, Scott. *Photoshop 7 Down & Dirty Tricks*. New Riders. Indianapolis. 2002.

McClelland, Deke. *Photoshop 7 Bible*. John Wiley & Sons. New York. 2002.

Internet Resource

Photoshop World Wide Users Group. Retrieved from: *http://www.wwug.com/forums/adobe_photoshop/*. November 2002.

On the CD-ROM

For those of you who would like to try out the procedures described in this chapter, a resource has been provided on the CD-ROM. Look in *exercises/ch15* for the following file:

doorcloser.eps Completed technical illustration geometry ready for line rendering practice.

Perspective Drawing

Axonometric Projection

Technical Animation

Rendering

Technical Illustration

3D Modeling

Chapter 11

Photo Tracing

In This Chapter

This chapter presents a fundamental method of creating technical illustration line work. You'll see later in the chapter that this line work can also become the basis for more sophisticated illustrations. This method is called *photo tracing* and is useful when dimensions, engineering drawings, or CAD data do not exist, but when photographic data does. Photo tracing is the first step in creating accurate line work for both line and continuous-tone illustrations.

Because photographs, unless taken with an extreme telephoto lens, have varying amounts of perspective convergence, it's important to have a solid understanding of perspective theory when approaching a photo tracing. If you are aware of where vanishing points and horizon lines are located, your photo tracings will be more accurate and effective.

Chapter Objectives

In this chapter you will understand how to:

▼ Analyze a photograph to determine whether or not manual tracing or software-based autotracing is appropriate

▼ Set parameters when scanning photographs for placing in PostScript drawing programs

▼ Find the underlying structure of horizon lines, vanishing points, and thrust lines that control the geometry of the photograph

233

- ▼ Use receding thrust lines to assure the correct alignment of ellipses

- ▼ Edit the construction for visibility

- ▼ Apply simple line weight rendering and selective block shading for added realism

- ▼ Understand the relationship of an accurate photo tracing to a finished continuous tone illustration

The Theory behind Photo Tracing

A complete technical illustrator is able to produce effective results from a variety of source materials. It is not uncommon, when assigned a technical illustration job, to be presented with photographs as the only source of information about the job. Possibly, the assembly or product you are illustrating is a custom prototype, and no drawings exist. Or such significant changes have been made that drawings, if they do exist, are worthless.

If you are used to working from CAD data or engineering prints, and neither is available while photographs are, you must be able to use the photographs. The photo tracing technique is usually faster and more accurate than starting a construction from scratch.

Photographs have two advantages. First, a photograph reflects the actual condition of the subject. If a technician chooses to disregard an installation diagram, the photograph will reflect how the product looks as it's shipped. Second, sizes and geometric relationships are correct in photographs. You don't have to worry whether you have the correct ellipse exposure or the right measurement.

But photographs also have distinct disadvantages. The photo has already been taken and there is much you don't have control over. The projection is perspective, and the distance from the subject and the angle of view have been fixed. The farther from the target or center of the camera's lens (the center of vision), the greater the difference from the axonometric view. Also, in tight spaces photographers tend to use very short focal length lenses, and the shorter the focal length of the camera, the more the perspective view deviates from an axonometric view. Consider any 35 mm camera lens less than 200 mm to include perspective; any lens less than 45 mm to introduce a degree of distortion; and any lens less than 25 mm to introduce a degree of *unacceptable* distortion.

A perspective view, by its very nature, is the exact opposite of an axonometric view. Similar distances cannot be compared across a perspective view because distances appear smaller the further the distances are from the viewer (this is called convergence).

The good news is that the tilt of an ellipse in a perspective view is controlled by its perpendicular thrust axis, as is an ellipse in an axonometric view. However, perspective ellipses vary in exposure as they move to the outside of the viewer's range of vision, and as you will see, the ellipses are not true ellipses at all.

You may want to return to Chapter 9 to make sure you are familiar with the terms and concepts presented there because they will be used in this chapter.

Autotracing a Photograph

PostScript drawing programs almost always have tools that will automatically trace raster images and produce Bezier curve outlines. A *Bezier curve* is a mathematical description that shapes a curve based on the position and attraction strength of control points on the curve. Autotrace tools work by comparing adjacent pixels and detecting edges of objects where values change. Even though parameters can be altered to tighten or loosen these comparisons, successful autotracing is highly problematic. You have to have a high contrast raster image of almost perfect subject matter. You will probably find, as I have, that these tools are worthless for the vast majority of continuous-tone photographic images you want to trace. You'll spend more time editing the autotrace than you'd invest in a manual tracing, plus, you won't have the geometric accuracy.

Stand-alone programs such as Adobe's Streamline give you more control over the autotracing process but only marginally improve results on all but the most favorable photographs. More industrial-strength applications such as Intercap rely on autotracing of existing paper drawings and photographs with increased success, but of course with increased computing requirements.

If you are able to manually trace a photograph with accuracy, you will be able to evaluate the effectiveness and appropriateness of automated tracing products. Because photo tracing relies almost totally on manually tracing features on a photograph with the **Pen Tool**, a short discussion on Bezier curves is warranted. Be advised, however, that skill in this area is only developed by practice; you have to trace several photographs before you will feel comfortable with the **Pen Tool**.

Pen Tool Options

The **Pen Tool** has several options that need to be understood before the tool is used.

▼ The **Pen Tool** starts a curve, continues a curve, and then ends the curve. Depress the **Ctl Key** and the pen turns into the last **Selection Tool** used. Click anywhere on the drawing to lose the tool and complete the line. You then can start a new discontinuous line.

▼ A single click creates a sharp corner; a click-and-drag action creates a curve. Figure 11.1a shows a line created by three individual clicks of the **Pen Tool**. It looks like a polyline. Figure 11.1b shows a line started with a single click, continued with a click-and-drag action, and finished with a single click. In Figure 11.1c all three points on the line have been created with a click-and-drag action.

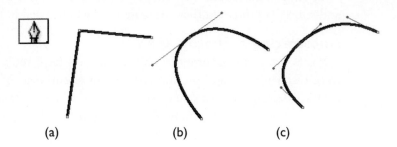

(a)   (b)   (c)

Figure 11.1   Pen actions determine the shape of the curve.

▼ The **Add Point Pen Tool** (Figure 11.2) adds corner points (click) or curve points (click-and-drag) to a selected curve. Because you have this tool you don't have to worry about having the optimum number of curve points when first creating the trace.

▼ The **Delete Pen Tool** (Figure 11.3) removes a point when the tool is clicked on it. Note that the curve segment becomes flatter and straighter because the curve point that rounded it (from Figure 11.2) has been removed.

▼ The **Change Point Type Pen Tool** (Figure 11.4) changes a corner point to a curve point and a curve point to a corner point. The curve from Figure 11.3 has been turned into a polyline because the middle point has been switched from a curve point to a corner point.

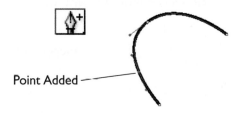

Point Added

Figure 11.2   The **Add Pen Tool** adds a corner point (click) or a curve point (click-and-drag).

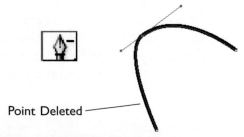

Point Deleted

Figure 11.3   The **Delete Pen Tool** removes a curve point.

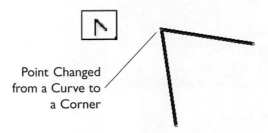

Point Changed
from a Curve to
a Corner

Figure 11.4   The **Change Pen Tool** switches curve points into corners and corners into curves.

Pen Tool Practices

In practice you want as few points as necessary to define the curve. Start with corner points at each end of the curve and drag a curve point at the apex (highest or lowest point) or any place the curve changes direction or is tangent to another shape. Remember, you can always add, delete, or change points.

▼ Begin with a preliminary tracing. Use corner points at each end and curve points in between (Figure 11.5).

▼ Using the **Filled Selection Tool** on the curve will move the entire curve. Using the **Open Selection Tool** on the curve will allow individual curve points and their handles to be edited.

▼ Click on the curve itself with the **Open Selection Tool**. The curve is highlighted, and the curve points are open boxes. This means the curve itself or individual points can be moved. Click and drag a curve point until the curve better approximates the desired tracing (Figure 11.6). The point becomes filled (selected), and its handles, if it's a curve point, become visible. Once the point is in the correct location, adjust its handles until the curve matches the photograph.

▼ Again with the **Open Selection Tool**, click on a curve point whose curve shape on either side needs to be reshaped (Figure 11.7). The point becomes filled and its handles become visible. If you click on the curve itself, and not on a curve point, only the handle on that side is displayed. Click on the point to display both handles.

▼ Move these handles by the little boxes at their ends to reshape the curve coming into and going out of the point. Bezier curve handles are independent. That is, the curves on either side of the point don't have to be symmetrical.

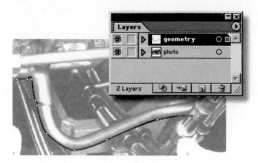

Figure 11.5   Preliminary tracing.

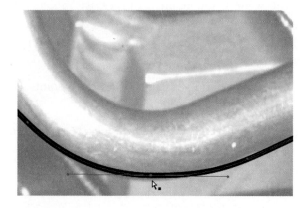

Figure 11.6   Move curve points to better approximate the tracing.

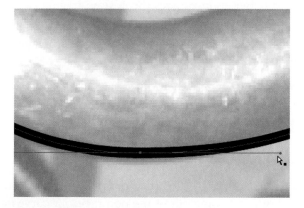

Figure 11.7   Move handles to reshape the curve between the points.

The Goofy Curve Point

Every so often you have a curve point that doesn't make sense. You pull on the handles and the curve doesn't respond the way you think it should. The reason for this may be that the curve handles are turned inside out.

▼ Observe the curve in Figure 11.8. The middle point has only one handle extending from it, and the curve appears twisted.

Figure 11.8   Curve with a suspicious point.

▼ Click on the suspect point to select it, and grabbing the handle (Figure 11.9), rotate it about the point until the curve straightens out (Figure 11.10).

Figure 11.9   Point rotated by one of its handles.

▼ The point's other handle has been collapsed on top of the point itself. With the **Open Selection Tool**, grab just to the left of the point and drag. You may be rewarded with the missing handle (Figure 11.10). If not, use the **Change Pen Tool**, change the point to a corner, and then change it back to a curve. You should have your two handles.

Figure 11.10   Point rotated and handles extended.

Circle Geometry in Perspective

Figure 11.11 demonstrates the effect of distance from the *center of vision* on a circle's orientation and exposure. A circle, directly in front of a viewer and perpendicular to the direction of sight, will appear as a true circle. As that circle moves horizontally away from this central position, it becomes less and less exposed (narrower), yet its minor axis remains aligned to its perpendicular thrust (the line from the center of the ellipse to the vanishing point).

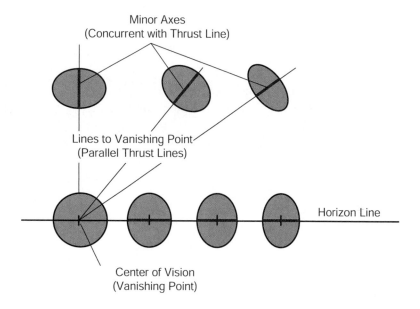

Figure 11.11   The relationship of the ellipse minor axis and the exposure in a perspective view.

At all times, the minor axis of the ellipse remains concurrent with the axis perpendicular to the circle (thrust axis). The ellipses on the horizon line in Figure 11.11 all are aligned such that their minor axes remain concurrent with the line from the circle's center to the vanishing point (the center of vision).

The ellipses above the horizon line show this more clearly. An ellipse directly above the center of vision is less exposed than the one centered on the vanishing point. As an ellipse is drawn out from the center of vision notice how, by keeping the minor axis aligned with this thrust axis, the ellipse tilts.

▼ Note

Perspective distortion is introduced at any point away from the center of vision and increases exponentially the greater this distance. As you might surmise, you'll be more successful with tracing detail close to this center of vision.

▼ Illustration in Industry

Knott Laboratory
Denver, Colorado

Images Copyright Knott Laboratory.

Software

AutoCAD 2002
3ds max 4.2
Adobe Photoshop
Adobe Premiere
Adobe After Effects
PC Crash

Hardware

Windows P4x2 PC
Rendering Farm

Vehicular accident reconstruction requires engineering precision, for physics-based motion and dynamics, as well as high-end material, lighting, and rendering procedures. In this project, for fun, we were interested in the forces needed to tip a semi-tractor trailer. It took several cars at highway speeds, impacting at different times, to do the job. Our texture and procedural maps were made in-house, combining real photographs and videos with computer graphics materials. A detailed survey was completed at the scene in Los Angeles to obtain accurate geometry of the road, buildings, and scenery. This was generated in three dimensions and materials representing signs, store fronts, and road markings were applied to the models. Physics-based sun systems and shadow systems were used to create realistic lighting ,and blur effects helped show the effects of moving vehicles.

The Team

William Neale, Director
Nathan Rose
Chris Hughes
Tom Reyes
Scott Mcfadden
Hailey King

Scanning Photographs

The quality of a source photograph (contrast, detail, grain, or digital resolution) contributes to the ultimate success of a photo tracing. Consider the following when scanning photographs for tracing:

▼ The greater the resolution and bit depth of the scan, the more trouble you'll have while tracing (zooming, panning, saving). Always use the *least* amount of scan data required to do the job.

▼ Though PostScript illustration programs are resolution independent, raster scans are resolution dependent. The larger the physical dimensions of the original scan, the greater the resolution and, thereby, its detail.

▼ If the photo is in color, give up color for greater resolution and grays.

▼ Photographs with a full range of continuous tones (256 grays) will almost never autotrace effectively because there are no definite changes from black to white. Always keep the original photograph handy for reference. You don't want to rely on the scan as your sole technical source.

Figure 11.12 shows a technical photograph scanned at 150 dpi, in gray scale, and placed on the default Illustrator layer. This layer has been renamed "photo" and turned into a template with a 75 percent image display. By making the photo layer a template, it can't be moved. By dimming the display, all black and color lines used in construction will be easily seen. To follow along with the instruction that follows, import *valves.jpg* found on the CD-ROM in *exercises/ch11*.

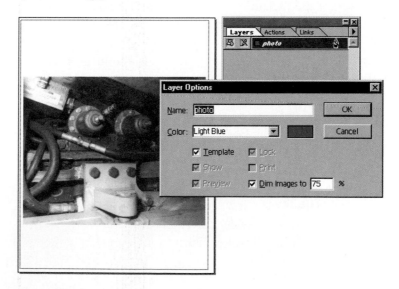

Figure 11.12   Place the scanned photograph on Illustrator's renamed default layer.

Perspective Geometry

Because the photograph in Figure 11.12 was taken using a rather short focal lens (25 mm in a 35-mm film format), a significant amount of perspective (and subsequent distortion) has been introduced. Figure 11.13 establishes the basic geometry of two of the photograph's most salient features: the two valves. These valves are parallel and horizontal, with axis lines nearly perpendicular to the plane of the photograph. When central axis lines through the valves are drawn on a "geometry" layer, they cross on the horizon line, at the left vanishing point (LVP). This left vanishing point is where all lines horizontal and parallel to the axes of the valves are directed. Compare Figure 11.13 with Figure 11.11. The circular features of the valves are represented by the right two ellipses above the horizon line in Figure 11.11. They are tilted so that their minor axes are concurrent with their central axes, and they have become thinner, or less exposed.

Figure 11.13　　Parallel features locate vanishing point and horizon line.

▼ Rule # 3

Parallel lines will always be directed to a common vanishing point. Two sets of parallel lines will then define two vanishing points and the horizon line for that plane will pass through those two points.

You can tell that the machine itself is almost parallel to the picture plane because horizontal lines above and below the horizon line are almost horizontal. This means that the LVP is very close to the location of the center of vision (CV). You can see then that a majority of the photograph is outside the area around the center of vision. This is not the best situation for a photo tracing because most of the illustration will be moderately to severely distorted.

The three large hexagonal fasteners (hex bolts) below the valves are aligned horizontally, as are the centers of the valves. If one were to draw lines through these horizontal points (Figure 11.14) and extend the lines until they cross, they should cross at a point on the horizon line. This vanishing point is where all lines horizontal and parallel to the axis of the machine are directed.

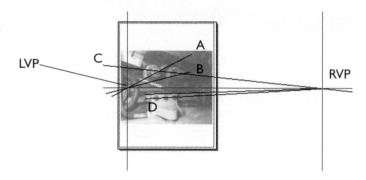

Figure 11.14   Other parallel features locate additional vanishing points.

▼ **Hint**

The accuracy of the vanishing point is a function of how far apart these lines are and how far apart the points are that determine the lines. A small amount of error on two points very close together can multiply as the line is extended to the horizon line.

In Figure 11.14 a fundamental perspective condition is established. The two parallel valve axes are horizontal and locate the LVP. Lines C and D are horizontal and parallel and locate the RVP. If you have been accurate and there isn't significant distortion in the photograph, the LVP and RVP should define a horizontal line, in this case, the *horizontal horizon line*.

Don't be surprised if parallel lines don't perfectly meet at the same point on the horizon line. Also, don't be surprised if two vanishing points don't line up perfectly on a horizon line. This is due, again, to a combination of *distortion* (the camera's lens) and *accuracy* (your ability to find centers and edges and to estimate angles). One solution (after checking for accuracy) is to spread the difference out between the points, until the desired condition is met.

Why is it important to locate these horizon lines and vanishing points when doing a photo tracing?

▼   Actual edges on the photo may be difficult to detect. With a correct vanishing point, you only need one point on the line to accurately find its direction.

▼   Distortion at the edges of the image may be unacceptable. By using consistent horizon lines and vanishing points you can *normalize* the perspective, taking the distortion out.

With the horizon line, and RVP and LVP established, the controlling geometry for parts on the photograph can be found. Figure 11.15 shows a detail of the photograph with axis lines for parts parallel to the valves. These axis lines are perpendicular to the projection plane and are all directed to the center of vision. They are the lines that will determine the orientation or tilt of ellipses.

Figure 11.15 Other parallel features such as axis lines are directed toward common vanishing points.

Create a Vignette Technique

Because you almost always want to feature or highlight a portion of an illustration, and because geometry becomes more and more distorted the greater its distance from the center of vision, a technique called a *vignette* is used. Simply stated, a vignette is an image that fades out from its center. But because we are doing a line tracing, we can't actually fade out our lines. But we can choose when to stop drawing, giving the illusion of a vignette.

Figure 11.16 shows the photograph in Figure 11.15 edited into an oval, centered about the major features. The line tracing will end at the edges of this oval, producing what for a line drawing is a vignette. Additionally, with the reduction of bit data in the placed photograph, the display will pan, zoom, redraw, and save more rapidly.

▼ Illustrator Tip

The **Closed** (black) **Selection Tool** is used to move a curve without changing its shape. The **Open** (white) **Selection Tool** is used to reshape the curve, move its anchors, and reposition the handles. If an acchor is filled in (black), it is selected; if an anchor is not filled in (white), it is ready to be selected.

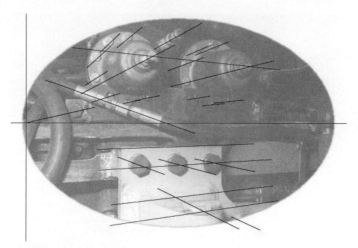

Figure 11.16   Define an oval area to create a vignette.

Ellipse Axes and Perpendicular Thrust

Figure 11.11 established the relationship between the minor ellipse axis and the thrust line perpendicular to the circle that recedes into the distance to the vanishing point. Figure 11.17 shows an ellipse defining the front of the left-hand valve body (a cone), correctly aligned with the thrust axis. Salient points (center and minor axis ends) are aligned on the thrust line. The exposure of the ellipse is determined by stretching its handles until the elliptical shape visually matches the photograph. This takes experience, and the more you work with ellipses, the more quickly you can match the photograph.

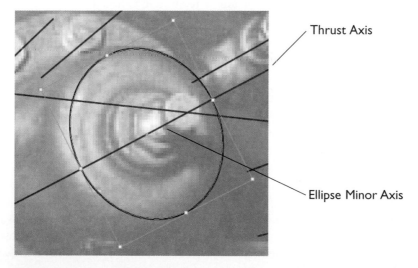

Figure 11.17   An ellipse is rotated so that its minor axis is concurrent with the perpendicular thrust.

Always find a large ellipse along a given thrust line. This first ellipse can be used (**Copy**, **Paste**, and **Scale**) for all subsequent ellipses along the axis. The exposure (width) may have to be increased slightly as the ellipse approaches the center of vision, but it is much more efficient than beginning each ellipse from a toolbox oval.

Figure 11.18 shows how this original ellipse has been enlarged or reduced to coincide with the valve's circular features on the photograph. A small amount of editing with the **Scissors Tool** results in a completed valve body (Figures 11.19 and 11.20).

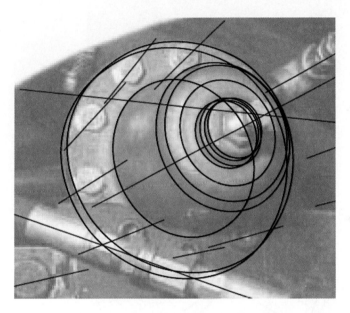

Figure 11.18   Additional circular features are created from the first ellipse.

By starting with a relatively large ellipse you have the best chance that subsequent ellipses, made from the first by duplicating and scaling, will be accurate.

▼ Select the ellipse with the **Filled Selection Tool**.

▼ Choose the **Scale Tool** and leave the origin at the center of the ellipse.

▼ Grab the ellipse on its perimeter and depress the **Shift** key.

▼ Turn off **Smart Guides** if they are on. Move the cursor along an imagined diagonal. If you move the cursor horizontally or vertically, the resized ellipse will tend to snap to the original height or width.

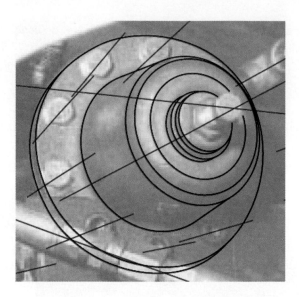

Figure 11.19 Ellipses are edited for visibility, resulting in a finished part.

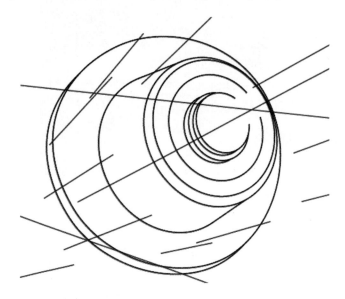

Figure 11.20 Valve body with photo layer hidden.

In an axonometric drawing, the second valve could easily be made by copying the first and moving it to its own axis. But in a perspective drawing, the second valve body has a central thrust line at a shallower angle (Figure 11.15), further removed from the center of vision. The ellipses defining this second valve body are rotated and narrowed and must be constructed separately.

Hexagonal Fasteners

There are several hexagonal fasteners in the photograph in Figure 11.15 holding the front and rear sections of the valve bodies together. Because the axes of all mounting holes are parallel to the central axis of the valve, all axes are directed to the same vanishing point. Because of the small size of the fasteners, a simplified construction will be conducted.

Figure 11.21 shows the beginning elliptical construction for one of the hex mounting bolts. Notice that all centers and minor axes are on, or coincidental with, the thrust line. Rather than constructing the hexagon from scratch, we will use the **Polygon Tool** to create a quick shape. This hex must fit the ellipses in Figure 11.21. But since we didn't use AxonHelper to determine the exposure, rather simply matched the exposure of the ellipse to the photograph, we don't know the major-to-minor axis relationship needed to scale the hexagon to match.

Figure 11.22 shows one of the ellipses from Figure 11.21 rotated to the horizontal so that its major-to-minor axis relationship can be determined. In this case, the minor axis is approximately 90.95 percent of the major axis. The hexagon is scaled identically.

Figure 11.23 shows this scaled hexagon, rotated so one of its sides (the middle side) is perpendicular to the fastener's thrust. This first hex is on the bottom of the fastener. Hex bolts will hardly ever be perfectly aligned like this, but if you force them to be, they will lie in perspective much better.

Figure 11.24 shows this first hex bolt edited for visibility. Figure 11.25 shows it correctly aligned on the mounting bolt's thrust line.

Ellipses Aligned

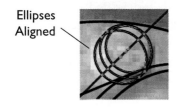

Scaled to Fit

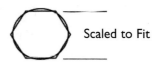

Figure 11.21 Hexagon approximation by ellipse.

Figure 11.22 The hexagon scaled for body.

Hexagon Aligned

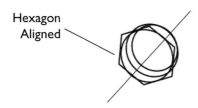

Figure 11.23 Scaled hexagon is rotated.

Figure 11.24 Completed and edited fastener.

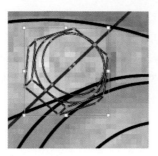

Figure 11.25   Completed hexagonal fastener aligned on photograph.

In Figure 11.11 we demonstrated how the exposure and tilt of an ellipse in a perspective view changes as ellipses are positioned away from the center of vision. This is one reason why a perspective view may be inappropriate for large subjects that have considerable detail. To see the detail, you have to be up close to the object. But the closer you are, the more of the subject is outside your cone of vision, introducing unacceptable distortion. In our valve assembly example, we have omitted detail outside the cone of vision by making use of the vignette technique.

Because the mounting bolts are fairly close together on each valve body, much time can be saved by using one hex bolt as the basis for the others on the same valve, even in a perspective view. If we keep the middle hex plane perpendicular to the thrust, and adjust the scale, credible (and highly efficient) results can be achieved. Figure 11.26 shows the completed left side valve with its mounting bolts. The original bolt (at the top) was **Copied**, **Rotated**, and **Scaled** as necessary for the other three positions. Additionally, other hexagonal features such as the adjusting screw and jamb nut on the front of the valve body were made from this original hexagonal fastener construction.

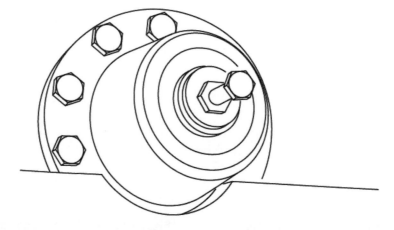

Figure 11.26   The original construction is duplicated, scaled, and rotated as necessary.

Features at Other Angles

The hose below the left valve in Figure 11.15 doesn't adhere to the two vanishing points we previously located. Still, a central axis or thrust line must be found to assure that features are properly aligned. Figure 11.27 shows the hose fittings with their minor axes properly aligned with the axis of the hose. Figure 11.28 shows the fittings completed and edited for visibility. As was done before, much of the geometry was created by **Copying**, **Pasting**, **Scaling**, and final editing.

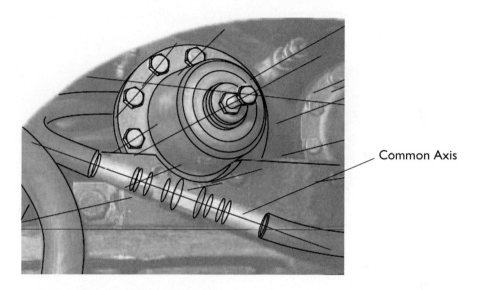

Figure 11.27   Other aligned cylindrical features are drawn by first locating the common axis.

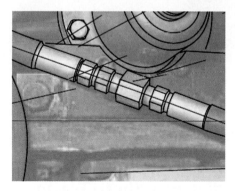

Figure 11.28   Completed hose fittings, edited for visibility.

Line Caps and Joins

As the details of the photo tracing were completed, maximum use was made of object snaps and **Smart Guides**. But without specifying how lines will be finished about their ends and corners, intersections—even with end points perfectly aligned like that shown in Figure 11.29—will yield less than desirable results. Breaks in continuous lines, like that in Figure 11.29, can be avoided by selecting all line work and turning on the **Rounded Cap** and **Join** settings in the **Stroke Window**. Figure 11.30 shows the effect of changing these two line settings. This produces a result similar to a traditional ink illustration where lines have a distinctive rounded termination.

Figure 11.29   Default end **Cap** and **Join** settings do not produce smooth, continuous lines.

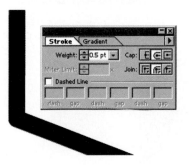

Figure 11.30   By extending the **Cap** and **Join** with the **Rounded** option, smooth lines can be made.

Line Weight Shading of a Photo Tracing

A final line tracing is shown in Figure 11.31. You can see the vignette effect as the lines stop at the edge of the oval photograph. For some purposes, this can be considered a finished technical illustration, and call-outs or other descriptors can be added to increase its effectiveness. But to make the illustration even more descriptive, selective lines need to be made bolder or thicker. This was covered in detail in the Chapter 10.

You don't want to overrender the line tracing, so start with the simplest approach. (Always copy the line tracing to a new layer called *render* so that the original geometry is preserved.) Figure 11.32 shows a simple two-line technique where the outside edges of each part are made heavier. The increase in readability is obvious.

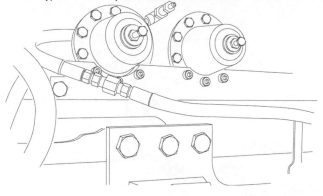

Figure 11.31   Completed photo tracing with all lines rendered in a single line weight.

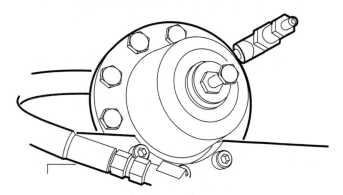

Figure 11.32   A simple two-line rendering scheme.

If parts still look flat, especially where one part overhangs another, consider a three-, four-, or possibly five-line rendering scheme. At all times, the lowest number of line changes needed to effectively communicate the illustration will result in greater profitability for you.

Selective Block Shading

You may need to add selective block shading to the illustration (Figure 11.33) to further add a 3D feeling. Block shading represents areas of geometry in shade or shadow. By applying selective shadows, greater depth can be achieved. This should be done carefully, on a separate layer, so that overzealous use of shadows can be easily removed.

In Figure 11.33 you can see that the shadow of the valve flange on the rear hose effectively sends the hose to the rear. Likewise, the shadow below the hose fitting establishes the fitting in front of the frame. Other small block shadows help fasteners appear more three dimensional.

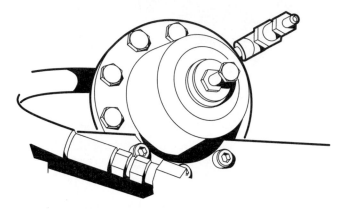

Figure 11.33   Block shading adds shadows to give greater depth and realism.

Continuous Tone from a Photo Tracing

Figure 11.34 shows the valve detail with continuous-tone gray scale applied to the line tracing. Because the original photo tracing probably wasn't done with gradient fills in mind, much of the line work will have to be joined into continuous paths so that predictable gradient rendering can be applied. Again, copy the line work to a new layer to preserve the geometry. You'll want to duplicate selected line work and move it off the photo tracing to join the lines and apply the gradients. When satisfied with the results, move the finished part back onto the line work.

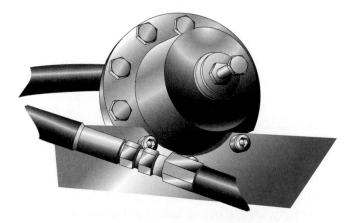

Figure 11.34   Continuous-tone rendering of photo tracing with thin line work displayed for emphasis.

When creating manual blends in Adobe Illustrator to fill a solid area, it is better to blend between shapes rather than lines. This isn't as critical if the area to be filled is small. However, if these shapes are enlarged as much as 200 percent, the potential problem can be seen immediately. In this example, when blending between lines and the **Scale Strokes & Effects** check box is unchecked (as it is by default), the lines may become separated and the smooth blend is destroyed. However, when shapes are enlarged, they overlap and no gaps are seen.

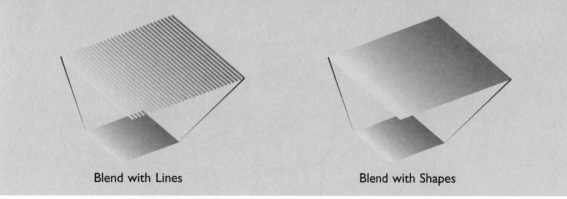

Blend with Lines   Blend with Shapes

Photo Tracing Examples

Large, complicated pieces of equipment are prime candidates for photo tracing. For example, Figure 11.35 shows a photo tracing of a valve installation so that individual components can be identified. It would take hundreds of engineering drawings to describe all its components. Construction in perspective might take a week or longer. But a photo tracing, if you have an acceptable photograph, might require only 8 hours from start to finish.

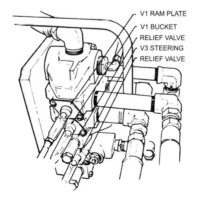

Figure 11.35   Photo tracing for component identification.

Photo tracing is effective for consumer industrial products like the saw shown in Figure 11.36. Contract illustrators can work more easily from photographs, especially if a company doesn't want its engineering drawings (drawings that can be considered legal documents) released to the public.

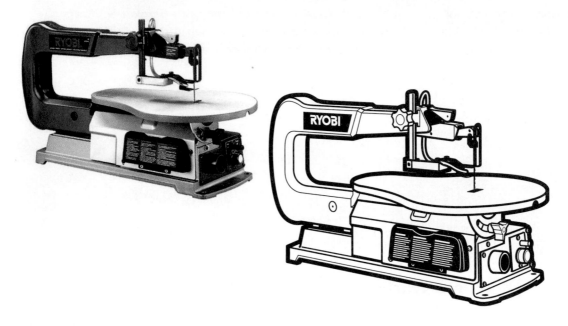

Figure 11.36   Photo tracing of a consumer product (Greg Maxson).

▼ Illustrator Tip

Another method for creating variable line weights makes use of Adobe Illustrator's ability to turn a line into a shape. Because a shape made from a line starts out with consistent thickness, the thicker portion can be formed by editing individual handles. For example the washer in (a) uses a single line weight for all lines. In (b), the lines have been duplicated and converted to outlines (**Object|Path|Outline Stroke**). You can see how the red outlines have been edited for thickness. In (c) the washer is shown as part of a fastener assembly.

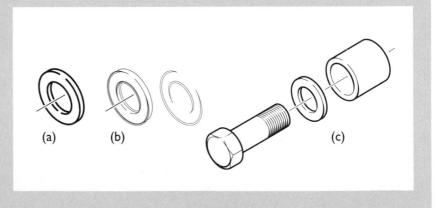

Taking It to the Limit

Chapters 12 through 14 are devoted to techniques that produce resolution-independent, near photorealistic results in color. Many times, a final color illustration begins as a photo tracing. Figures 11.37 through 11.39 show illustrations that are fundamentally different. One uses a line drawing of an antiquity as the basis for the final illustration, while the others present photo tracings of commercial products.

Review

Photographs make excellent resources for technical illustrations because all scaling, foreshortening, and perspective convergence are accurately represented. However, photographs also capture information that you may not want in the illustration, necessitating decisions on your part as to what should be omitted and the choice of rendering method.

Analog photographs must be scanned at a resolution sufficient for illustration purposes, a resolution through which the underlying structure of the perspective can be found. The behavior of lines, curves, and ellipses in perspective views follow strict geometric rules and, once learned, control the appearance of all elements in the perspective view.

Figure 11.37   Product illustration based on an effective photo tracing (Michael Bridgman).

Figure 11.38   Nonmechanical illustration
                    (Yunhe Bao).

Figure 11.39   Product illustration
                    (Takahiro Ikeda).

▼ Note

These examples may be only partially rendered and show the photo tracing line work necessary to support final rendering. Remember that the final illustrations are resolution-independent PostScript and can be printed on a variety of output devices with optimum results.

Figure 11.40   Industrial illustration (Sam Steinborn).

Text Resources

Ang, Tom. *Digital Photographer's Handbook.* DK Publishing. London. 2002.

Freeman, Michael. *The Complete Guide to Digital Photography*. Silver Pixel Press. Rochester, New York. 2001.

McClelland, Deke and Eismann, Katrin. *Real World Digital Photography.* Peachpit Press. Berkeley. 1999.

Internet Resources

The following Internet sources cover the use of Adobe Illustrator techniques. Several also feature applications suitable for technical illustration.

Portsort portfolio site. Retrieved from: *http://www.portsort.com/home.cfm.* November 2002.

Illustrator Answers. Retrieved from: *http://www.illustratoranswers.com/.* November 2002.

ErgoDraw Tutorials. Retrieved from: *http://ergodraw.looktwo.com/e/tutorials/ tutorials.shtml.* November 2002.

Go to Graphics. Retrieved from: *http://www.2ginc.com/tutorials/illus.html.* November 2002.

On the CD-ROM

For those of you who would like to follow along with the photo tracing procedures in this chapter, several resources have been provided on the CD-ROM. Look in *exercises/ch16* for the following files:

valves.jpg   Import this raster file into Adobe Illustrator to follow the steps in this chapter.

photo1.jpg through photo1.jpg   Use these raster images as the bases for additional photo tracingexercises.

Chapter 12

Emphasis with Color

In This Chapter

Up to this point we have dealt with technical illustration constructions and the line rendering that naturally develops from accurate geometry. Effective technical illustrations *must* be accurate; projections and constructions must be chosen that are appropriate for the intended use of the illustration.

Different areas of an illustration can be emphasized by selective changes in line weight, as was discussed in Chapter 10. In many cases, this is sufficient to distinguish between parts or features. But there are instances where additional contrast is necessary—short of realistic coloration—and this involves two fundamental techniques: grayscale shading and spot color.

Of course, the use of grays isn't the same as the use of color, but as you will soon see, using spot color to create emphasis is, at least partially, based on value (lightness and darkness), so understanding grays is a good place to start.

As was pointed out earlier from Horton's *Illustrating Computer Information*, black and white plus one additional color is optimum for identification in an illustration. But as was also revealed, many individuals have trouble distinguishing subtle differences in hue. Indeed, a greater impact is made by varying *value* than in making changes in the same or similar hues. So when using gray or color to highlight components in an illustration remember that color or tone must be really different to be effective.

Chapter Objectives

In this chapter you will understand:

- ▼ The selection of color space appropriate for illustration reproduction
- ▼ The selection of color models appropriate for the illustration task
- ▼ The arrangement of areas of gray tone emphasis, both in terms of front to back, and layer order
- ▼ How spot color swatches are specified in PANTONE colors, added to the color palette, and assigned to selections
- ▼ The relationship of PANTONE and CMYK process spot colors

Illustration Color Space

Adobe Illustrator makes use of two color space definitions: CMYK and RGB. Each attempts to define color in terms of the technologies used for reproduction. Each reproduction technology has certain limitations and is capable of reproducing a different *gamut,* or range of colors. Every printer, scanner, camera, or monitor has its own gamut, and the coordination and matching of these dissimilar gamuts is one of the challenging aspects of reproduction. You may wish to review Chapter 2, where additional concerns of color reproduction are discussed.

CMYK Color Space

CMYK defines color in terms of the cyan, magenta, yellow, and black inks used in the four-color printing process. This is a *subtractive* color model in that the colors we see are the result of allowing only certain wavelengths of light to reflect off the printed surface. So when we see yellow on a printed sheet, that's the color wavelength that the yellow ink reflects. In a subtractive color model, white is the absence of color and black is the combination of all colors. This is important to understand. White is not ink. It is the absence of ink, allowing the paper's white surface to be seen. CMYK is also called *process color* and is used when pigment (ink, toner, or dye) is placed on some medium (paper or film). CMYK color space is defined by percentages of each of the four colors and when applied to PostScript, effectively results in color whose limitation is not the data, but rather the reproduction method. CMYK color is described in 8 or 16 bits each of cyan, magenta, yellow, and black data, resulting in 32- or 64-bit color.

▼ **Note**

In most cases 16-bit CMYK color (64 bits total) is unnecessary or even unusable. Remember, 24-bit color produces a gamut of 16.7 million colors, many of which are outside the range of human perception *and* outside the gamut of many reproduction devices. CMYK color with 16 bits of data per channel produces a file that cannot be used in many graphics programs.

RGB Color Space

RGB defines color by the red, green, and blue phosphors used to illuminate display monitors (or the color gels placed in front of theatrical lighting). RGB is an *additive* color model in that the color we see is the result of combining projective light. In RGB, white is the combination of all colors, while black is the absence of all color. RGB is used when color is displayed on monitors or projection screens. Also a few digital printing devices make use of red, green, blue, and black inks. RGB color is described in 8-bits each of red, green, and blue data, resulting in 24-bit color (16.7 million colors).

The selection of color space is made when creating a new file (Figure 12.1) and should be made with the end use of the illustration in mind. Of course, you may not know an illustration's intended purpose when you begin, and indeed, illustrations tend to be repurposed more and more. A general guideline is to create illustrations as if they were going to be reproduced by CMYK printing. That way you can always export (rasterize) the PostScript to various formats where color space can be reduced. Put another way, create your illustrations with the broadest color space, knowing that you can sample this space down if necessary.

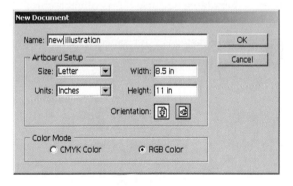

Figure 12.1   Selection of color space model.

You can combine elements from different color spaces in Illustrator but when saving the file, you will be notified of this problem. When opening an Illustrator file with both RGB and CMYK elements, you will be forced to choose one color space or the other. The file will then be CMYK; however, elements may still be RGB and present problems when the file is RIP'd. If you have combined RGB raster data (from PhotoShop, for example) with CMYK PostScript in an EPS file, the solution is to go back and change the original raster color space and place the raster file again in Illustrator.

The results of mixing color spaces can range from trivial to fatal. At best, the file is successfully RIP'd, but colors shift slightly as one color space is converted to another. At worst, the printer, multimedia engine, or Web browser simply chokes and the offending image isn't printed or displayed.

Color Models

Once the color space is selected, you have the opportunity to select the color model (mode) that best fits your needs. A *color model* is a method whereby all colors can be described. RGB and CMYK are two device-dependent models, one for display and one for printing. You may want to use a device-independent model that makes more sense to you as an illustrator. The Adobe Illustrator **Color Window** provides methods whereby you can smoothly move back and forth between color models depending on your needs.

Gray-Scale Color Model

This model defines color only in terms of lightness and darkness. In PostScript, CMYK grays are represented as a percentage of black ink. In RGB color space grays are represented as equal amounts of red, green, and blue light. An RGB value of 115, 115, 115 will result in a gray value because an equal amount of red, green, and blue are in the mix. Figure 12.2 shows this gray-scale color model in the color window.

CMYK files can be very large because 8 bits of each color channel must be saved. A black-and-white (or gray scale) raster illustration should be saved in gray scale mode. That way, when the raster is later saved in EPS format, only black channel data is saved. When placed in Illustrator and saved in EPS, the gray scale raster only contributes to the black separation.

Figure 12.2   Gray-scale color model.

▼ Tech Tip

There is an inverse relationship between screen pitch [*lines per inch* (lpi)] and the number of grays you can display or print with digital data. For a given ppi, the more lpi, the finer the detail that can be printed but with fewer grays. Reduce the lpi and you can print more grays but the screen will be coarser.

▼ **Note**

It is usually easier to visualize gray as a percentage of black rather than equal RGB values. In CMYK, all inks other than black can be set to zero, and black ink adjusted as with gray scale. In HSB, hue and saturation can be set to zero and brightness adjusted as black.

RGB Color Model

This model uses 256 values (8 bits of information, 0 to 255) for each color component. The combination of two RGB primaries results in one of the CMYK primaries. For example, 8 bits of red and blue produces magenta (Figure 12.3), 8 bits of green and red produces yellow and 8 bits of blue and green produces cyan. Although many technical illustrations will be displayed in RGB, most illustrators will use this mode only to check for color matching between print (CMYK) and display (RGB).

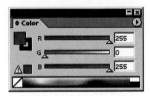

Figure 12.3   RGB color model.

HSB Color Model

The HSB (hue, saturation, brightness) color model defines color using a 360-degree color wheel. In Figure 12.4 the hue is at 30.97 degrees around the wheel. A *hue* is a particular color (red as opposed to reddish orange, for example); saturation is the intensity of the color when combined with white (a tint); *brightness* is the hue and intensity when mixed with black (a shade). This is the preferred color model for technical illustration because once a hue is chosen, saturation can be decreased to arrive at highlights while brightness can be decreased to create shadows without affecting the color or hue.

Figure 12.4   HSB color model.

CMYK Color Model

Figure 12.5 shows the same brown color specified in the previous example, but this time in CMYK. You can easily see the increased complexity of describing brown in terms of four separate components. Illustrators use this color model for matching known CMYK values. The model can then be switched to HSB for determining the full range of highlight, shade, and shadow colors.

Figure 12.5   CMYK color model.

Web-Safe RGB

This color model specifies only those colors that are produced by Web browsers without dithering. This limits the palette to 256 colors (fewer on Windows because a number of colors are reserved for system display). This doesn't mean that you can't display more colors. It means that displaying two or more Web-safe pixels next to each other (dithering) will approximate colors outside the Web-safe palette. This gives the visual impression of the missing color. Web colors are specified in *hexadecimal*, three sets of numerals and letters (Figure 12.6).

Hexadecimal is the mode used for specifying color in a Web browser using HTML, XML, or Java. The only time you would concern yourself over whether a color is Web-safe or will be dithered is if color fidelity is critical.

Because EPS data isn't naturally displayed on Web pages without specialized plug-ins, some graphical gymnastics have to be performed to get images intended for print publications into a form suitable for the Web. Two issues are in play here:

▼ **Resolution**. Without specialized viewers Web images have fixed raster resolution. CMYK vectors must be rasterized at an appropriate resolution.

▼ **Color Depth**. CMYK color must be converted to RGB color and in a data format appropriate for the Web (JPG, GIF, PNG).

The ultimate solution to this dilemma is to use Adobe's *Portable Document Format*. This way, all color and resolution issues are taken care of by the Adobe Distiller. The technical documents are then either viewed in the browser, if a PDF plug-in is available, or viewed outside the browser in the Acrobat Reader.

Figure 12.6   Web-safe RGB color model.

Emphasis with Gray Scale

Let us first explore how the addition of grays can add emphasis to a technical illustration. Consider Figure 12.7, a line illustration of a hydraulic roof support used in underground mining operations.

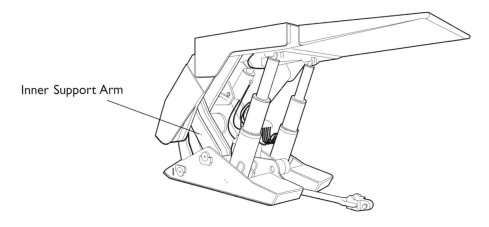

Inner Support Arm

Figure 12.7   Hydraulic roof support.

We could simply use a call-out to identify a part or component. However, as you can see in Figure 12.7, this may be somewhat ambiguous. The actual extent of the inner support arm is not clearly shown by the call-out and leader.

Compare Figure 12.7 to Figure 12.8 where the component has been further identified with a gray panel. Here, the actual extent of the component is easily discernible. We call this a spot color or, more accurately, a *spot tone*.

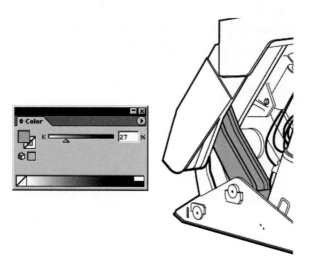

Figure 12.8   Emphasis with gray scale.

Copying the geometry that defines the shape from the line drawing and joining
elements of the outline using **Object|Path|Join** can form an accurate gray spot
tone. However, because the gray area will have no stroke, it is usually easier to lock
the underlying geometry and simply trace over the shape with the **Pen Tool**, creating a new panel that can be shaded. The boundary of the gray panel only needs to
be somewhere on the black line defining the shape, and this can often be done
freehand.

Several approaches are available to position the gray panel:

▼ Create the gray panel on the same layer as the geometry and move the
panel to the *back* so that the line work is on top. However, if any shapes
require a white fill for visibility, you'll have additional editing to remove
the fills. Figure 12.9 shows the panel moved to the back, revealing a shape
that has a white fill.

▼ Create the panel on a *lower* layer (Figure 12.10) so that the lines are on
top. This gives you added flexibility in displaying or not displaying the
gray emphasis, but the problem with filled shapes persists.

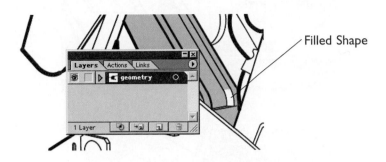

Figure 12.9   Gray panel at back on same layer.

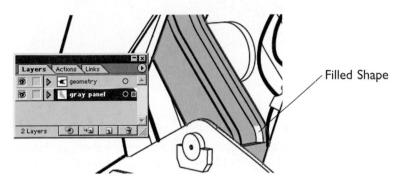

Figure 12.10   Gray panel on lower layer.

▼ Place the gray spot tone layer *above* the geometry layer (Figure 12.11) and adjust the gray panel's transparency to allow the lines to show through. Because you are making the panel transparent, it needs to be darker to start with to arrive at the same gray as in previous examples. This solves the filled shape problem and places the spot tone on a separate layer where it can be selectively displayed and edited. However, the gray tone not only darkens the inner support arm, it *lightens* the inner support arm line work. This may or may not be a significant handicap because it is the area of tone, and not the geometry alone, that now describes the desired feature.

▼ Tip

Spot color areas will never be stroked. In other words, they will never have an outline because it is the line illustration that provides the line work. You will hardly ever have a single shape from the line rendering that you can use for the spot color unless the shape you are trying to highlight is very simple. An added benefit of a separate spot color panel is that it can easily be hidden or displayed on a separate layer.

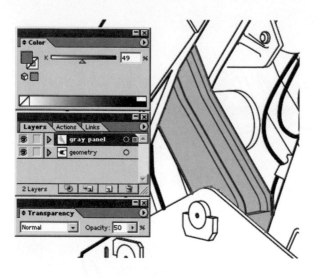

Figure 12.11   Gray panel on higher level with transparency.

▼ Note

This concern for lightening the lines is avoidable when printing from plates. The gray can be separated and printed first as a spot color; the black lines are then overprinted. However, if the illustration is sent to a one-pass digital printer, all colors are flattened and rasterized, making lines behind the spot tone lighter.

Spot Color

A spot color is a second color added for emphasis. In our previous example, the color wasn't a color at all, but rather a gray tone. Historically, illustrators had to be very careful about using color because of its increased cost in analog printing. Costs increased as the number of plates increased, as the number of times negatives and halftones had to be shot increased, and as the number of times each printed sheet had to be run through the press. This remains the case today in direct-to-plate and direct-to-press printing environments.

However, many documents that contain technical illustrations are printed on high-speed dry process or ink-jet printers where the cost impact of color and its separations is minimized. The following discussion will assume that you really don't know on what kind of press your work will be printed, so you have to assume it will be traditional analog.

As a review of what was presented in Chapter 2, spot color can be specified as a single color or as a process color. As a process color, four printing plates (one each for cyan, magenta, yellow, and black inks) will have to be made just to print a second color. As a spot color, only two plates will have to be made, one for black and one for the second color. If you would happen to have black and four spot colors, you'd be better off specifying spot colors as process colors. That way, you could get by with four rather than five plates. When you use process colors for spot color applications, you never actually print the spot color. It is the combination of cyan, magenta, yellow, and black halftone dots that gives the *impression* of the spot color.

Spot Color Swatches

A spot color is specified as a swatch in the **Swatches Palette** in Adobe Illustrator. Spot colors can be custom mixed from CMYK inks, but to achieve a high level of predictability and reproducibility, the PANTONE Matching System (PMS) is used almost exclusively. Both solid colors and process colors are specified in terms of PANTONE's matching system. To use a PANTONE color as a spot color in an illustration:

▼ Open the **Swatches Window**. You may choose to delete swatches for a particular illustration so you can deal with only those colors you need.

▼ Choose **Windows|Swatch Libraries|PANTONE Uncoated**. Select **List View** from the **Options Menu**. This way, you have access to both the visual color and the PANTONE numeric specification.

▼ Find the color you want (in our case PANTONE 1915U) and click on it. The swatch, with a small identifier in the lower right corner, is added to the **Swatches Palette**.

▼ Double click on the swatch to bring up the **Swatch Options** dialog box. In Figure 12.12 two tints of the spot color, 20 percent and 40 percent have been added using the HSB color mode.

Assign the Spot Color

Select the area for the spot color. Simple shapes that have been correctly bound and stroked can be immediately filled. However, most spot color applications are over multiple shapes, so a new spot color shape must usually be created.

Click on the swatch and the color is assigned to the area. Note in Figure 12.12 that a **Global** option is selected by default. This means that any change in the spot color will automatically be reflected in all shapes that have that color, without the shapes being selected. This allows you to easily experiment with spot colors.

Tints (lighter colors) of the spot color can be handled in two ways. First, if a special ink is mixed for the spot color, all tints can be printed on a separate plate at the same time. Black lines are printed last. If the tints are CMYK halftones, they are printed along with the other color, again, with black printed last.

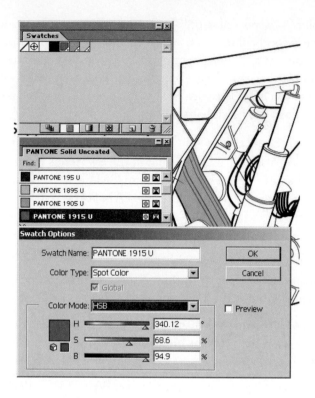

Figure 12.12   PANTONE spot color assigned to area of emphasis.

PANTONE and Process Spot Color

The **Swatch Options** dialog box allows you to switch between color modes. Because PANTONE provides an easy way to use paper color swatches, it is often easier to simply identify the spot color you want as a PANTONE color and then let Illustrator determine the CMYK components if process color plates are already being made. Figure 12.13 shows how PANTONE 1915U is matched with CMYK color 0, 71, 20, 0. PANTONE also provides matching swatches in CMYK, so you *could* start in CMYK if you so desire.

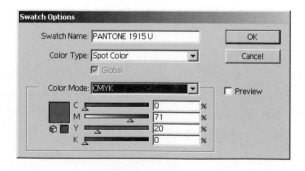

Figure 12.13   PANTONE to CMYK color matching.

Spot Color and Overprinting

Unless you tell it not to, when the PostScript is RIP'd, objects on top cut away bottom objects that they overlap. This can be problematic when spot color fills are on top or when lines are on top and can cause misalignment of spot color fills as individual plates are made. When spot color areas are on top of lines, set them to overprint. When lines are on top of spot color, set them to overprint (Figure 12.14).

Figure 12.14   Spot color overprinting.

▼ Note

If you use the **Snap to Points** option as you create spot color areas, these areas will follow the centerlines of the black lines. Turn off the stroke on the spot color areas, and you can almost be guaranteed that areas bounded by object lines will be totally filled with spot color. This makes trapping less problematic.

Spot Color and Digital Presses

The rasterization process that converts PostScript to printable digital dots removes many of the concerns of traditional spot colors. For one thing, dry-toner and ink-jet printers remove much of the problem of spot color registration because multiple plates, stripping, and make-ready are either eliminated or automated. The cost advantage of spot over process color is also reduced. Other than colorant consumption, it's just as easy to print in four color, as it is in black and white.

And, of course, technical illustrations that are displayed as part of Web or online documentation and training bear no financial burden whatsoever for color, and the order that spot color appears in the file is of no consequence. A display monitor or RGB projector is just as happy displaying 16.7 million colors as it is displaying black and white.

Review

Even in an environment of free color, there remains an argument for spot color. Spot color accomplishes something that full color cannot: it draws attention to specific components in an illustration because of contrast—contrast of color and value. Your eye is easily drawn to the spot color in Figure 12.12 because the spot color doesn't have to compete with other shades and colors. Changing the color or saturation in a full-color illustration (dimming the rest of the hydraulic roof support, for example) may not be as effective as using spot color. Even in an environment where full-color holds no financial disadvantage, there are compelling reasons to use spot color.

Text Resources

The following resources provide additional information about using color for emphasis.

Adams, J. Michael, and Dolin, Penny Ann. *Printing Technology*. 5th Edition. Delmar/Thompson Learning. Albany, New York. 2002

PANTONE Color Formula Guide One Thousand. PANTONE Corporation. Carlstadt, New Jersey. 1993.

Golding, Mordy, and White, Dave. *PANTONE Web Color Resource Kit*. Hayden Books. Book & CD-ROM edition. Indianapolis. 1997.

Internet Resources

About Color Blindness. Retrieved from: *http://www.delamare.unr.edu/cb/*. November 2002.

Marvin's Corner: Printing with Spot Color Inks. Retrieved from: *http://marvin.mrtoads.com/spot.html*. November 2002.

Spot vs. Process Color. Retrieved from: *http://graphicdesign.about.com/library/weekly/aa060399*. May 2002.

On the CD-ROM

For those of you who would like to try out the procedures described in this chapter, a resource has been provided on the CD-ROM. Look in *exercises/ch12* for the following file:

support.eps   The hydraulic support used in this chapter.

Perspective Drawing

Axonometric Projection ▶

Technical Animation

Technical Illustration

3D Modeling ▶

Rendering ▶

▶ Technical Illustration

Chapter 13

Color Rendering

In This Chapter

By now you are aware that Adobe Illustrator is probably the most popular vector graphics application currently being used to create technical illustration. One aspect of the program's popularity is the relative ease with which you can render an illustration.

In Chapter 12 we pointed out the importance of gray-scale shading and spot color to better illustrate form and function and instances where these techniques would be appropriate. The next logical step beyond these basic rendering techniques is realistic colorization. Oftentimes realistic colorization is necessary to bring additional attention or clarity to an illustration. Realistic colorization would be appropriate and useful in areas such as product sales sheets and brochures, architectural renderings, and in news and infographics.

Color illustration requires both the technical skills of axonometric and perspective projections and constructions as well as an understanding of color, material, surface finish and how light affects each of them. To use color effectively, you need to understand color theory and be a keen observer of color in everyday life. You should collect photographs of everything; you should clip examples from magazines and scour businesses for color brochures.

If you want to be a color illustrator, you might start by visiting several of the Web sites listed at the end of this chapter. If you see an effective technique, analyze it using the knowledge you have gained in this text, and try it. Remember, many of the great artists and illustrators in history began by copying the masters. Out of this copying their own unique styles developed.

Chapter Objectives

In this chapter you will understand:

- ▼ The mixing of custom colors in Adobe Illustrator
- ▼ The creation of custom color **Swatches**
- ▼ The use and modification of color **Stroke** and **Fill**
- ▼ The making and use of **Blends** and **Gradients**
- ▼ How to use the **Gradient Mesh** effectively
- ▼ The application of **Clipping Mask** and **Pathfinder** to rendering

Color in Adobe Illustrator

Color in Adobe Illustrator can be specified using the PANTONE Matching System or by creating custom colors using CMYK, RGB, or HSB values. In many situations, it can be cumbersome to search through hundreds of colors in the PANTONE system to find just the right color. For print applications, it makes sense to specify color in terms of CMYK, where the final output will be in the form of CMYK inks. In this chapter we will focus on specifying color in terms of CMYK.

▼ Note

Adobe Illustrator's color picker allows easy switching between CMYK and other color models. Adding to the difficulty of color matching is the fact that no matter which model you use, you will always be viewing RGB color space. Unfortunately, Adobe Illustrator doesn't give you the option to work in device-independent LABxyz color space.

Custom Color Mixes

Figure 13.1 shows the **Color Palette** in Adobe Illustrator's **Color Window**. CMYK color is specified by the percentage of each cyan, magenta, yellow, and black ink in each halftone spot.

Figure 13.1   Adobe Illustrator's **CMYK Color Palette**.

You will enter specific values for each of the process colors in this window, or you can select a color by clicking on the color bar at the bottom of the window. It may be more efficient to click on the color bar to get an approximate color and then adjust the process color sliders to arrive at the final color (Figure 13.2). Any selected object in your file will take on these color attributes as you select the color. Or you can simply drag your selected color from the **Fill** or **Stroke** box and drop the color onto the object.

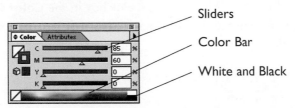

Sliders
Color Bar
White and Black

Figure 13.2   Process color values in the **CMYK Color Palette**.

When you *tint* a color, you add white. When you *shade* a color, you add black. This is problematic in CMYK. However, if you switch to the **HSB Color Palette** the color started in CMYK can be tinted by lowering the saturation (**S**) and shaded by lowering the brightness (**B**). When you are satisfied with the results (Figure 13.3), return to the **CMYK Color Palette** where the sliders reflect the change.

Hue (Color)
Saturation (Lightness)
Brightness

▼ Tip

As saturation is decreased, increase the brightness to lighten the hue even further.

Figure 13.3   Tinting and shading in the **HSB Color Palette**.

Custom Color Swatches

For a project that requires the use of a specific group of colors within a single illustration or across several illustrations, you can easily create and save color **Swatches**. This assures a much faster and more consistent color application within an illustration and between families of illustrations. To make use of Adobe Illustrator's custom color swatches, follow these steps.

▼ Create your color in a **Color Palette** (Figure 13.4). In this case we will be creating a CMYK process color swatch. Note: Adobe Illustrator lets you create CMYK colors in an RGB file. You have the option of changing the color space to CMYK PostScript when you save the file in EPS format.

Fill Box

Stroke Box

Color Box

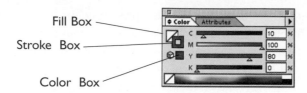

Figure 13.4   Color created in **Color Palette**.

▼ Drag this new color from the **Color** box in the **Color Palette** to the **Swatches Palette** (Figure 13.5) and drop it into an open area beside the last color.

Figure 13.5   Color as it appears in the **Swatches Palette**.

▼ Double click on this red swatch in the **Swatches Palette** to access the **Swatch Options** dialog box. Here, you can name the color (called CTI red), and select the **Color Type** and **Color Mode** (Figure 13.6). If the color will be part of blends or gradients, choose **Process Color** as **Color Type**. If you are going to use the color as an accent, choose **Spot Color**.

▼ Tip

Use descriptive names for your color swatches. You can use object-oriented names (like *wagon red*) or more rendering-oriented names (like *wagon mid-tone*).

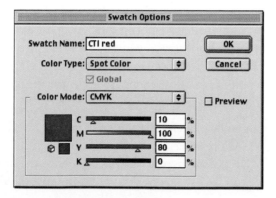

Figure 13.6   The **Swatch Options** dialog box.

▼ After giving the color a name, click **OK**, and then click on this color in the **Swatches Palette**. In Figure 13.7 you will notice that the red color now displays as a single spot color in the **Color Palette**.

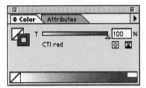

Figure 13.7   Red spot color displayed in the **Color Palette**.

Tints of the CTI red can now be specified in terms of a percentage, rather than the individual CMYK values that make up the color. Because CMYK colors will be printed as halftones (even as spot colors), and most printing will be done on white paper, dialing down a percentage of a color decreases the halftone spot size, revealing more white paper. This lightens the tint. This method enables the illustrator, designer, or client to specify colors of objects or components of objects in terms of how light the color is.

▼ Illustrator Tip

To achieve the impression of depth in this example, three Adobe Illustrator techniques must be employed: arranging **Front to Back** within a layer, the ordering of **Layers**, and the use of **Transparency**. In this refrigerator, all three techniques are used. In order to see detail through the left side panel, a transparency setting of 30 percent is used. To place the open door to the front, its layer is higher on the layer stack. And to keep certain elements on the door visible, they are moved to the front within the layer.

Color Stroke and Fill

Color **Stroke** and **Fill** in Adobe Illustrator is also selected using the **Color Palette**. At this point you have noticed two boxes in the upper left corner of the window. As shown in Figures 13-8a and b, toggling between these two boxes enables you to select the color fill and stroke values, and these values will display in the **Color Palette**. The **Fill** color is a 30 percent tint of the **Stroke** color.

Fill Box

Stroke Box

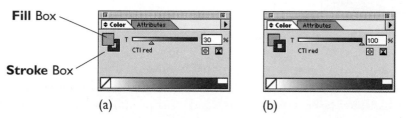

(a)   (b)

Figure 13.8 (a) The **Fill** box toggled to the front. (b) The **Stroke** box toggled to the front.

Color Gradients

It is very common as a technical illustrator to be required to accurately render various shapes and surfaces in color. For example, representing a simple cylinder is certainly a task that illustrators are expected to routinely accomplish. The **Gradient Tool** in Adobe Illustrator makes this task quick and efficient.

▼ To begin the cylinder, draw a simple rectangle with the **Rectangle Tool**. Paint this rectangle with a black stroke and no fill. This is the orthographic front (rectangular) view of a cylinder in vertical position (Figure 13.9).

Figure 13.9   A simple rectangle represents a cylinder.

▼ With the **Gradient Tool Window** displayed, select the type of gradient you want to create, in this case a **Linear Gradient** (Figure 13.10). The **Linear Gradient** forms parallel reflections on the surface of the cylinder that will, in turn, be parallel to the cylinder's thrust axis.

Figure 13.10　Linear Gradient option selected.

▼ Click below the gradient bar to add color keys at the approximate locations where the gradient will change values (Figure 13.11). This will create another stop in the gradient where color can be applied via the **Color** or **Swatches Palette**.

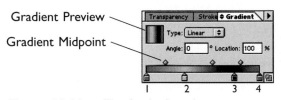

Color Key Added

Figure 13.11　Modify the gradient by adding color keys.

▼ Continue to fine-tune the gradient, adding and adjusting stops and mid-points (the diamond icons above the slider), and applying various colors, until the gradient will sufficiently describe a cylindrical surface. The mid-point diamonds determine how rapidly one color key blends into its neighbors. The finished gradient is shown as a preview in Figure 13.12.

Gradient Preview

Gradient Midpoint

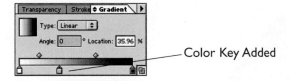

1　Ambient (Base) Color
2　Highlight Color
3　Diffuse (Shade) Color
4　Reflected Color

Figure 13.12　The finished gradient.

▼ Click on the finished gradient in the upper left corner of the **Gradient Palette**, and drag it onto the rectangle we created earlier (Figure 13.13). The direction and length of the gradient can be adjusted using the **Gradient Tool** from the **Toolbox**.

Highlight　　　　　　　　　　　　　　Diffuse

Ambient　　　　　　　　　　　　　　　Reflected

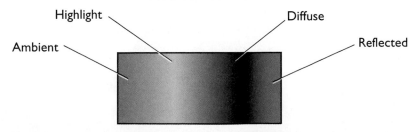

Figure 13.13　The finished gradient applied to the object.

The direction of the **Linear Gradient** will be perpendicular to the direction line drawn by the **Gradient Tool**. Use the **Shift** key to constrain the direction of the gradient to the angles set in **Edit|Preferences|General**. Where you start and stop the direction line determines how much (or how little) of the gradient is applied to the selected object.

Figure 13.14 shows how the judicious use of gradients can really add impact to an illustration. The key to using gradients effectively is to not get carried away. Don't make them overly complex or use gradients to describe every surface. The gradient will remain oriented to the object when rotated.

Figure 13.14   Illustration of a shock absorber created by making use of cylindrical gradients.

Realistic pictorial views (axonometric or perspective projections) don't always do the best job at communicating the technical aspects of a product. Many times you will combine pictorial, principal orthographic, and diagrammatic views in the same illustration. In this shock absorber example, the actual functioning of the product is best described by a combination of a pictorial view, for part identification, and diagrammatic views, that describe the design function.

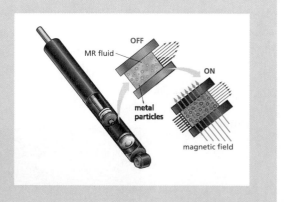

Blends

In its most basic form, Adobe Illustrator's **Blend Tool** takes a beginning line or shape, and an ending line or shape, and creates a number of intermediate objects that make a smooth transition between these base objects. Not only is the physical shape blended, but **Fill** and **Stroke** parameters are also blended. You can blend between any number of base objects and Illustrator will provide the transitions. The primary application of this technique is to create a gradient with greater flexibility than the **Radial** and **Linear** options of the **Gradient Fill Tool**. Additionally, other objects such as lines, shapes, text, or other blends can be grouped together to form even more detailed blend objects.

A blended gradient is used in conjunction with **Pathfinder** operations (2D Booleans) and a **Clipping Mask**. First you create the gradient you desire, and then the shape that the gradient will go into, and then use that shape to mask the blend.

The options for the **Blend Tool** are set in **Object|Blend|Blend Options**. Each blend is based on the current setting in **Blend Options**. Start with the **Smooth Color** option, and Illustrator will determine the number of transitions necessary to make a smooth blend based on the base objects. Too many steps and the blend is unnecessarily dense. Too few steps and you will have gaps in the blend. After you get the hand of blending, you can try the other two options, **Specified Steps** and **Specified Distance**.

Blend objects can be easily edited. The base objects can be edited using the **Open Selection Tool**. This provides a powerful interactive feature that lets you adjust the blend (you don't have to know the correct colors before you start). The color of a base object can be adjusted, changing the transitions on either side in real time. You can immediately see the results of your changes. The transition objects cannot be edited unless you choose **Object|Blend|Expand** or **Release**. This allows each transition object to be selected with the **Open Selection Tool** and adjusted individually.

You can also add to either end. By using the **Open Selection Tool**, you can select the first or last base object and another base object. Adobe Illustrator will continue the blend with the current **Blend Options** settings.

▼ Note

Although you can blend closed and unclosed shapes (shapes and lines), you probably won't get the results you want. It's better to blend lines with lines and shapes with shapes.

Line Blends

A **Line Blend** uses straight or curved lines to define the beginning and ending elements of the blend. If you use straight lines, it may appear identical to a **Gradient Fill**. A **Gradient Fill** is a call to the PostScript interpreter at print time and because it contains no geometric elements, requires considerably less file space than does a blend. For this reason, if given the choice between the two for simple linear or radial gradients, choose **Gradient Fill**.

The simplest use for blends is to create and distribute lines evenly between two base object lines. This process can be applied to technical charts or diagrams to create equally spaced rows and columns or to arrange a repeating element evenly. This technique could be used to create knurls, splines, or any other repeating surface feature. You could also blend between two lines to create a scale guide or grid for a technical illustration. A technical diagram, like that in Figure 13.15, can aid in gaining an understanding of simple line blends and their benefit.

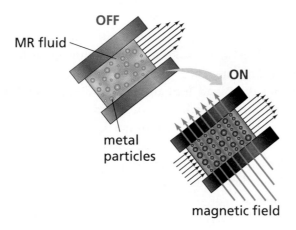

Figure 13.15   A technical diagram having repeating elements.

To demonstrate how the **Blend Tool** can be used to space repeating elements, we will apply it to the green arrows in Figure 13.15. Note that there are five evenly spaced steps between the first and last green arrows (the base objects). In Figure 13.16 the arrows have been isolated.

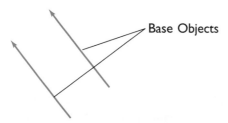

Figure 13.16   The base objects for blending repeating elements.

▼ Select both the green arrow shafts. Click on the **Blend Tool** in the **Illustrator Toolbox**.

▼ Click on corresponding anchor points on each of the two lines, holding down the **ALT** or **Option** keys when clicking the second anchor point (Figure 13.17). This will open the **Blend Options** dialog box.

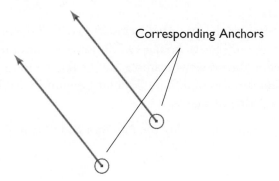

Figure 13.17   Corresponding anchor points circled in red for emphasis.

▼ Specify the number of blends, or intermediate objects, between the lines. Select the **Specified Steps** spacing option (Figure 13.18), enter **5**, and click **OK**. The finished distribution of intermediate objects is shown in Figure 13.19.

Figure 13.18   Enter the number of intermediate steps in the **Blend** dialog box.

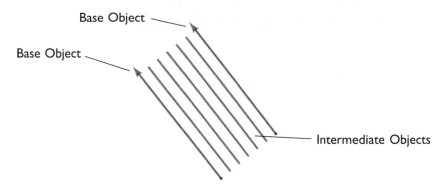

Figure 13.19   Finished distribution of blend objects.

As you can see from Figure 13.19, five equally spaced intermediate lines have been created between the first and last green arrow base object. The ends of the shafts can then be adjusted independently after you expand the blend from the **Object|Blend|Expand** pull-down menu. **Expand** allows each blend object to be adjusted with the **Open Selection Tool**, but maintains the integrity of the blend object. **Release** destroys the blend object and creates independent intermediate objects. Repeat this process to create intermediate arrowhead shapes as well.

Shape Blends

A **Shape Blend** uses shapes (usually closed) to define the beginning and ending elements of the blend. This technique can be used to distribute shapes, just as lines were distributed in the previous example. But by far, the most useful application is to create gradient surfaces of infinite flexibility. Figure 13.20 shows the line art for the eventual rendering of a cone.

▼ Create a simple cone using the **Pen Tool** and **Ellipse Tool**.

Figure 13.20   Line art of conical shape.

▼ Generate two blends: one blend to transition smoothly from a base color to a highlight color, and a second blend to transition smoothly from a base color to a shadow color.

▼ Use the **Pen Tool** and create a wedge shape that covers half of the cone. Duplicate and reposition this original wedge, adjusting the anchor points as needed (Figure 13.21).

▼ Place the narrower highlight and shadow wedges in front of the base color wedges. Figure 13.21 shows you what the wedges might look like. The green wedges represent the sunlight side. The blue wedges represent the side of the cone away from the sun.

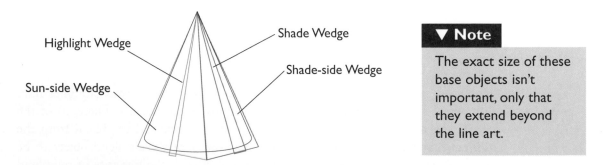

▼ **Note**

The exact size of these base objects isn't important, only that they extend beyond the line art.

Figure 13.21   Base object wedge shapes.

▼ Fill the two larger wedges with an appropriate base color. In this case, a blue-gray to suggest metal. Fill the two narrower wedges with appropriate highlight and shade colors to complement the base color (same hue, just lighter or darker).

CMYK Base Color

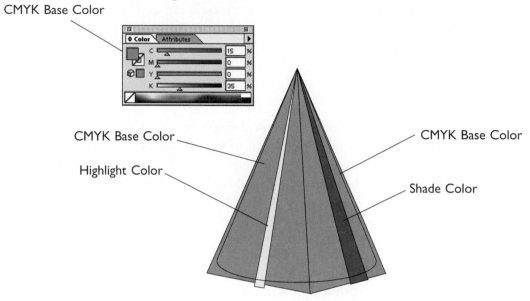

CMYK Base Color

Highlight Color

CMYK Base Color

Shade Color

Figure 13.22   Base color specified in CMYK color space.

Note

The two larger wedges will smoothly blend together because they share the same color where they abut.

▼ With the wedge shapes filled with color, repeat the blending process previously discussed. In this case, use **32** as the number of blend steps. Blend the highlight wedge with its background and then blend the shade wedge with its background (Figure 13.23). The red cone shape will be used as a **Clipping Mask**. This process is covered in the next section.

Figure 13.23   The completion of the blended wedges with cone shown in red.

A note regarding spacing options: You could use the automatic **Smooth Color** option rather than the **Specified Steps** option in this case. However, the automatic option will generate copious blend steps, often resulting in more complex blends than are necessary. On the other hand, be careful that your step values are enough to avoid any color banding when the illustration is output. The optimum number of blend steps is something you will learn with experience, based on the colors used and the amount of area between base objects.

Clipping Mask

It may be possible to create blends out of shapes that exactly define parts in a technical illustration. But more often, you will create a blend and use it with many shapes. You wouldn't want to have to create a new blend for every shape. To avoid this, you will be using a shape as a **Clipping Mask**. When a shape is used in this manner, you essentially create a window through which the blend will be seen. This way, you have a single, general-purpose blend that can be used for almost any shape. However, there is a downside to this method. The entire blend remains in the file; only the portion contained within the mask is displayed. This can pose problems when the illustration is rasterized, and unpredictable results may occur when the file is printed on a PostScript printer.

Figure 13.24a shows how the same blend can be used across three (or any number of) axonometric projections. Use AxonHelper to determine the correct rotation of the blend so that it is parallel to the thrust axis.

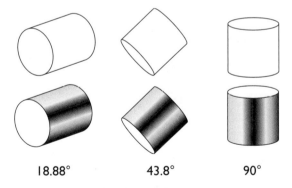

| 18.88° | 43.8° | 90° |

Figure 13.24   Three axonometric **Clipping Masks** using the same but rotated **Blend**.

In the case of the cone in Figure 13.23, the red outline serves as the **Clipping Mask** through which the conical blend will be seen.

▼ To complete the cone, move the outline of the cone to the front of the blends, and paint it with no fill and no stroke.

▼ Select the cone and the blends together (Figure 13.25). Choose **Object|Clipping Mask|Make**.

The finished cone can be further enhanced by the addition of a thin highlight edge at the bottom and a thicker outline to create even more dimensionality (Figure 13.26).

Perhaps the most important point to keep in mind when rendering an object using shape blends is that the shapes are not only controlling the color of the object. The shapes also serve as contour lines that describe the geometry of the object. This is very important to the technical illustrator as he or she attempts to accurately mimic the shape and surface of objects around them.

Figure 13.25   Result of **Clipping Mask**.

Figure 13.26   Enhanced and finished cone.

Pathfinder

The **Pathfinder Window** displays options available for combining paths (union, difference, intersect, plus several others). Because you may want a blend to go all the way across a surface, you don't want to break it up. Figure 13.27 shows a cylinder and keyway paths with the **Pathfinder Window** open. The keyway is the front object, so it will be subtracted from the cylinder.

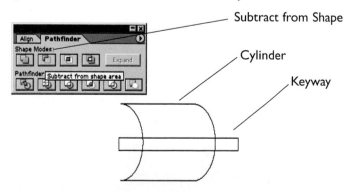

Figure 13.27   Use **Pathfinder** to perform 2D Boolean-type operations.

Figure 13.28 shows the result of the **Pathfinder|Shape Mode| Subtract** operation. Note that the keyway has not broken the cylinder into two separate parts, rather it has created a mask based on its shape (Figure 13.28b). This is important for how the **Clipping Mask** will display the **Blend** in the new object.

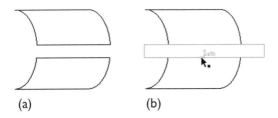

(a)   (b)

Figure 13.28   Subtraction doesn't remove a shape; it only masks.

The blend is positioned behind the path (Figure 13.29). Both objects are selected and the command **Object|Clipping Mask|Make** is executed.

Figure 13.29   Position the blend behind the **Clipping Mask**.

The blend is displayed within the **Clipping Mask** as defined by the **Pathfinder|Shape Modes|Subtract** command. You can see in Figure 13.30a that the chrome blend is displayed across the cylindrical surface, uninterrupted, with the keyway simply an opening in the blend. However, the entire blend and the keyway objects remain in the file as shown in Figure 13.30b.

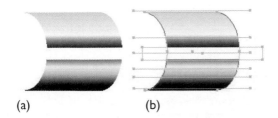

(a)   (b)

Figure 13.30   (a) The blend is uninterrupted across the **Clipping Mask.** (b) All objects remain in the file.

Gradient Mesh

Usually, simple linear or radial gradients will sufficiently describe most surfaces that technical illustrators will encounter. However, there are times when you will be asked to render shapes or surfaces that are more organic than machined parts. The backpack illustration in Figure 13.31 is a good example of an ideal use for the **Gradient Mesh**.

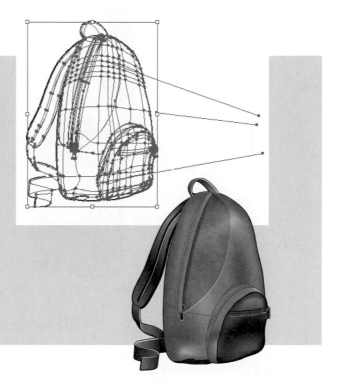

▼ Illustrator Tip

Along with **Blend** and **Gradient Fill**, **Gradient Mesh** provides the ultimate flexibility in harnessing Adobe Illustrator's PostScript rendering power. Review Figure 2.4 for an overview of how this technique is based on a surface mesh and how highlight, midrange, and shade values are based on HSB color space. This is also the chance for you to blur the boundaries between 2D and 3D illustration. As you can see in this mesh detail, the surface mesh is closely akin to a 3D wireframe. By editing the mesh so that it follows the pictorial projection, the blends better match the surfaces.

A **Gradient Mesh** is created by generating a series of rows and columns across a shape. The shape does not have to be closed, but generally it is convenient and more practicable to have a closed shape. It is also helpful to have a **Fill Color** loaded into the **Color Palette** before you begin. Unique colors can be assigned to the intersections of mesh lines (nodes); the colors are impacted by the length and position of the Bezier handles associated with the nodes. Internal nodes have two sets of handles because they are connected on both sides. External nodes have one set of handles.

In Figure 13.31 you can see how this is used to give the backpack a natural, soft, nonmechanical look. This would be extremely difficult if you used line or shape blends. Also notice how the same mesh is used on two different parts of the backpack, necessitating the use of a **Clipping Mask**.

To better see how the backpack was constructed, observe Figure 13.31. You can see that along with **Gradient Mesh**, the use of **Layers** is critical. The original outline shape of the backpack is used 4 times: as a black outline, as a brown outline, as the basis of the orange **Gradient Mesh**, and as the black portion of the pack with a **Clipping Mask**. Details such as seams, zippers, the strap, and the front pocket are constructed separately.

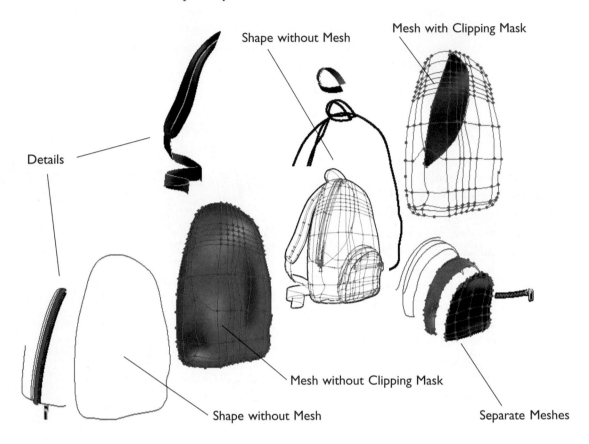

Figure 13.31   An example of the **Gradient Mesh** in use.

Gradient Mesh Example

In practice, it may be productive to sketch your idea of a possible solution in order to plan the alignment of the mesh and the placement of the colors. If you are working from a photograph, place the image on a background layer for reference.

Figure 13.32 shows a sketch of a racing vehicle cowl with preliminary contour lines defining the surface. As was mentioned in the previous Illustrator Tip, take a 3D approach to mapping the mesh to the object. The cowl appears to be defined by five columns and six rows (counting the cells, not the mesh lines).

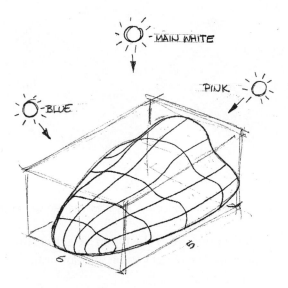

Figure 13.32   Preliminary **Gradient Mesh** planning sketch.

▼ Create the shape. In this case the sketch was scanned and traced in Adobe Illustrator. Choose **Appearance|Flat** and enter **Columns:5** and **Rows:6** (Figure 13.33). The mesh will be evenly applied across the shape.

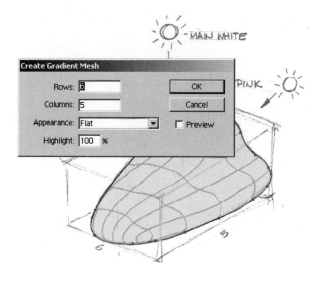

Figure 13.33   **Gradient Mesh** based on planning sketch.

▼ Edit the gradient mesh with the **Open Selection Tool**. The mesh that Adobe Illustrator creates won't define the surface the way you need (Figure 13.34a). Move the nodes so that the mesh approximates the surface (Figure 13.34b). Use the scan as a guide.

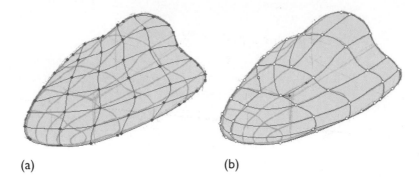

(a)   (b)

Figure 13.34   (a) Mesh as created by Adobe Illustrator.   (b) Mesh edited to better describe the surface.

▼ Select all the nodes with the **Open Selection Tool**. Assign a single color that represents the general ambient color of the object (not a shade and not a highlight).

▼ Pick a highlight tint of the general ambient body color. Apply this tint to nodes on the top edges of the shape (Figure 13.35). You can select multiple nodes by holding down the **Shift** key.

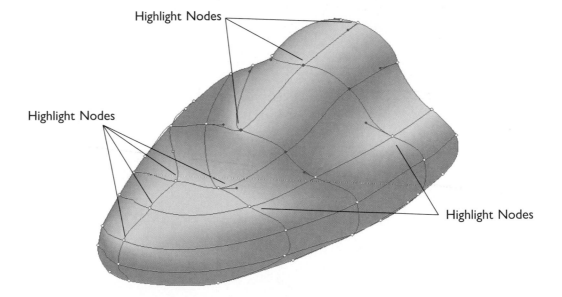

Figure 13.35   Highlight tint applied to nodes on top of curves.

▼ Pick a shade of the general ambient color. Apply this darker color to the nodes located where the surface bends underneath (Figure 13.36).

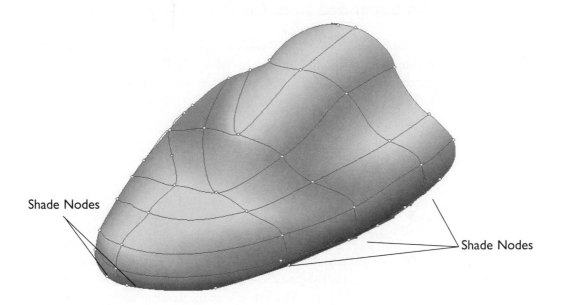

Shade Nodes

Shade Nodes

Figure 13.36   Shade applied to the nodes on the bottom of the curves.

▼ Pick a reflected color. This may be a blue light reflected from the sky or a pink light reflected from the sun. Apply these colors to the outside of the shape, to nodes pointing toward these light sources (review Figure 13.32). Start with the existing color and add the reflected color. The completed surface is shown in Figure 13.37.

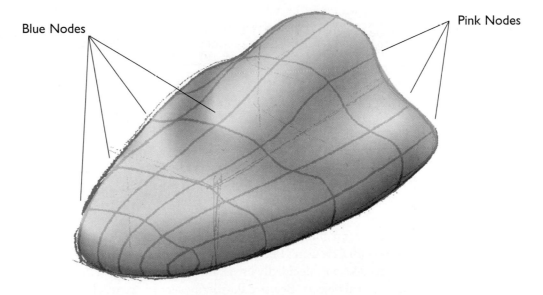

Pink Nodes

Blue Nodes

Figure 13.37   Completed **Gradient Mesh** surface.

An Expanded Role for Photoshop

Up until now Photoshop has taken a back seat to Illustrator as the tool of choice for technical illustrations. But as many technical illustrators have found, and as Adobe Systems constantly promotes, the two graphics tools are actually complementary, rather than competitive. Some of this has to do with the tools themselves.

▼ Interfaces, terminology, and tools in both applications now share a common design philosophy.

▼ Both have become *metatools*. That is, there are vector aspects of Photoshop and raster aspects of Illustrator.

▼ Technical illustrations are repurposed more often. Vector illustrations need to be rasterized, resampled, and colorized for Web and multimedia.

▼ And finally, the necessity that large, expensive, dedicated computers be used to manipulate the large raster files required for high resolution has been mitigated. Modern affordable desktop computers have the processor speeds and memory to handle almost any raster illustration task.

Take the following Illustrator Tip for example. It features the traditional use of a raster image in technical illustration. A scanned photograph forms the background for a resolution-independent vector overlay.

▼ Illustrator Tip

Technical illustrations may combine fixed-resolution raster images with resolution-independent PostScript. Considerable time can be saved by using a digital photograph or scan, . Always match the resolution and format of the raster image with the output resolution. For example, if the output is process color lithography at 1200 ppi, your image should be a CMYK TIF file at a resolution evenly divisible into 1200, say 300 dpi.

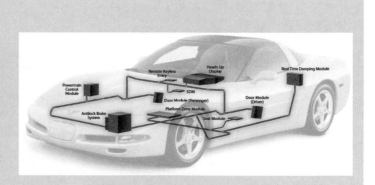

However, a quick review of the tasks that Adobe Illustrator *doesn't* do well reveals the potential for using Photoshop. Illustrator uses vector strokes and gradients somost objects take on a decidedly linear, mechanical feeling. This is totally appropriate for mechanical objects. Because PostScript vectors are resolution independent, fine detail isn't dependent on pixel size. And because PostScript drawing programs are beginning to take on some of the characteristics of CAD programs (snaps, trims, mathematical constraints, and parametric features), they have become even more appropriate for maps, diagrams, technical drawings, and data representation.

Figure 13.38 displays the type of subject matter perfect for raster technical illustration. It is evident that the hand sketch technique would be difficult, if not impossible ,to produce using vectors. True, this technique requires more artistic ability than the mechanical constructions we have covered up to this point, but if you look carefully at the development, you'll recognize many common technical considerations such as ellipse alignment and perspective convergence.

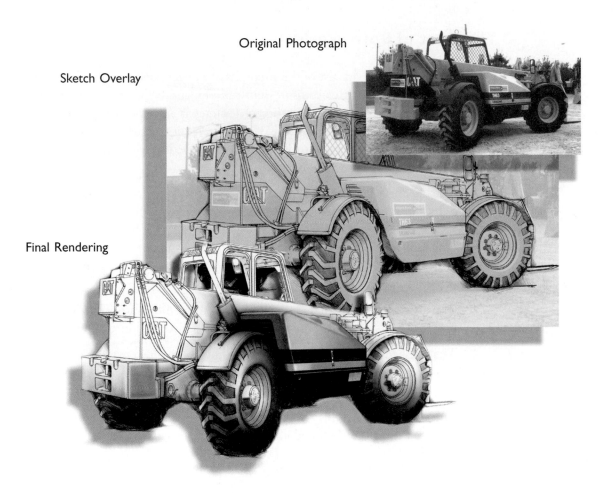

Figure 13.38   Resolution-dependent raster illustration (Daniels & Daniels).

The photograph contains much more information than is needed for the illustration but removes the necessity of doing a complete pictorial construction. The sketch could be done directly in Photoshop or Painter or, if you are confident with a soft pencil, a hand sketch could be done on a vellum overlay and then scanned. If you carefully study Figure 13.38, you'll notice that the final rendering actually contains three distinct rendering styles:

▼ The rough sketch technique

▼ A transparent colorization of the chassis and parts of the structure. This is much like watercolor or marker.

▼ A more detailed rendering of the side panel. This is actually the subject of the illustration and is rendered in a photorealistic technique so that it is in its environment, but stands out for inspection.

The Best of Both Worlds

The previous example made use of the strengths of a raster illustration tool, namely, the freedom (especially with a pressure-sensitive pen and tablet) to sketch expressive lines and shading. But what if technical accuracy is an important factor in the final illustration? In the case of Figure 13.39, the mechanical accuracy of a PostScript drawing program is combined with the freedom and flexibility of a raster program to create an effective technical illustration.

▼ The client supplied the original line art, in this case a low-resolution bitmap. (This is not unusual. You might wonder why the client didn't supply a vector file making the illustrator's job easier. The answer may be that the client had no idea who had the file, or where it was stored. Or, the client never had the vector file in the first place, only the publication file it was placed in.)

▼ The artwork was scanned and placed into Adobe Illustrator. The line art was created using the axonometric techniques discussed in this text.

▼ The finished line art was rasterized in Photoshop where extensive selections were created and saved. These selections were expanded by one or two pixels to cover the linework (called a color spread).

▼ Color textures were created and saved in a separate file. These textures were copied to the clipboard and using **Paste Into**, individual selections were filled.

Review

In this chapter you were presented with several fundamental techniques for color illustration. All PostScript color illustrations begin with paths and **Stroke** and **Fill** techniques. These paths can be filled with solid colors, tints, or shades. For additional realism, **Gradient Fill** provides an almost unlimited range of potential surface finishes.

Figure 13.39   Raster illustration from vector underlay (Daniels & Daniels).

Adobe Illustrator allows custom gradients to be formed from line or shape blends, further increasing the range of potential rendered surfaces. To make use of this technique, a **Clipping Mask** displays the blend within a shape. But for ultimate control of color rendering, a **Gradient Mesh** allows multiple colors to smoothly blend across any surface shape you can imagine.

Raster and vector rendering can complement each other. Use the accuracy and resolution-independence of PostScript vectors to assure accurate constructions and projections. Then, rasterize the vectors at your illustration's target resolution and create a more freehand-style illustration in Adobe Photoshop or Fractal Painter. The complete technical illustrator is able to match the tool to the job, moving between vector and raster graphics as necessary.

Text Resources

Alspach, Ted, and Murdock, Kelly. *Illustrator 10 Bible*. John Wiley & Sons. New York. 2002.

Blatner, David, and Fraser, Bruce. *Real World Adobe Photoshop 7*. Peachpit Press. Berkeley. 2002.

Johnson, Harald. *Mastering Digital Printing: The Photographer's and Artist's Guide to High-Quality Digital Output*. Muska & Lipman. Cincinnati. 2002.

Kieran, Michael. *Photoshop Color Correction*. Peachpit Press. Berkeley. 2002.

McClelland, Deke. *Real World Adobe Illustrator 10*. Peachpit Press. Berkeley. 2002.

Steuer, Sharon. *The Illustrator 10 WOW! Book*. Peachpit Press. Berkeley. 2002.

Internet Resources

Adobe Illustrator Tips and Techniques. Retrieved from *http://www.desktoppublishing.com/tipsillustr.html*. February 2003.

Adobe Illustrator Tutorials. Retrieved from: *http://graphicssoft.about.com/cs/illustratorart/*. February 2003.

Adobe Illustrator Users Group. Retrieved from: *http://groups.yahoo.com/group/Adobe-Illustrator/*. February 2003.

Adobe's Corporate Tips & Techniques. Retrieved from: *http://www.adobe.co.uk/products/tips/illustrator.html*. February 2003.

Analytical Graphics, Inc. Retrieved from: *http://www.analyticalgraphics.com*. February 2003.

CAD Tools for Adobe Illustrator. Retrieved from: *http://www.hotdoor.com/CADtools/*. February 2003.

Daniels and Daniels Illustration. Retrieved from: *http://www.beaudaniels.com*. February 2003.

Kevin Hulsey Technical Illustration. Retrieved from: *http://www.khulsey.com*. February 2002.

Nidus Technical Illustration. Retrieved from: *http://www.nidus-corp.com/cutp.html*. February 2003.

On the CD-ROM

For those of you who would like to try out the procedures described in this chapter, several resources have been provided on the CD-ROM. Look in *exercises/ch13* for the following files:

backpack.eps.	Completed backpack illustration to study the gradient mesh.
shock.eps.	Completed shock illustration to study gradients.

Perspective Drawing

Axonometric Projection ▶

Technical Animation

Rendering ▶

Technical Illustration

3D Modeling ▶

Chapter 14

PostScript Materials

In This Chapter

Technical illustrations may be completed with simple line weights, as was discussed in Chapter 10. Or, individual colors can be used to highlight areas of interest. This was covered in detail in Chapter 12. In Chapter 13, you were shown how color rendering positively impacts an illustration's effectiveness. In this chapter, you will find a library of PostScript materials that, when combined with accurate constructions, can produce the highest-quality technical illustrations.

PostScript materials are, by their nature, more powerful and flexible than are raster materials. Raster materials are of a fixed resolution. You have to be careful to create them at the intended resolution and not enlarge them. When you enlarge a raster material, you decrease the resolution. And you are well aware by now that PostScript materials make use of the printer's resolution. The material itself is described mathematically.

Raster materials can be scanned or captured by a digital camera, and at a high-enough resolution, can be combined effectively with vectors. With PostScript materials, you may still want to start with a scan or digital photograph on a background layer and create the PostScript material using it as a guide.

Building upon the color rendering basics you learned in Chapters 12 and 13, this chapter will take you one more step toward accurately rendered technical illustrations. An important aspect of doing so is to be able to simulate the materials that make up objects—such as wood, various metals, plastic, glass, and rubber. Accurately simulating materials is important and will lend credibility to your illustrations. A good place to begin is to simply be a good observer—visually absorbing all the materials around you. Make mental notes about what makes these different materials unique and how light interacts with them. With these in mind, you will understand how to replicate a number of materials common to technical illustration.

Chapter Objectives

In this chapter you will understand:

- ▼ How to make use of **Swatch**, **Symbol**, **Custom Brushes**, and **Blend** libraries to render a technical illustration
- ▼ How PostScript materials for basic geometric forms are created
- ▼ That sophisticated materials are many times comprised of several basic materials
- ▼ How light and reflected light impacts different materials
- ▼ How to make a list of basic materials that you may encounter in technical illustration
- ▼ That color and value are used together to differentiate between various materials
- ▼ How appropriate colors are used to render materials effectively.

Building Libraries

It would be a shame to have to create a new texture from scratch each time you render a technical illustration. Some materials (colors and gradient fills) can be saved in **Swatch** libraries. You'll find a number of **Custom Brushes** on the accompanying CD-ROM in *tools/brushes*. You can also create libraries of blends as **Symbols** to be used again and again. Although two files can't share the same symbol library, a symbol dropped into one file, copied, and then pasted into another file becomes a symbol in the new file.

   Swatches, **Custom Brushes**, **Blends**, and **Symbols** are saved in individual Adobe Illustrator files. For example, you'll find the file *geo_shapes.eps* in *tools/blends*. This Adobe Illustrator file contains a library of basic geometric shape blends that can be copied and pasted from the **Symbol Palette** to an illustration. Once in the illustration, the blend can be scaled and the colors of base objects in the blend adjusted as necessary.

Light and Reflected Light

It is important to recognize how light interacts with different materials when you are rendering technical illustrations in color. Materials around you are affected differently by direct or indirect light, and even by the time of day. Generally, more direct or intense light on an object will generate stark, higher-contrast highlights and shadows. Indirect or diffused light will generate smooth, softer highlights and shadows. Reflected light on the back side of objects will also vary with lighting conditions. Figure 14.1a and 14.1b show the effects of different lighting conditions on the same object.

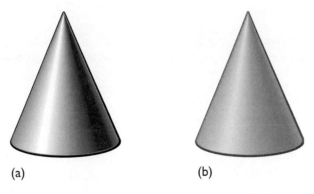

(a)   (b)

Figure 14.1   (a) Cone in bright, direct light. (b) Cone in low, indirect light.

Material Characteristics

A surface or material with a flat or dull finish may have a slight highlight or even none at all. Materials such as cast metal and raw wood are good examples because of their rough or unfinished surfaces. On the other hand, materials such as glass, chrome, and stainless steel typically reflect much more light due to their smooth or highly polished finishes. The differences between their highlights and shadows are extreme.

A technical illustrator must possess a basic knowledge of the manufacturing processes involved in creating specific materials. For instance, a machined steel shaft is manufactured and finished quite differently than a cast-iron engine block. A firm understanding of these processes will result in more accurately rendered objects.

Basic Materials

As a technical illustrator, you will encounter and be required to render hundreds of different materials in your career. Materials pertaining to technical illustration generally fit into three categories:

▼ Natural materials

▼ Man made materials

▼ Processed materials

Natural materials include human skin and hair, concrete, wood, stone, fluids, and the sky. Human-made materials include fiberglass insulation, cloth, carpet, leather, and building materials such as shingles. Generally, processed materials are rubber, plastic, glass, chrome, iron, steel, aluminum, bronze, and brass.

▼ Note

Always save an effective material rendering so it can be used again. Gradients, blends, and transparent overlays can be edited and rescaled, and their colors adjusted, for use over and over again.

The basic materials listed can be broken down further. Various woods, metals, and different types of glass and plastics are all around us. A technical illustrator involved in the engineering or manufacture of a machinery may routinely be called upon to render materials from a particular list, while another illustrator may render materials unique to architectural or medical illustrations. Over time, you will build a list of your own materials that will best suit you and the project at hand.

▼ Illustrator Tip

Repetitive features can be efficiently constructed by creating a **Custom Brush** as in this example of a bicycle chain. You can specify the (a) beginning, (b) middle and (c) end of each brush stroke. (d) You can see that each component tiles seamlessly. (e) When the chain stroke is joined, the beginning and end overlap for a continuous pattern.

Brush strokes, however, lie in the plane of the paper, and although you can shear the stroke path into axonometric projection, the brush that paints the stroke remains orthogonal. The solution is to **Distort** the brush stroke so that the brush shapes are in correct projection.

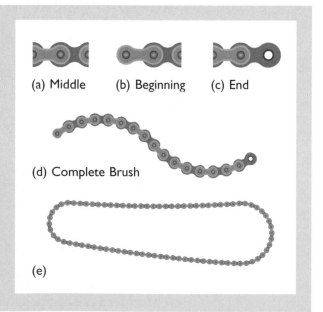

(a) Middle   (b) Beginning   (c) End

(d) Complete Brush

(e)

Describing Materials with Color

In addition to understanding the effects of light and the manufacturing processes, the color component of rendering materials is equally as important. Experienced illustrators know that a different material can be suggested by simply changing a color or intensifying a highlight. The results of a change like this can be realized in the comparison of the siding illustrations in Figure 14.2.

(a) Steel   (b) Aluminum   (c) Vinyl

Figure 14.2   Modify the color and highlights to suggest different materials.

Pulling It All Together

Let's take a closer look at the shock absorber shown in Figure 14.3. This particular illustration contains several of the more common materials associated with automotive or industrial technical illustrations. Because the shapes are generally cylindrical, a common **Gradient Fill** can be used to represent polished metal, painted metal, and rubber.

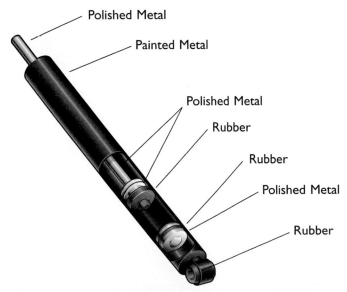

Figure 14.3   Cut-away of a shock absorber.

Figure 14.4   Chrome, aluminum, and rubber.

Later in this chapter you will be shown exactly how to represent various metallic materials, but from Figure 14.4 you can quickly see that with a slight change in ambient color as well as the projected and reflected light, dramatic differences in material and finish can be shown.

Chrome Piston

Chrome is a material that is used throughout a variety of industries and therefore is rendered frequently in technical illustrations. (See pages 316 through 319 for a description of both indoor and outdoor chrome.) At first glance, chrome's appearance may seem identical to highly polished steel. However, steel can be rendered to have a polished or dull surface, while chrome is always characterized by a high contrast, polished surface. Another characteristic of chrome is the presence of a horizon line being reflected back onto the surface of the object. This horizon line can be seen in the piston shaft gradient in Figure 14.5, where blue meets dark gray in the gradient color bar. The abrupt change in color is fine-tuned by adjusting the color keys and midpoints. Chrome is further distinguished by a bright, almost white highlight, contrasting dark horizon line, and the reflection of colors from surrounding parts or objects. The shaft of the piston has been isolated in Figure 14.6 to show you the application of the chrome gradient. Note that the reflections are parallel to the axis of the shaft. This is a characteristic of cylindrical materials.

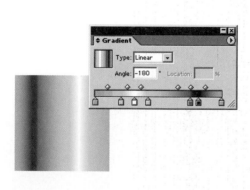

Figure 14.5   Chrome gradient as it appears in the **Gradient** window.

Figure 14.6   Chrome gradient applied.

Painted Aluminum Shell

Aluminum is another material that is very common to technical illustration. It exists in a wide variety of colors and finishes—from dull, matte finishes to glossy, and painted. For the shock bsorber, a warm, gray color was used in various tones to create the gradient. To achieve the glossy surface, a narrow highlight was built into the gradient, as well as a wide band of reflected light to better delineate the cylindrical surface. The aluminum gradient is shown in an interior detail in Figure 14.7, and the gradient is applied to the exterior shell in Figure 14.8.

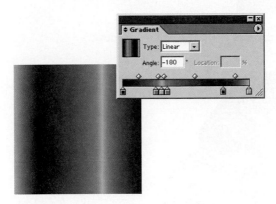

Figure 14.7   Aluminum gradient as it appears in the **Gradient** window.

Figure 14.8   Aluminum gradient applied to the exterior shell.

Rubber Bushing

Rubber usually is found in items such as bushings, grommets, weather stripping, moldings, hoses, O-rings, and tires. It generally exists only in a handful of colors—black, and perhaps the occasional deep red or green for hose applications. For the bushing on the shock absorber, a cool, gray color was used in various tones to create the gradient. To achieve a softer, dull surface, a wider highlight was built into the gradient, as opposed to the narrow highlight used in the aluminum gradient. The rubber gradient is shown in detail in Figure 14.9, and the gradient is applied to the bushing in Figure 14.10.

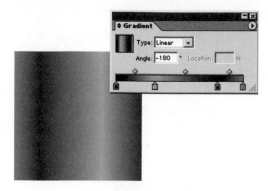

Figure 14.9   Rubber gradient as it appears in the **Gradient** window.

Figure 14.10   Rubber gradient applied to the bushing.

▼ Illustrator Tip

There are situations where a simple linear or radial gradient will not accurately describe a surface. For example, to render the end of a machined shaft, (a) begin with an ellipse filled with a medium base color, like gray. (b) Create two wedge shapes—one large and one small. Place one on top of the other, fill the large shape with the same medium gray as the ellipse, and fill the small shape with a lighter gray. (c) Blend between these shapes to create a smooth transition. (d) **Copy** this blend group and **Rotate** it 180 degrees about the center point. **Mask** the blends with a copy of the ellipse, and place the masked blends directly on top of the medium gray ellipse (d).

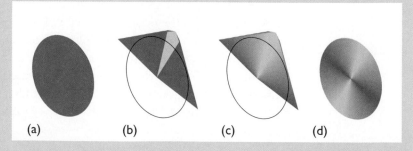

Material Gallery

The materials rendered in the shock illustration are just a sampling of the materials you will encounter when faced with rendering technical illustrations. Some additional materials are shown in Figures 14.11 to 14.31.

Additional realism can be obtained when rendering some materials by establishing the base geometry and color in Adobe Illustrator first, exporting it to Photoshop, and adding textures and other fine detail. The figures in Chapter 13 demonstrated this. This multi-application approach is well suited for rendering natural or organic objects, when nonmechanical details are not easily achieved in Illustrator.

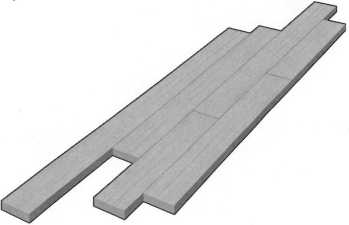

Figure 14.11   Wood flooring.

▼ Illustrator Tip

Illustrator brushes can be used to simplify the construction of more complex objects or surfaces by creating a texture. For instance, (a) a **Pattern Brush** is created from two simple shapes, and (b) then applied to multiple, overlying strokes. The colors are adjusted on random strokes and the brushed strokes are expanded. (c) The resulting pattern uses a **Mask** to match the shape of the object or surface to be filled and **Distort** can twist the shapes into pictorial projection. In this example, this expanding brush technique was used to simulate a commonly used home construction material.

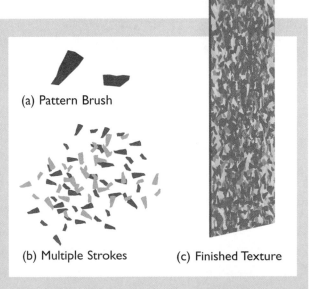

(a) Pattern Brush

(b) Multiple Strokes   (c) Finished Texture

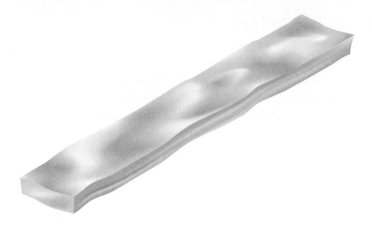

Figure 14.12   Fiberglass insulation.

Often it is easier to construct an object or surface in a plan view first, and then rotate and scale or skew the surface into the desired plane. To render the insulation, rows and columns were added to a simple rectangle using the **Gradient Mesh Tool**. The rows and columns were colored using the three color swatches. Use **Distort** to project this rendered surface onto virtually any plane, and the gradient mesh will go along for the ride.

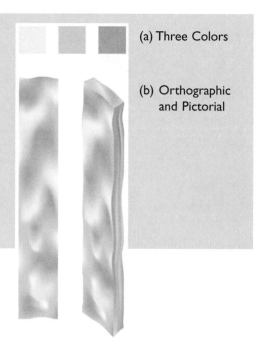

(a) Three Colors

(b) Orthographic and Pictorial

Round Metal Panel with Joint

Many industrial products require enclosures, packaging, or cases that are comprised of closely fitting panels. If you look at refrigerators, stoves, furnaces, electrical panels, computer cases, or electronic components, you'll see examples of the surface detail similar to Figure 14.13.

Figure 14.13   A rounded panel with joint.

1. Create the panel profile using any of the axonometric or perspective construction techniques covered in this text (Figure 14.14a). Duplicate this profile and move it to the side to be eventually used as joint lines.

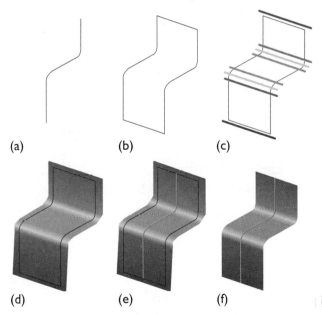

(a)   (b)   (c)

(d)   (e)   (f)

Figure 14.14   Developmental steps for rounded panel with joints.

2. Duplicate the original profile along the receding axis and connect the tops and bottoms with a straight line. Join these corners into a contiguous shape (Figure 14.14b). This will serve as the shape for a **Clipping Mask.**

3. Create lines at the receding axis angle that are slightly wider than the panel shape (Figure 14.14c).

▼ Determine a base object color. In this case we are using a blue metallic color with HSB values of 200, 40, 40.

▼ Place a line with the base color at each transition position. Assign highlight color (almost white) to the lines at the center of the fillet and round.

▼ The two vertical surfaces of the panel are essentially the same color, with a slight color transition between the top and bottom.

▼ The horizontal surface's color is between the highlight and the base color.

▼ Use **Blend Options|Specified Steps** because of the significant difference in blend distances between the flat areas of the panels and the fillet and round. The fillet and round require approximately a quarter the number of steps. The finished blend is shown in (Figure 14.14d).

4. Move the panel profile to the middle of the blend and bring it to the front of the stack. This is your joint line (Figure 14.14e).

▼ Align the top of the joint line to the top of the mask shape.

▼ Duplicate the joint. Offset so that both are visible and aligned to the top of the mask shape.

▼ Select the right joint line and pick up the deepest shade color in the blend with the **Eyedropper Tool**. This is the joint shadow.

▼ Repeat this process for the left joint line with the lightest color in the blend. This is the joint highlight. Group the blend and the panel joints.

5. Move the mask shape to in front of the blend and select both. Choose **Object|Clipping Mask|Make**. The blend is clipped to the panel (Figure 14.14f).

Window/Interior Detail

Many large industrial products such as automobiles, trucks, aircraft, and construction equipment include operator compartments that feature glass windows. To show these effectively, you must be able to render the glass windows and the interior (Figure 14.15). The glass will have more or less reflections, depending on the surrounding environment and whether or not the window is tinted. Of course, this also depends on how clean the window is. The interior will have more or less detail depending on window tinting, reflections, the color of interior objects, and the scale. Figures 14.16 through 14.21 record the development of this rendering technique.

Figure 14.15   Automotive windows with interior in background.

1. Create paths that describe the exterior of the subject, the window openings, and the interior. Place each of these components on separate layers for selective display as shown in Figure 14.16.

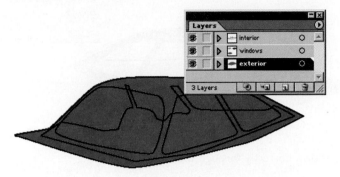

Figure 14.16   Exterior (filled), interior, and window paths.

2. Move a copy of the windows off to the side for later use. Duplicate this again.

3. Create a **Compound Object** that opens the windows in the exterior.

   ▼ Hide the interior layer.

   ▼ Assure that the windows layer is above the exterior layer. This will subtract the windows from the exterior.

   ▼ Choose **Object|Compound Object|Make**. The windows are subtracted from the exterior (Figure 14.17).

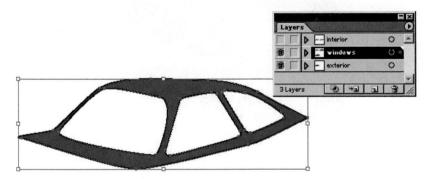

Figure 14.17   Windows are subtracted from the body.

4. Create the visibility of the interior by moving the interior layer to the bottom of the layer list. The interior is now correctly masked by the **Compound Object** (Figure 14.18).

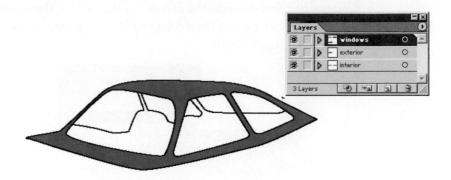

Figure 14.18   Interior is shown correctly through the window openings.

5. Using the duplicate copies of the window outlines as **Clipping Masks**, create blends for the windshield and the side glass (Figure 14.19).

▼ Create base objects for the blends and position them appropriately. The windshield is pointed more toward the sun.

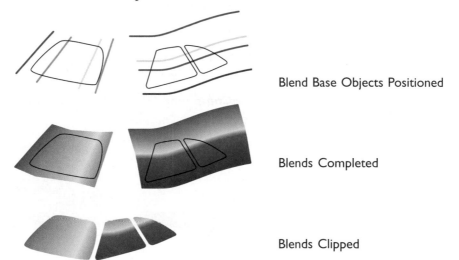

Blend Base Objects Positioned

Blends Completed

Blends Clipped

Figure 14.19   Window blends use window shapes to clip to the outline.

▼ Create the blends with the **Blend Tool** and **Blend Options** set to **Specified Steps**. Choose a number of steps that produces a smooth blend.

▼ Use the windshield as a **Clipping Mask** for the windshield blend.

▼ Select the two side windows and make them a **Compound Path** so they will function as a single **Clipping Mask**.

6. Turn the stroke off for the exterior and interior. Assign a 60 percent black to the interior as shown in Figure 14.20.

▼ Make sure that the windows are on the window layer.

▼ Pick an anchor to match on both a window blend and the compound path of the car body.

▼ Adjust the transparency of the windows to get the level of reflectivity and opacity you desire (Figure 14.21).

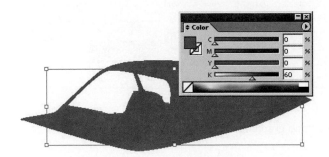

Figure 14.20   Interior is filled with shadow color.

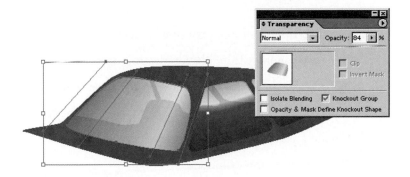

Figure 14.21   Transparency of window blend is adjusted.

Metallic Textures

A large percentage of products shown in technical illustrations are metal. This includes polished chrome, brushed aluminum, painted metals, and metal that has been rough cast. There are several techniques that can be used to create metallic textures: **Gradient Fill**, **Line** and **Shape Blends**, and **Custom Brushes**. But because metallic surfaces are very predictable, and some illustrations may have hundreds of chrome pieces, you want to use the most efficient method. When these are linear—as is the case with cylinders and flat plates—the chrome reflections will be linear also. For these reasons we will demonstrate metallic textures using **Gradient Fill**.

Figure 14.22 shows the **Gradient Fill** for an inside chrome texture. Chrome that is indoors has a hard, cold appearance and is usually completely devoid of color.

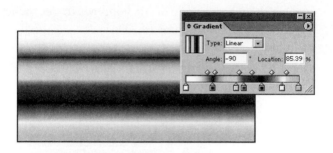

Figure 14.22   Indoor chrome gradient.

1. Begin by creating a shape that will hold the chrome gradient. In this case, we use a rectangle, a shape that might be considered the front view of a cylinder.

2. Double-click the **Gradient Tool** and click below the color bar to place the color keys.

   ▼ The inside chrome texture will have a centered dominant black reflection.

   ▼ Place another thin, black reflection on the upper half.

   ▼ Place a thin, white reflection on the lower half.

   ▼ Adjust the midpoint diamonds to sharpen the transitions (Figure 14.22).

3. Create a path to hold the gradient (Figure 14.23a).

   ▼ Duplicate the cylinder.

   ▼ Trim the elliptical ends.

   ▼ Connect the elliptical ends with limiting elements.

   ▼ Join the ellipses and limiting elements.

4. Fill the path with the gradient (Figure 14.23b) .

   ▼ Set **Edit|Preferences|General|Constrain Angle** to 90 + the axis angle. *Note*: A cylinder facing to the left will require a 90 - the axis angle. This constrains the **Gradient Tool** to an angle perpendicular to the thrust axis.

   ▼ Click on the cylinder shape, and then click on the **Gradient Tool**.

   ▼ Hold down the **Shift** key and drag across the cylinder. If you start on the top limit and drag to the bottom limit before releasing the mouse button, the gradient will be distributed across the area. If you start outside or inside the selected path, the path will be filled differently.

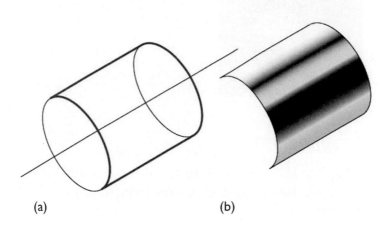

(a)   (b)

Figure 14.23   (a) Path to hold the gradient. (b) Gradient constrained to the thrust axis perpendicular.

5. Complete the end of the cylinder (Figure 14.24). In this case we have added a small chamfer.

▼ Drag the **Gradient Tool** across the chamfer so that a light portion of the gradient is next to the dark part of the cylinder. Adjust the color keys and midpoint diamonds so that dark is next to light and light is next to dark.

▼ Repeat the same operation for the cylinder's flat end. Change the angle so that the gradient is roughly parallel to the cylinder's elliptical major axis.

Figure 14.24   Gradient applied to chamfer end.

This base chrome texture can be used as the basis of other metallic finishes. Figure 14.25 shows how the color keys of the indoor chrome can be edited to render outside chrome. The changes include:

▼ Blue reflection from the sky

▼ Brown-yellow reflection from the ground

▼ White reflection at the horizon

Figure 14.25   Outdoor chrome material.

Brushed aluminum is similar to indoor chrome but does not have the hard sur-face finish that causes stark black and white reflections. The changes (Figure 14.26) include:

▼ Dark reflections are lightened.

▼ Light reflections are made light gray.

▼ A small amount of blue is added in light and dark reflections.

Figure 14.26   Brushed aluminum material.

Painted metal (Figure 14.27) has the softest of the surfaces:

▼ Light and dark reflections approach the same value.

▼ There are fewer color keys.

▼ Keys are added at the top and bottom to frame the edges.

Figure 14.27   Painted metal material.

Transparent Blends

Transparent blends are easy and convenient to make in Adobe Illustrator. Figure 14.28 shows a direction arrow that blends from transparent to opaque. In this example, a direction arrow overlays a panel with text. Were the arrow to be opaque, it would have to be repositioned or the text behind the arrow would be unreadable. The solution is to create an arrow that smoothly blends from transparent to opaque.

Figure 14.28   Transparent blend.

▼ **Tech Tip**

Transparent blends can also be used to make more complex **Clipping Mask** backgrounds. By overlaying a series of more or less transparent blends, you can create greater texture depth than can be achieved by a single blend.

Figure 14.29 shows the development of a transparent blend and its use as the background for a mask.

▼ The arrow will be used as a clipping mask (Figure 14.29a). Duplicate the arrow and place it on the paste board so that it can be edited later for shadow edges.

▼ Create beginning and ending base objects for the blend and assign both the same color (Figure 14.29b). This should be the ambient color of the arrow.

▼ Set the transparency of the base object on the tail of the arrow to zero. The transparent base object is shown in red.

▼ Set **Blend Options** to an appropriate **Specified Steps** value based on the width of the base objects and the blend distance.

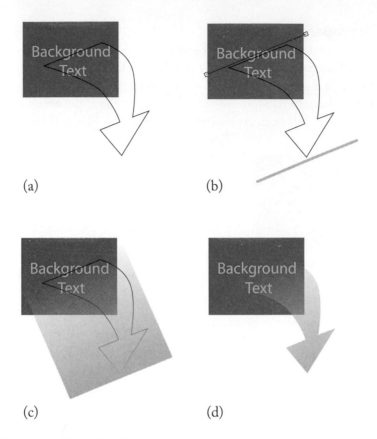

Figure 14.29 Transparent blend development.

▼ When you make the blend, it will begin opaque at the arrow head and blend smoothly to transparent (Figure 14.29c). Duplicate this blend and set it aside to use for the arrow's thickness.

▼ Move the arrow path to the front. Select both the arrow and the blend object and choose **Object|Clipping Mask|Make**. The arrow masks the blend, and the text on the panel is readable through the direction arrow (Figure 14.29d).

A nice finishing touch is the shadow edge thickness shown in Figure 14.28. Because the arrow blends to transparent, you can't simply trim the arrow path you created at the start of this exercise and assign a shade color. This is because the edge has to blend from opaque to transparent just as the arrow does. Lines, unfortunately, cannot be stroked with a gradient. There is a way around this:

▼ Trim the duplicate arrow path and keep the portion that shows thickness on the left side.

▼ Select the line and choose **Object|Path|Outline Stroke**. The line is converted to a closed path. Turn the stroke off. This path can serve as a

Clipping Mask for a blend that starts out as the shade color and ends transparent.

▼ Position the duplicate blend behind the thickness line and adjust the opaque base object's color by dialing down the brightness using the HSB color model.

▼ Bring the line to the front and position it on the arrow. The arrow and its edge now blend out to transparent.

Wires, Hoses, and Tubes

One of the most simple yet critical components of a technical illustration are wires, hoses, and tubes. Any deviation from parallelism in the opposing sides of such objects will be easily detected and take away from the effectiveness of the illustration. Luckily, there are several techniques in Adobe Illustrator that assure accurate geometry. Figure 14.30 records the development of a wire.

▼ With the **Pen Tool**, create a path that defines the centerline of the wire (Figure 14.30a).

▼ Select the path and choose **Object|Path|Offset Path**. Enter a value equal to one-half the wire's diameter as the offset distance. As shown in Figure 14.30b, a new path is created with the top and bottom parallel to the original line.

▼ **Trim** the ends from the offset path. These lines can be assigned different line weights for a simple rendering treatment (Figure 14.30c).

▼ To render the wire in continuous tone, use the top, centerline, and bottom as base objects for a blend (Figure 14.30d). Choose colors and values that describe the wire material.

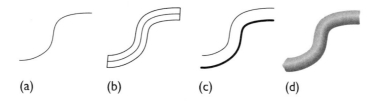

(a)   (b)   (c)   (d)

Figure 14.30   Construction and rendering of a simple wire.

As long as objects such as wires are relatively flat and don't loop around on themselves, rendering with blends is straightforward. However, when a wire loops around in front of itself, the rendering becomes much more complex. Figure 14.31 shows the construction necessary to create a looping wire and render it.

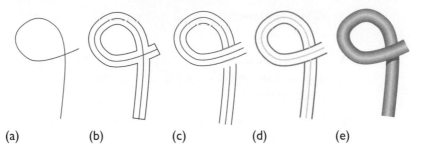

(a)   (b)   (c)   (d)   (e)

Figure 14.31   Transparent blend development.

▼ Create the wire centerline as shown in Figure 14.31a.

▼ Select the path and choose **Object|Path|Offset Path**. Enter a value equal to one-half the wire's diameter as the offset distance. As shown in Figure 14.31b, a new path is created parallel to the original line. Note that the wire us simply outlined and the spatial positioning of the loop has been lost.

▼ This is a **Compound Path**. Choose **Object|Compound Path|Release** in order to edit the shape.

▼ **Trim** the ends of the wire away. **Trim** (break) the shapes and add the green connectors shown in Figure 14.31c. Make sure that these green connectors actually complete the paths.

▼ Assign color and stroke to the outside base objects. Assign a lighter color to the centerline which becomes the highlight base object (Figure 14.31d).

▼ Perform a blend between these three base objects (Figure 14.31e). Add a small shadow to further give the impression that the wire bends around in front of itself.

Review

Second only to solidly built geometry, accurately rendered object materials can make or break technical illustrations. We have only touched on the infinite number of materials possible in technical illustrations. However, armed with the approaches covered on these pages, you'll quickly find yourself creating your own materials and textures. You will start to take notice and make mental notes of the materials that exist in the real world, so they can be recalled to render a great variety of subjects. Determining the placement of a highlight on an object and its corresponding shadow, selecting the proper base color to represent a steel shaft, or applying the appropriate texture to fiberglass insulation will require you to visually absorb the materials around you.

Text Resources

Alspach, Ted, and Murdock, Kelly. *Illustrator 10 Bible*. John Wiley & Sons. New York. 2002.

Blatner, David, and Fraser, Bruce. *Real World Adobe Photoshop 7*. Peachpit Press. Berkeley. 2002.

Johnson, Harald. *Mastering Digital Printing: The Photographer's and Artist's Guide to High-Quality Digital Output*. Muska & Lipman. Cincinnati. 2002.

Kieran, Michael. *Photoshop Color Correction*. Peachpit Press. Berkeley. 2002.

McClelland, Deke. *Real World Adobe Illustrator 10*. Peachpit Press. Berkeley. 2002.

Steuer, Sharon. *The Illustrator 10 WOW! Book*. Peachpit Press. Berkeley. 2002.

Internet Resources

Adobe Illustrator Tips and Techniques. Retrieved from: *http://www.desktoppublishing.com/tipsillustr.html*. February 2003.

Adobe Illustrator Tutorials. Retrieved from: *http://graphicssoft.about.com/cs/illustratorart/*. February 2003.

Adobe Illustrator Users Group. Retrieved from: *http://groups.yahoo.com/group/Adobe-Illustrator/*. February 2003.

Adobe's Corporate Tips & Techniques. Retrieved from: *http://www.adobe.co.uk/products/tips/illustrator.html*. February 2003.

Analytical Graphics, Inc. Retrieved from: *http://www.analyticalgraphics.com*. February 2003.

CAD Tools for Adobe Illustrator. Retrieved from: *http://www.hotdoor.com/CADtools/*. February 2003.

Daniels and Daniels Illustration. Retrieved from: *http://www.beaudaniels.com*. February 2003.

Kevin Hulsey Technical Illustration. Retrieved from: *http://www.khulsey.com*. February 2002.

Nidus Technical Illustration. Retrieved from: *http://www.nidus-corp.com/cutp.html*. February 2003.

On the CD-ROM

For those of you who would like to try out the procedures described in this chapter, several resources have been provided on the CD-ROM. Look in *exercises/ch14* for the following files:

arrow.eps	Transparent arrow example
flooring.eps	Wood flooring illustration
geo_shapes.eps	Basic geometric shape blends for your use
insulation.eps	Fiberglass insulation illustration
panel.eps	Rounded and filleted panel illustration
shock.eps	Shock absorber illustration
shock_grad.eps	Material gradients used in the shock absorber illustration
siding.eps	Varying color and highlights
window.eps	Automotive window illustration
wire.eps	Straight and looping wire exercise

Perspective Drawing

Axonometric Projection

Technical Animation

Rendering

3D Modeling

Technical Illustration

Chapter 15

Text and Technical Illustrations

In This Chapter

Although a picture may be worth a thousand words, technical illustrations at times still require text in the form of call-outs, notes, annotations, or, in some cases, text on actual parts (in axonometric or perspective projection). Such text should always take on a supporting role; if you need to laboriously explain a part's function using text, you probably have an ineffective illustration.

To understand how text can be integrated into technical illustrations, you need to understand how textual and graphic data are integrated into data formats. You must understand the relationship of text as it appears on a display monitor with text as it is printed on resolution-independent PostScript printing devices, and the outlines are rasterized, or RIP'd, into a pattern of printer dots.

Chapter Objectives

In this chapter you will understand:

▼ Digital typography and its terms, strengths, and weaknesses

▼ The difference between fixed-size screen fonts, outline display fonts, and outline printer fonts

▼ The impact color and value have on the readability of type

325

▼ The effect of turning font descriptions into vector outlines

▼ The effect of rendering outline fonts into a bitmap

▼ The limitations of bringing text into Adobe Illustrator from CAD programs

▼ How to place typography correctly into axonometric or perspective position

▼ The correct placement of call-outs and notes for maximum readability

Typography Primer

A *font family* is a distinct typographic letter style. For example *Helvetica* and *Gill Sans* are sans serif font families. Their strokes are uniform and devoid of serifs—small extensions at their ends. *Times* and *Caslon* are serif font families. Their strokes are variable and do have serifs (Figure 15.1).

Sans Serif   Serif

Figure 15.1   Sans serif and serif font families.

A *font* is the collection of all characters of a given example at a certain style and size. Not all fonts have the same number of characters or even identical characters. When someone asks, "What font is that?" they usually mean, "What is that font's family?"

Type is measured in *points*. One point is equal to 1/72 of an inch so a capital letter "M" in 72-point type would be nominally 1 in tall.

▼ Note

There's a reason why a 72-point Helvetica M isn't the same size as a 72-point Times M and why neither are exactly 1 in tall. This measurement refers not to the actual size of the letter but to the size of the slug, or piece of material, that an M sits on in traditional mechanical printing.

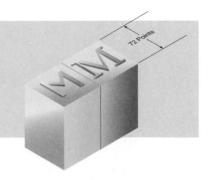

Leading is the spacing from the baseline of one line of type to the baseline of the next line of type (Figure 15.2). This is also referred to as line spacing. The term *11/13* refers to 11-point type with 13 points of spacing between successive baselines. Type set too *tight* is difficult to read. Type set too *loose* is difficult to recognize as continuous lettering.

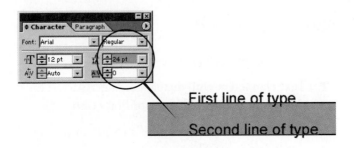

Figure 15.2   Increasing leading increases spacing between lines.

A type designer builds a certain amount of natural leading into the font so successive lines of type don't bump into each other. This is not standard and depends on the form of the letters, the intended use of the font, and the sense of the designer.

Type can be *monospaced* or *variable spaced*. Standard typewriters produced monospaced type; you can find computer fonts that are monospaced and look just like they were done on a manual typewriter. Monospaced fonts are difficult to read for long periods of time.

Variable-spaced fonts adjust spacing between letters based on which two letters are next to each other. For example, study Figure 15.3. The mono-spaced letters "V" and "O" have a totally different spacing than do the variably-spaced letters "V" and "O." Intraletter spacing is called *kerning* (also called letter spacing), and like leading, the amount of kerning in variable-spaced fonts is decided on by the font designer. However, selecting text and increasing or decreasing kerning can override this spacing, usually as a percentage of the normal amount. Note that the fonts in Figure 15.3 are the same point size but have different letter heights.

Figure 15.3   Spacing between successive letters is called kerning.

> ▼ **Note**
>
> The banes of typesetting are the space bar and the **Return** key. Avoid using the space bar to position type along a horizontal line. Instead, use indents and tabs. Avoid separating lines of type vertically with a number of returns. Instead, use space before or space after the line.

Digital Typography

Digital typography exists in one of three possible forms:

▼ Text characters with textual characteristics. This means that you can edit the text as you would in a word processor.

▼ Text characters comprised of curved outlines but lacking text characteristics. In this case, the text is *resolution-independent* vector artwork. Each character can be reshaped as artwork, but all text properties have been lost.

▼ Text characters comprised of a pattern of picture elements (pixels). This text is *resolution dependent* and exhibits no text or vector art characteristics. It is essentially raster artwork.

In order to display text on a raster monitor (like the one you probably have unless you have a specialized vector stroke monitor of which there aren't very many around), the characters must necessarily be rendered to the screen as a pattern of dots. This can be done in several ways.

Fixed-Size Screen Fonts

Both the Windows and Macintosh operating systems (as well as Unix and Linex) display most of their system textual information at predictable sizes. To do this efficiently a fixed-size bitmap *system font* is used and that font's characters are defined in a set size (10 point, 12 point, etc.) bitmap. These system fonts should *never* be used for text on technical illustrations because they will print at a low fixed resolution instead of at a high resolution.

PostScript Screen Fonts

A set of fixed-size screen fonts corresponding to common PostScript type sizes can be used to accurately render on the monitor what will be printed on a PostScript printer. Because a majority of publication-quality technical illustrations will be printed on PostScript devices, using PostScript screen fonts produces the most predictable results. But because PostScript printer fonts reside in the printer (or are downloaded to the printer as necessary), a set of fixed-size display fonts are used to correspond to the desired printer fonts.

Though PostScript output produces the quality necessary for technical illustrations, viewing a fixed-size PostScript screen font on the screen (like Helvetica) can be problematic, especially if you choose odd-sized text. Say, for example, you need 17.5-point Helvetica. The chances that you have 17-point Helvetica screen font is nil, so the program will make 17.5-point Helvetica out of the closest size you have, in this case, probably 14 or 16 point. The result is small differences in alignment and type fitting from screen to output. The solution is to use outline display fonts corresponding to the printer fonts.

Outline Screen Fonts

The solution to the fixed-size screen font problem is to use an outline screen font, such as TrueType or OpenType. Outline fonts use a mathematical description of letterforms so that dimensionally correct letters and numbers can be rendered to the screen at the correct size and style in real time. In other words, the bitmap representation of 17.5-point Helvetica is created as the letterforms are displayed on the screen. Adobe's Multiple Master font technology allows entire families of type to be developed from a basic outline description, and their OpenType fonts minimize problems that might arise between PostScript Type 1 and TrueType technologies.

Using outline display fonts means that you will get good-looking screen display of any character, at any size, and any style. This would be entirely appropriate for technical illustrations that live on display monitors. However, technical illustrations containing TrueType fonts may cause problems when sent to PostScript printers. In fact, many print shops will not accept—or take responsibility for—output containing TrueType. Why? Because TrueType was originally created to solve the problem of displaying readable type at any size on a monitor, not for printing; when rendered by PostScript printing engines, variable results may occur.

Adobe's Portable Document Format

Much of the concern over typographic issues in technical illustrations has been addressed in Adobe's Portable Document Format (PDF). PDF functions as a container for metafile data: text, raster graphics, and vector graphics. It solves the problem of how to simultaneously deliver technical training (containing technical illustrations) on-line and by traditional printing.

PostScript Type 1 fonts are rasterized directly to the screen by the PDF viewer, much as are TrueType fonts. So when displaying a technical illustration on a display monitor, PostScript and TrueType outlines are accurately rasterized and displayed—PostScript by Adobe Acrobat and TrueType by the computer system.

When the PDF technical illustration is printed, other factors come into play. A raster image processor (RIP) must handle converting text and graphics into printer dots. PostScript fonts are RIP'd directly, but TrueType outlines must first be converted to Type 1 outlines and then rasterized. Though this process is transparent to the operator, subtle differences do occur, especially in copy fitting. Because the metrics of Ariel TrueType and Helvetica PostScript are slightly different, a body of copy may change. A passage that was designed to be 27 lines long may spill over onto line 28 after TrueType is converted to PostScript and printed (or be one line short). Figure 15.4 demonstrates the variability resulting from converting TrueType to PostScript. Text that should clear an illustration (Figure 15.4a) may print over the figure (Figure 15.4b).

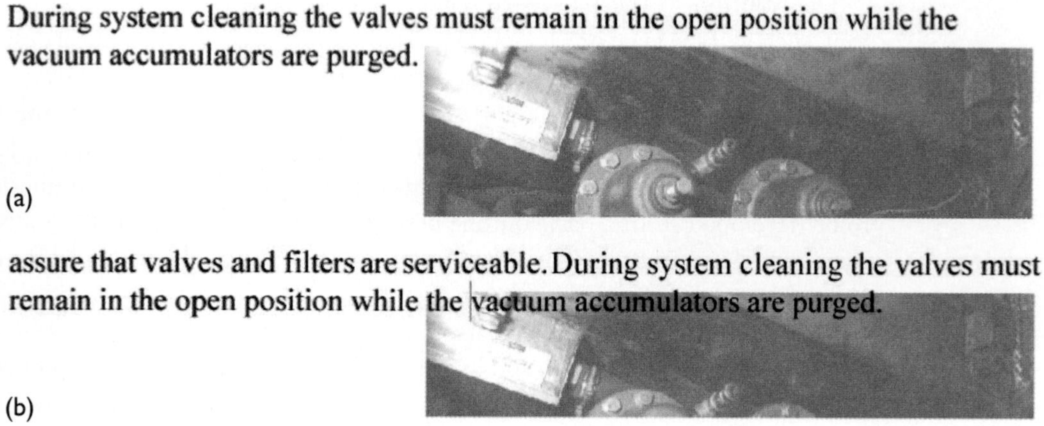

During system cleaning the valves must remain in the open position while the vacuum accumulators are purged.

(a)

assure that valves and filters are serviceable. During system cleaning the valves must remain in the open position while the vacuum accumulators are purged.

(b)

Figure 15.4   Printing TrueType may lead to unexpected results.

Converting TrueType to Outlines

Because technical illustrations generally use typography to *supplement* the image, there is little reason to use anything other than a basic sans serif font for call-outs, and a basic serif font for longer passages of body copy. If you use TrueType in technical illustrations, consider converting the type to outlines after the final editing. The text becomes Bezier EPS artwork (just like the rest of the illustration) and will RIP accurately. An additional feature of converting text to outlines is the ability to modify characters (Figure 15.5). When converted, the art that looks like text will be both filled and stroked.

The stroke may be turned off so that the fill extends to the middle of the stroke, providing for crisper text outlines (Figure 15.6). This slightly opens up the inside of letters (called *counters*) and makes them more readable, especially in small point sizes.

Once turned into artwork, there will be no conversion process at the RIP, as has been previously discussed, and any substitution problem (where Acrobat substitutes a Type 1 font for something you used that it doesn't recognize) is eliminated.

Figure 15.5   Convert fonts to outlines for printing on PostScript devices and character modification.

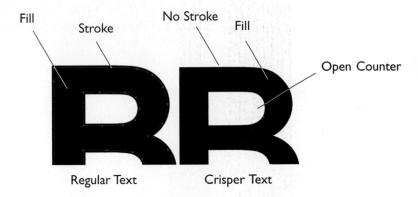

Figure 15.6   Remove stroke so that converted text appears crisper.

▼ **Note**

Always convert type to outlines if you have any question whether someone who will later use the file might not have the correct fonts installed. Distribute the EPS file with converted type for printing and review, and keep the Adobe Illustrator file with unconverted typography for potential revisions.

Text From Raster Files

On occasion you may find text on raster files that end up as technical illustrations, or at least as part of technical illustrations. Such text can be disastrous because of the resolution dependence of the raster file. Most PDF files are RIP'd at 1440 dpi and above. Text in a 600-dpi raster file will appear much coarser and, possibly unreadable at small point sizes. Follow these steps if you need to increase the quality of text you find in raster files:

▼   Make a print of the file for reference.

▼ In a raster editor, remove the text. This can be a difficult process if the text is on top of detailed artwork (Figure 15.7a). Even after selective editing (Figure 15.7b), the results may be unacceptable.

▼ If removing the text effectively proves impossible, consider placing a solid color panel over the text (Figure 15.8).

▼ Save the edited raster file in EPS format.

▼ **Place** the raster file in Adobe Illustrator.

▼ Set outline text in the correct position. If the text is TrueType, convert it to outlines.

▼ Save the illustration in EPS format. EPS is a container that holds the fixed-resolution raster data and the resolution-independent text or text art (Figure 15.9).

(a)

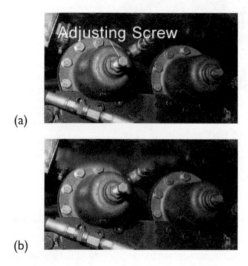

(b)

Figure 15.7   Effective and ineffective removal of text.

▼ Photoshop Tip

The detail of the background behind the text probably isn't all that important. If you can duplicate the surrounding detail, you probably can fool the observer into thinking the image continues behind the text. The easiest way is to select an area of interesting detail with a **Feathered Lasso Tool** and duplicate-move (**ALT** key down) the "patch" over the raster text. Or, you can use the **Rubber Stamp Tool** to "paint" target pixels over the text.

Panel Ready for PostScript Text

Figure 15.8   A solid panel may be the only solution if a raster pattern cannot be effectively edited in.

Adjusting Screw

Figure 15.9   Outline text added in a vector program. Compare results to the text in Figure 15.7a.

Text from CAD Files

Using geometry, or even entire illustrations from CAD programs, is highly problematic, as is further discussed in Chapter 17. Even if geometry is usable, text may not be, and editing or reworking geometry and text from a CAD program may not be worth the time. It may be more effective to use the CAD data to create new geometry.

Wide-format ink-jet and high-resolution laser printing have generally replaced pen plotting for technical drawings. Still most CAD programs create graphical output, including text, as vector strokes, optimized for pen plotting. Windows CAD programs may make use of TrueType fonts for screen display, and these fonts may look fine when printed to a laser printer. But when vector stroke text is brought into a PostScript illustration program, the results are less than professional. As you might imagine, CAD programs generally know nothing about PostScript or PostScript fonts.

Figure 15.10 shows the result of copying Ariel TrueType text from AutoCAD and pasting directly into Adobe Illustrator. The text remains Ariel and can be edited normally in Illustrator. However, the inherent problems of printing TrueType to PostScript printers remains. The solution is to convert the Ariel to outlines.

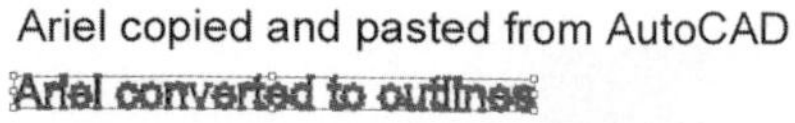

Ariel copied and pasted from AutoCAD
Ariel converted to outlines

Figure 15.10   Font characteristics are preserved when copying and pasting between AutoCAD and Illustrator.

AutoCAD files can be saved in EPS format. Unfortunately, the Ariel is lost (Figure 15.11) as the font outlines are converted to triangular polygons. This is the worst of all worlds: the text can't be edited because it has been converted to vectors and the conversion wasn't to Bezier curves, but rather to polygons.

Ariel saved as EPS and opened in Illustrator

Figure 15.11   Converting Ariel to EPS in AutoCAD results in text being converted to triangular polygons.

AutoCAD's intermediate DXF format handles text differently that the first two methods (Figure 15.12). When the AutoCAD file is saved in DXF format, the text continues to be Ariel and editable, but it is now contained in a text frame.

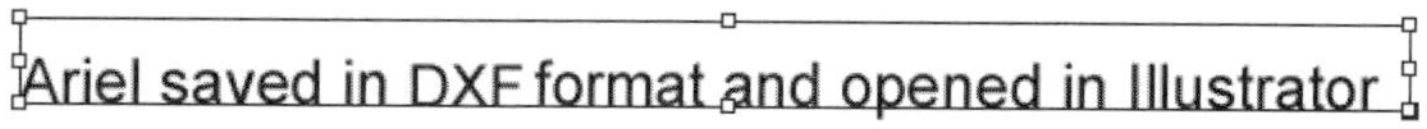

Ariel saved in DXF format and opened in Illustrator

Figure 15.12   AutoCAD's DXF format preserves font characteristics when opened in Adobe Illustrator.

Text and Color

We have mentioned previously that there is very little overhead in creating multi-color or full-color digital artwork. Add to that the fact that much technical illustration is repurposed for other functions—illustrators may not know in what forms their work will eventually be used.

Text becomes readable only when the characters and the background have sufficient contrast. In color illustration this contrast can be in hue or value. Unfortunately, when color illustration is converted to gray scale, all hue contrast is lost and readability is totally dependent on value contrast. Study Figure 15.13 where text has been overlaid on a color panel. Saturation and value are identical between the text and background. The text is only readable because there is a difference of hue.

Unfortunately, the text and the background have identical value (brightness) settings, so when the illustration is converted to gray scale (Figure 15.14), the type is unreadable.

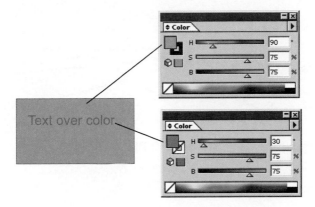

Figure 15.13   Text readability is a function of both hue and value contrast.

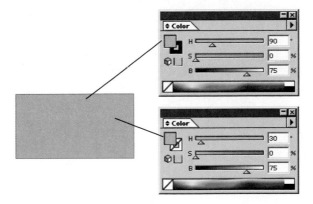

Figure 15.14   Stripped of hue, the text and background become the same gray.

Call-Outs

A *call-out* is a short passage of text that identifies a part in a technical illustration. Care should be taken not to clutter the illustration with text because the visual communication quality of the image may be adversely affected. Call-outs can be in the plane of the paper or in the plane of the drawing. Figure 15.15 shows call-outs in the plane of the paper. This is the easiest and most direct way to place the text. In Figure 15.16 the call-outs are in the plane of the drawing. By doing this, the pictorial space is maintained, though the text may be harder to read and spacing the call-outs becomes more problematic.

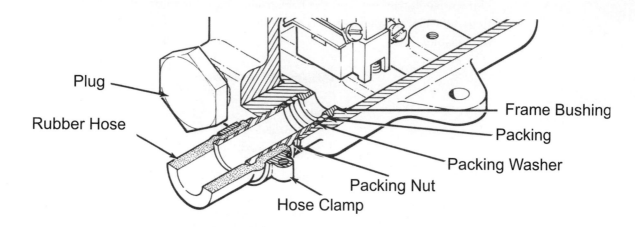

Plug
Rubber Hose
Frame Bushing
Packing
Packing Washer
Packing Nut
Hose Clamp

Figure 15.15 Text in the plane of the paper.

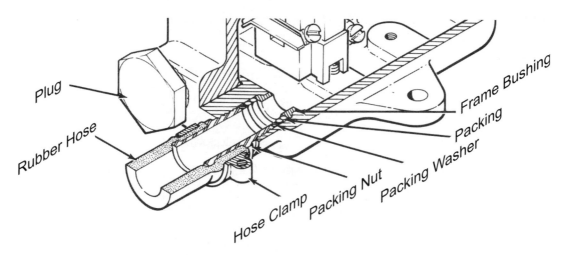

Plug
Rubber Hose
Frame Bushing
Packing
Packing Washer
Hose Clamp
Packing Nut

Figure 15.16 Text in the plane of the drawing.

Another method of identifying parts in a technical illustration assembly is to use numbers in balloons (Figure 15.17). This is standard engineering practice. The numbers are coordinated with a *bill of materials* where detailed information about each part is listed.

Call-outs are often used to identify components on a technical illustration. Generally these identifiers are either *aligned* or *unaligned*. Aligned call-outs follow an underlying guideline and are appropriate when the subject of the illustration does not have features that might confound reading (Figure 15.18). In an aligned design, the call-outs are aligned along vertical and horizontal guides and the leaders are set at identical angles. If the illustration has features parallel to the desired alignment, you should consider an unaligned arrangement (Figure 15.19).

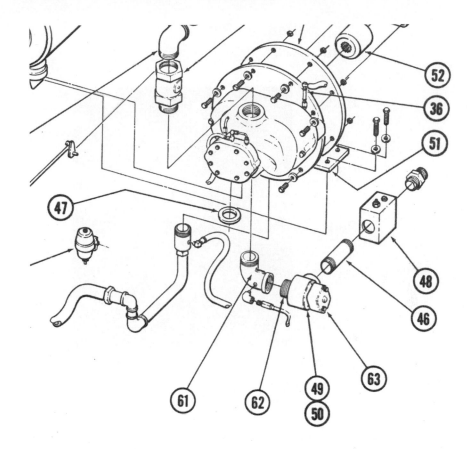

Figure 15.17   Balloons as call-outs.

▼ **Copyright Law**

Copyright law has an interesting interpretation of typography. The Copyright Office has historically ruled that the *shape* of letterforms exhibit *common industrial utility* and are, by themselves, not copyrightable. This means that the design of typography has not been deemed to have sufficient creativity or uniqueness to pass an intellectual property test. (To be copyrightable an expression must exhibit unique intellectual creativity.) What *is* copyrightable is the type family's name. So you can design a font that looks like Helvetica but you can't call it Helvetica, because that is a copyrighted type name. A way around this is to copyright a computer program or patent a process that's required to make the typography work (like Adobe's multiple master fonts). So the type shape can only be used if you have the copyrighted program.

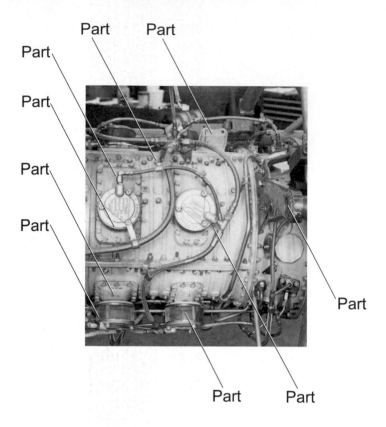

Figure 15.18   Aligned call-outs follow an underlying guideline.

It's helpful to create a separate layer in Adobe Illustrator for textual information, including call-outs. That way, the underlying graphics can be selectively displayed for additional uses. You can duplicate the text layer, **Select All**, and **Convert to Outlines**. That way, you have editable text and graphic text in the same file.

▼ Note

The line from a call-out to its part is called a *leader*. When the leader points to parts on photographs, as is the case in Figure 15.18, it is helpful to back the black leader with a thin white leader. That way, there is sufficient contrast between the various shades of gray on the photograph. This technique is also effective on line renderings where the white line breaks the object line, sending the object to the rear of the leader. Sometimes a black leader gets lost altogether, as was the case in Figure 15.6. Where the black leader crosses the black letter it is changed to white.

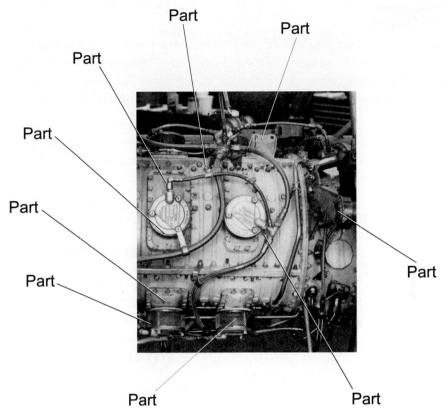

Figure 15.19   Unaligned call-outs are randomly placed around the illustration.

Text in Illustrations

Occasionally text actually appears in an illustration. When it does, you must correctly align the text to the projection axes. An otherwise effective illustration can be totally ruined if text is not correctly in projection.

Axonometric Text

In an axonometric illustration, text can be sheared into correct position either before or after conversion to outlines. It may be more convenient to wait until the illustration is completed to do this outline conversion for several reasons.

▼ If left as text, the sheared text can be swiped with the **Text Tool** and edited. Once converted to outlines the text must be typed again and resheared.

▼ Text that has been correctly sheared can be copied, pasted, and then edited into other text passages without reshearing. This shortens the process of shearing a number of text call-outs into axonometric position.

▼ **Note**

You can multiply select (**Shift** key down) individual text objects and perform axonometric shearing operations. The objects will shear correctly. However, because the operations are performed about the center of the multiple objects, their final position isn't easy to determine and the individual text objects will have to be moved to final positions.

To shear text into axonometric projection, simply apply the rotation-scale-rotation data from AxonHelper for the desired face. The text will be correctly projected onto the axonometric plane (Figure 15.20).

▼ Open AxonHelper to the correct shearing view. Determine the plane that you want the warning plate projected onto.

▼ Select the text with the **Closed Selection Tool**.

▼ Enter the shearing information in order.

▼ The text will be correctly sheared into axonometric projection.

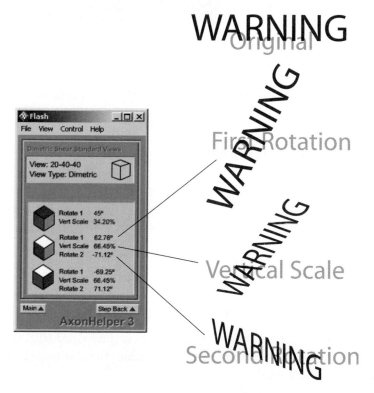

Figure 15.20   Axonometric text can be sheared into correct position using information from AxonHelper.

In Figure 15.21a, a warning plate is shown in orthogonal position. Without shearing, the construction of the letterforms in the axonometric view would be extremely difficult, but by using the same AxonHelper shearing data used to create the axonometric view (Figure 15.21b), the plate, and its lettering, are correctly projected.

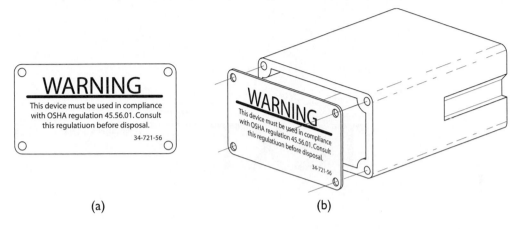

(a)   (b)

Figure 15.21   Axonometric text can be sheared into correct position using information from AxonHelper.

Perspective Text

In perspective, getting text into correct projection is more problematic because of convergence (refer to Chapter 9). Follow these steps as displayed in Figure 15.22:

▼ Create the perspective layout in which the text will be created. Locate vanishing points, measuring points, and the starting corner on the picture plane (Figure 15.22a).

▼ Create your text at the desired size and place on the common baseline/ground line at the starting point (Figure 15.22b).

▼ Enclose the text in a box, and using a magic point at the right measuring point, find the enclosing box in perspective (Figure 15.22c).

▼ Locate critical points of text characters (left, right, center, top, bottom) using measuring lines (Figure 15.22d). This is a perspective version of the *boxing-out method*.

▼ Copy text and scale to generally fit the perspective guidelines. Convert the text to outlines. Or, if the letters are generally straight strokes, complete with perspective construction.

▼ Using the **Open Selection Tool**, edit text points to align with perspective guidelines (Figure 15.22e).

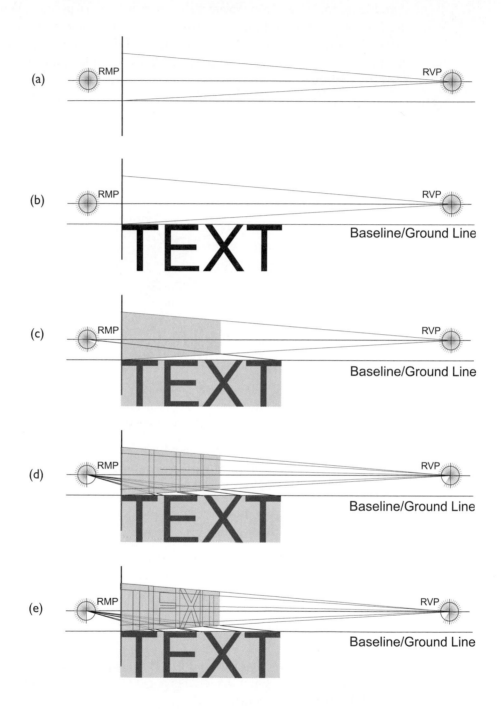

Figure 15.22   Fitting text into perspective requires finding the enclosing perspective rectangle.

▼ **Illustrator Tip**

Adobe Illustrator provides a method for adding perspective convergence to text outlines. In fact, this technique can be used to distort any Illustrator geometry into perspective. In the case of text, begin by converting the characters to outlines (**Type|Create Outlines**). You won't interactively see the type distort, so you will have to match perspective guide lines (a). Choose **Effect|Distort & Transform|Free Distort**. Move the handles of the distortion box (b) until the shape roughly matches the perspective (c). Successive free distortions revert back to a rectangular distortion box, so you can't use the shape of the box to guide your actions.

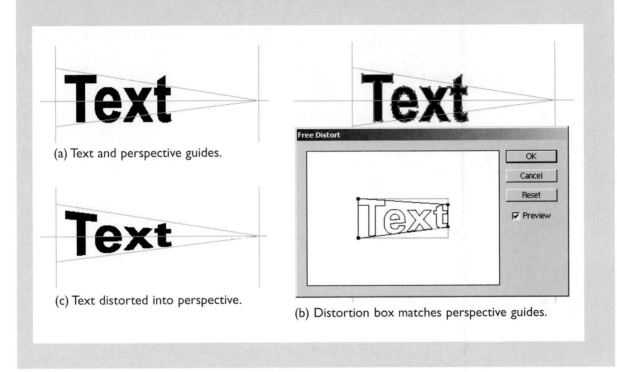

(a) Text and perspective guides.

(c) Text distorted into perspective.

(b) Distortion box matches perspective guides.

Review

Typographic elements are often found as parts of technical illustrations. When used to add textual explanations, they are referred to as call-outs. Call-outs can be placed in the plane of the paper or in a plane of the drawing and can be either aligned or unaligned. When text is part of the subject of the illustration, typography must strictly adhere to the system of projection.

Knowledge about digital typography will allow you to choose the most effective font technology for your illustration. With this knowledge you can recognize strengths and limitations of textual data from raster and vector applications and from CAD programs as well.

Text Resources

Baines, Phil, and Haslam, Andrew. *Type & Typography*. Watson-Guptill Publishers. New York. 2002.

Bringhurst, Robert. *The Elements of Typographic Style*. Hartley and Marks Publishers. Vancouver. 1997.

Heller, Steven, and Meggs, Phillip B. *Texts on Type: Critical Writings on Typography*. Allworth Press. New York. 2001.

Strivzer, Ilene. *Type Rules!* North Light Books. Cincinnati. 2001.

Internet Resources

Adobe Portable Document Format. Retrieved from: *http://www.adobe.com/products/acrobat/adobepdf.html*. August 2002.

Adobe PostScript. Retrieved from: *http://www.adobe.com/products/postscript/main.html*. August 2002.

Adobe Typography. Retrieved from: *http://www.adobe.com/print/main.html*. August 2002.

Planet Typography. Retrieved from: *http://www.planet-typography.com/*. December 2002.

TrueType Typography. Retrieved from: *http://www.truetype.demon.co.uk/*. December 2002.

Web Page Design for Designers-Typography. Retrieved from: *http://www.wpdfd.com/wpdtypo.htm*. December 2002.

On the CD-ROM

For those of you who would like to try out the procedures described in this chapter, several resources have been provided on the CD-ROM. Look in *exercises/ch15* for the following files:

callouts.ai	Assemble the callouts in both aligned and unaligned orientations. Use arrows or simple lines for leaders.
push.ai	Use this file to shear the letters into axonometric projection. Either engrave or emboss the letters.
raster_text.tif	A digital photograph with text. Replace the raster text with outline text in Adobe Illustrator.

Part Four

Modeling, Animation, and Technical Illustration

▼ **Turning Engineering Drawings into 3D Illustrations**

▼ **Using AutoCAD Data in Illustrations**

▼ **Modeling for Illustration**

▼ **Raster Materials**

▼ **Animation and Technical Illustration**

Perspective Drawing

Axonometric Projection

Technical Animation

Technical Illustration

Rendering

3D Modeling

Technical

Perspective Drawing

Axonometric Projection

Technical Animation

Technical Illustration

Rendering

3D Modeling

Chapter 16

Turning Engineering Drawings into 3D Illustrations

In This Chapter

In the preceding chapters you have seen how 2D technical illustrations can be made using a variety of techniques. In fact, Chapter 11 was devoted to *photo tracing*, the method of bringing a raster image into Adobe Illustrator to serve as the basis of an accurate orthographic tracing. It is important to have command of these 2D techniques because technical illustrations will, for the most part, be reproduced using traditional print processes, and much of the information from which illustrations are made may be two dimensional. Still, there will be times that you will want to make 3D technical illustrations—either because the information is already in three dimensions, or because you want the power and flexibility of 3D models.

For example, were you to need multiple unknown views (you can't predict ahead of time which views will be needed), it would be wise to model the subject and then extract the views as needed. Were you to rely on 2D constructions, you would have to re-create the geometry for every view, a process that in the end would be more time-consuming and less profitable.

You might also choose to model the illustration when you have access to existing 3D geometry. It will probably be more efficient to edit existing 3D geometry into something usable for an illustration than to make a 2D drawing from scratch. But this is predicated on having accurate and usable 3D data.

In this chapter you will learn how to turn 2D paper engineering drawings into 3D technical illustrations. That way, you are not dependent on having the 3D CAD geometry; all you need are accurate scale drawings. In Chapter 17 you will take the next step: turning digital 3D CAD geometry into technical illustrations.

Chapter Objectives

In this chapter you will understand how to:

▼ Determine which types of engineering drawings are appropriate for 3D conversion

▼ Scan engineering drawings into appropriate raster format

▼ Align and trace the scan into the necessary and sufficient vectors for modeling

▼ Save the vector tracing in a file format appropriate for your modeling program

▼ Choose appropriate procedures including naming conventions, hiding, coloring, and modeling processes

▼ Arrange the parts and camera for the desired technical illustration view

▼ Render the illustration in line, tone, and photorealistic modes

The Process

An effective illustration is always based on accurate geometry. This means that the drawing on which your 3D illustration is based *must* be drawn to scale or at the least, proportional. A drawing made *Not to Scale*, but which has correct numerical information attached to it, is appropriate for other 2D or 3D illustration techniques. So the first task is to secure a scale drawing.

We will be using a sectional assembly drawing as the basis of our illustration. Figure 16.1 shows the sectional assembly of a fan pulley. This assembly is tailor-made for 3D modeling because the parts are mostly cylindrical and can be formed by revolving a consistent cross-sectional shape about an axis. You will find that the assembly can be modeled very easily from these shapes, called *profiles*. We will show you the general procedure for creating the illustration, leaving several techniques either for you to explore on your own in 3ds max or another software program. In later chapters, such as in Chapter 19, you'll see how to add photorealism to the illustration. If you want to follow along, the necessary files can be found in *exercises/ch21* on the CD-ROM.

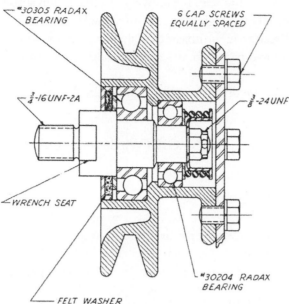

Figure 16.1   A scaled sectional assembly drawing appropriate for 3D technical illustration. (Source: Earl, *Engineering Design Graphics*, 1995.)

Scan the Drawing

The drawing itself does not have to be large. The dimensions of the scanner will usually determine the size of the drawing that you can use. The typical flatbed scanner will accept 11 x 17 sheets, so you may need to use a photocopier to reduce a large-scale drawing so that it can be scanned. Or, you can scan the large drawing in sections, *stitching* the pieces together in Photoshop. This scan of the engineering drawing is traced in Illustrator, providing accurate profiles.

The vector profiles you will be creating in Adobe Illustrator will be resolution independent so the size of the raster, if scanned from an 8.5 x 11 sheet, will generally be sufficient.

▼ Note

The reproduction from optical copiers (those that use traditional lenses) can be highly variable, especially as you move away from the center of the image. Lines may tend to "bend" as you approach the edges of a copy. You may have to correct this "stretching" of the geometry in Illustrator by making use of parallel, concentric, and tangent conditions. Digital copiers are generally more accurate across the width of the *charge coupled device* or scanning head.

It is important to scan the drawing at a resolution sufficient for tracing but if the resolution isset too high, the scan simply picks up unwanted artifacts (dirt, smudges, etc.) and results in an image that slows operations down in your vector program (like zoom and pan). To make an appropriate scan you should:

▼ Scan the drawing in black and white (1 bit).

▼ Scan at 150 dpi for an 8.5 x 11 sheet.

▼ Output to BMP or GIF (Windows); PICT or GIF (Mac).

Align the Scan

In Adobe Illustrator, open a new file and set the canvas size and orientation appropriately for the scan you just made. Pick **RGB Color Mode** because you'll be bringing the vectors into 3ds max, an RGB color space computer program.

Place the Scan

Choose **File|Place** and select your scan. The scan will be placed on the background layer, *Layer 1*, which is just fine, because you will be creating your vector profiles on another layer (Figure 16.2).

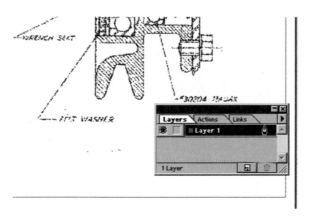

Figure 16.2   The scanned sectional assembly drawing on Layer 1 in Adobe Illustrator.

Lock the Scan

To keep the background from inadvertently being moved, you will want to lock the raster image. Double-click the layer name in the **Layers Window** and check the **Lock** button (Figure 16.3).

Dim the Scan

You can choose to **Dim** the scan at this time also. If the drawing becomes difficult to read, you can always take off the dim feature (Figure 16.3).

Figure 16.3   **Lock** and **Dim** features in the **Layer Options** control box.

Use Ruler Guides

You will want to use ruler guides to carefully align your profiles, so if rulers are not visible, choose **View|Show Rulers**. The guides that you will be pulling from the rulers are automatically locked, so you'll want to choose **View|Guides** and deselect **Lock Guides**.

Depending on the complexity of the object you are illustrating, create one or more additional layers. You may want to create a layer to hold geometry and guides for a particular part.

Layers and ruler guides are related. That is, when you drag ruler guides from the rulers, they are placed on the active layer. If you hide that layer, you hide the guides on that layer. If you try to create all the profiles on one layer, you may have so many guides displayed that you can't tell what you are doing. So for this example, we will start with an additional layer named pulley and add additional layers as needed.

▼ Illustrator Tip

Lines and curves can be made into guides (**View|Guides|Make Guides**), and guides can be made into geometry (**View|Guides|Release Guides**). Because the guides you pull out of rulers are either horizontal or vertical, this is the way to create angular or curved guides and then to turn the guides into geometry. These guides can be easily locked or hidden.

Pull the Centerline Guide

I prefer to use a solid guideline in blue. This is changed in the command **File|Preferences|Guides&Grid**. Begin by pulling a horizontal ruler guide for the centerline of the entire assembly. You can do this on the *pulley layer*. This centerline is critical once you get the profiles into 3ds max. It is this line that controls the axis of revolution for lathed (revolved) parts in max.

Align the Scan

Your scan probably won't be perfectly square (Figure 16.4a), but you can easily unlock and rotate the scan until the centerline of the scan is aligned with the horizontal ruler guide. Rather than try to rotate the scan manually, double-click the **Rotation Tool** and enter a small rotation number—0.1 degrees to start. Our scan was rotated 0.27 degrees (Figure 16.4b) but each scan will be different. Remember that positive rotation is counterclockwise. **Lock** the scan again when you are done.

Create a new layer called "pulley" to hold the guides for that part. After the scan is aligned, continue pulling guides from the horizontal and vertical rulers, boxing-out the general shape of the pulley (Figure 16.5).

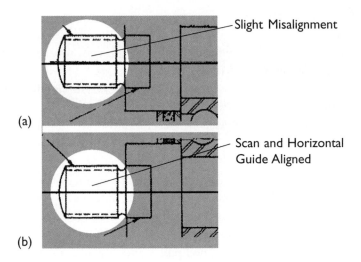

Figure 16.4   Assembly centerline before and after alignment with a horizontal ruler guide.

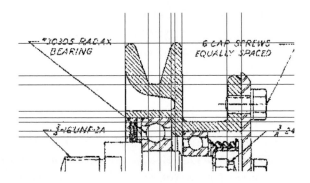

Figure 16.5   Ruler guides set for the construction of the pulley.

Create Profile Geometry

Because the pulley is symmetrical about the assembly centerline, only the top profile, in its correct position relative to the axis of revolution, is needed. This will be true for all symmetrical parts. There are several approaches to creating the profiles, but keep in mind, if you bring a valid *closed* profile into 3ds max, you will have greater success, without a lot of editing—such as welding of vertices.

Use the Pen Tool

Use the **Pen Tool** to create the profile. The first profile should be the centerline. You might try a 0.5-point red line for profile lines as this red contrasts nicely with the gray scan and blue guides. Choose no fill.

With a black 0.5-point line, follow the shape of the pulley, noting the **Smart Cursor** (Figure 16.6), indicating that guide intersections have been located. Try to *click* at each sharp corner and *click and drag* at each point of tangency. To make sure the path is joined, create the last point directly on top of the first point. You can add, subtract, or edit the profile (Figure 16.7) so that it matches the background scan. Profiles with a minimum number of vertices produce the best results in 3ds max. If you have unnecessary vertices, delete them and fine-tune the profile in Adobe Illustrator. Figure 16.8 shows profiles for the entire assembly with the background raster image hidden.

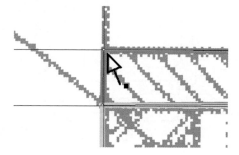

Figure 16.6   The **Smart Cursor** shows where guide intersections have been located.

▼ Tech Tip

How do you know if you have a closed profile? If you select the first and last points with the **Open Selection Arrow** and choose **Object|Path|Join**, you can tell if you have two points or one. If you get the option of making the points a corner or smooth joining, you know the points *aren't* joined. Join them. If you get a warning about needing two open end points, the profile is already joined.

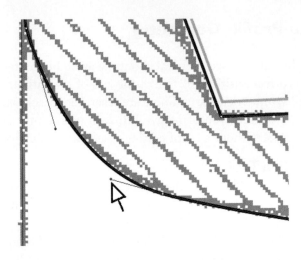

Figure 16.7   The profile is edited to coincide with the background scan.

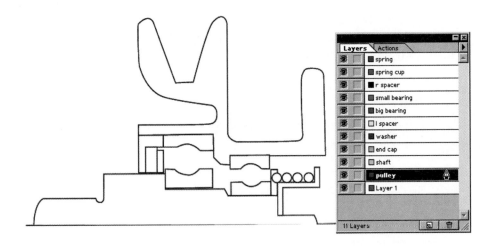

Figure 16.8   Finished profiles ready to be imported into 3ds max

Save the File

Because you don't need layers or the background raster image in 3ds max, save two versions of the file: one with layers and background image, and one with a single layer containing all profiles. Make sure you delete the **Ruler Guides** in this last file because they will appear as unnecessary long straight lines when the file is imported into 3ds max. Save the files in Adobe Illustrator (.ai) format, version 9. Macintosh illustrators should always use the file extension (.AI) even though it's unnecessary on the Mac because this is the way the Windows operating system identifies the file's creator type.

Import the Profiles into 3ds max

The creators of 3ds max knew that many illustrators, already familiar with products such as Adobe Illustrator, would create their 2D shapes in drawing programs and then bring them into Max for modeling. You don't *have* to do it this way. Scanned engineering drawings can be brought into Max as textures and traced over just as is done in Illustrator. The tools are just a little easier to use in a PostScript drawing program than in a CAD modeling program. Additionally, the Illustrator profiles are PostScript and can be sheared into an axonometric projection if need be.

▼ Note

Raster images can be brought directly into 3ds max and used for tracing. We prefer to use Adobe Illustrator to create the profiles because of Illustrator's more flexible drawing tools, layers, and guides. Additionally, by using Illustrator, you have a separate vector file that can more easily be edited and repurposed.

Import the Adobe Illustrator Vector Profiles

In 3ds max, choose **File|Import** and select the Adobe Illustrator (AI) import filter from the pop-down menu. Select the Adobe Illustrator file containing the pulley profiles. Because you will want to use the profiles individually to create the different parts of the assembly, choose **Import Shapes As: Multiple Objects** when prompted (Figure 16.9).

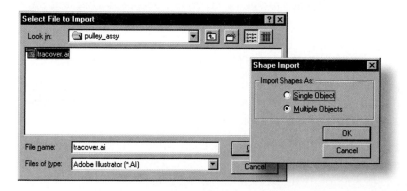

Figure 16.9   Import the Adobe Illustrator file into 3ds max.

Alignment with Axes

The XY geometry of the Adobe Illustrator file is mapped to the XY world coordinates of 3ds max, placing the profile geometry at Z 0 in the top view port (Figure 16.10). Depending on the units and level of zoom, the profiles may be very small. To fill all view ports with the profile geometry, click on the **Zoom Extents** icon at the bottom right of the screen. If the result is a line running across the view port (horizontally or vertically), you have imported ruler guides along with the profiles. Select these lines, delete, and redisplay.

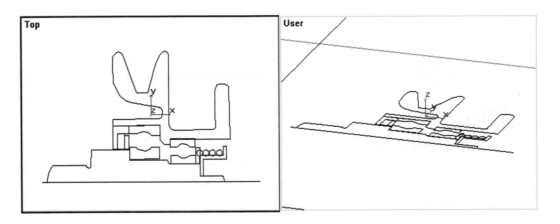

Figure 16.10   Profiles are normal in the top viewport of 3ds max.

▼ **3D Modeling Tip**

Spline Geometry from Illustrator

Spline geometry from Adobe Illustrator needs to be as simple as possible to yield the best results in max. Additionally, each profile needs to be *continuous*. That is, it must be one curve from start to end.

▼ Use the minimum number of anchors in Illustrator necessary to describe the profile.

▼ Make sure that Bezier handles are not "turned inside out." This happens when you drag a curve anchor the wrong direction. The curve may appear correct, but the anchor is actually rotated 180 degrees.

▼ Use a continuous curve in Illustrator. Even when you join individual curve segments in Illustrator, the curves are imported as separate segments into 3ds max. The vertices of these segments must be welded for correct results.

▼ It may be more efficient to **Freeze** imported profiles and trace over in 3ds max than to go back to Adobe Illustrator and fix problems.

Assign Intelligent Names

By default, Max has named the profiles Shape1, Shape2, ... , Shape11. Click on the pulley profile in the top view port and change its name from Shape1 to pulley. Do this for all the profiles. (You may have created your profiles in a different order, so be careful with the names.) See Figure 16.11 for the list of names.

Hide Everything But the Pulley and Axis

Click on the **Display** icon in the **Command Panel** and then choose **Hide by Name**. Choose the pulley and **CTL-click** to also select the axis. Choose **Invert** to quickly deselect the pulley and the axis and select all other profiles. Click **Hide** to hide all parts other than the pulley and the axis (Figure 6.11).

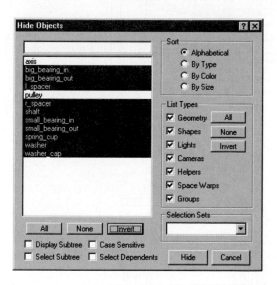

Figure 16.11   Hide a large number of parts by using the **Hide by Name** option.

Lathe the Shapes

Select the pulley and choose **Lathe** from the **Modifier List** in the **Command Panel**. Choose **32** as the number of segments. You should get a shape like that shown in Figure 16.12.

Set the Correct Axis

In the **Parameters** section of the **Lathe Modifier Rollout**, click on the **X-Axis** button to change the direction of the lathe. This is the correct direction (Figure 16.13), but the axis is in the middle of the profile and doesn't coincide with the axis of the symmetry. Because the axis of symmetry is the same as the assembly axis (remember, we said that it would be important!), we can use that.

Position the Axis

Click on **Lathe** in the **Modifier Stack** and then **Axis**. In the top view pull the lathe axis by the Y-axis arrow toward the front (down) until the X-axis arrow coincides with the axis of the assembly (Figure 16.14).

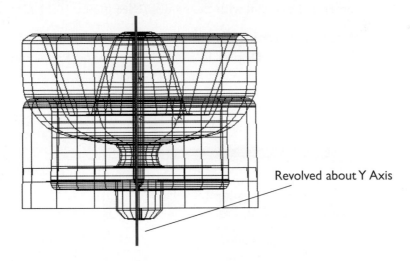

Revolved about Y Axis

Figure 16.12   Result of lathing the pulley profile about the Y axis.

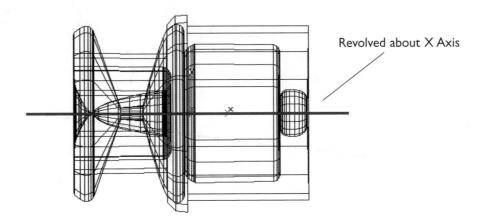

Revolved about X Axis

Figure 16.13   Correct the lathe axis to align in the X direction.

Complete the Lathed Parts

Use the same procedure to lathe the remaining parts of the assembly, hiding objects and shapes as necessary. Because the profiles were drawn in their assembled position in Adobe Illustrator, they will be correctly aligned when lathed in 3ds max. Remember to realign the axis of each lathe to the axis of the assembly as was done in Figure 16.14.

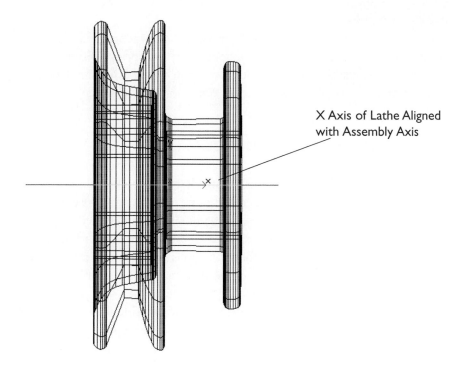

X Axis of Lathe Aligned with Assembly Axis

Figure 16.14 Reposition the lathe axis to coincide with the assembly axis.

Parts That Can't Be Lathed

Three features of the assembly cannot be modeled using the lathe feature: the balls in both of the bearings, the spring, and the castellated hexagonal nut. These small standard parts often make the difference between an effective and ineffective technical illustration, so they should be constructed carefully. The *Thomas Register* publishes "CAD Blocks," a CD library of standard parts and products from hundreds of vendors that can be brought into your modeling program, saving countless hours of drawing and modeling on your part.

▼ Note

You may have access to libraries of standard parts such as fasteners, springs, and bearings. Even if you don't, every time you create a part that you suspect will be used again, create a separate .max file with a descriptive name and store the file in a library directory. Because of Max's flexible **Merge** function, you can create a single file called *fasteners*, into which you merge all nuts, bolts, washers, and screws as you make them. Make sure you group the fasteners and call them by an identifiable name.

In the Introduction we mentioned that it is important to practice TGE (that's good enough) in technical illustration. Parts such as screws, bolts, nuts, and springs may be so small in an illustration that it is a waste of time to model them in detail. Knowing when a rough approximation of a part is good enough saves time and keeps files as small as possible.

The Balls of the Ball Bearings

The balls that separate the inner and outer races of the ball bearings may not be required if the final illustration view is generally from the front. But we will assume that they will be needed and must be included in the ball bearing model.

Begin by hiding all but the inner and outer race profiles of the larger bearing. Choose the icons for **Create|Geometry|Sphere** from the **Command Panel** and specify **32** as the number of sections of the sphere in **Parameters.** Click and drag a sphere that fits between the two bearing race profiles (Figure 16.15). The bearing profiles are at Z 0 as will be the sphere when it is constructed in the top viewport.

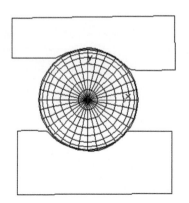

Switch to the left side view. With the ball selected, choose the icons for **Hierarchy|Adjust Pivot|Affect Pivot Only** and pull the ball's pivot to the axis line (use the **Y-Axis Lock** to make sure you move it straight). Deselect **Affect Pivot Only**.

With the ball still selected, choose **Array** from the toolbar and specify the number of copies (**8** and **2D**) and degrees about the X axis (**Rotate X=360**). You should have an array of bearing balls and if you were careful with the alignment, they should fit right into the races. Figure 16.16 shows the arrayed balls and bearing profile as well as the bearing with inner and outer races completed. Finally, group the races and balls into a single object and give the group an identifiable name so you can easily move it around. Repeat these steps for the small bearing.

Figure 16.15   The first ball is created at twelve o'clock, between the two race profiles.

Figure 16.16   The completed bearing after arraying the ball.

The Spring

Use *lofting* to create the spring by first establishing a *helical spline* to serve as its path. In our case, it is probably not critical to represent the spring's unsprung length or its ground ends. You do know the diameter of the wire, the diameter of the centerline of the spring, and the number of turns. This should be enough to make an accurate spring. Hide everything except the spring wire profile, the spring cup, right spacer, and the assembly centerline (Figure 16.17).

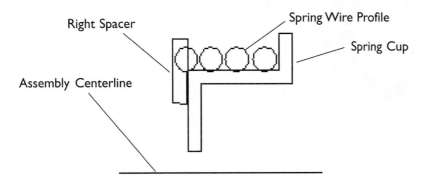

Figure 16.17   Profiles needed to create the spring.

The spring wire profile will serve as the loft *shape* and the helix as its *path*. In the **Command Panel** choose **Create|Spline|Helix** and draw out a diameter starting on the assembly axis and ending at the wire profile's center. The next click will establish the length of the helix. This is relatively unimportant because the helix can be scaled later to its unsprung length in the fields in the **Parameters** rollout. In the rollout, enter the same value for radius 2 as you made for radius 1. This makes a cylindrical as opposed to a conical spring.

With the helix still selected, enter the number of turns (**3**) in the **Turns** data field. Finally rotate the helix 90 degrees about the view Y axis. It should be in a position similar to Figure 16.18a. After scaling the helix along the X axis to a length between the spacer and the lip on the spring cup, it should look like Figure 16.18b.

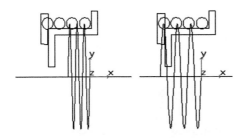

Figure 16.18   (a) The spring shape and path after rotation and scaling. (b) The spring ready to be lofted.

Select the wire profile (the **Shape**, one of the circles), and choose the **Create|Geometry|Compound Object|Loft** from the **Command Panel**. Choose **Get Path** and click on the helix. Choose **Move** to actually use the path and shape and not leave a copy. The finished spring is shown in Figure 16.19.

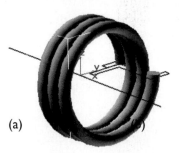

(a)

Figure 16.19   The finished spring. Note the profiles in place that control its size and position.

The Castellated Nut

The castellated (slotted) nut can certainly be accurately modeled (Figure 16.20a), and for those of you completing this assignment, it is included on the CD-ROM in *exercises/ch16/max*. Unless you are going to show an extreme closeup of that part of the assembly, you should represent the part with tubes—one with 6 sides and one with 16 sides (Figure 16.20b). Not only is modeling time reduced, but the display and rendering is more efficient and the file is ultimately smaller.

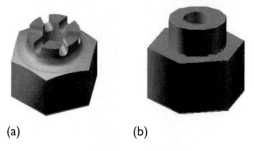

(a)   (b)

Figure 16.20   (a) An accurately modeled slotted nut. (b) A more practical solution for viewing at a distance.

Arrange Parts on the Axis

When all the parts are completed, use the **X-Axis Lock** to constrain front or top view movement along the assembly axis. Position the parts comfortably along the axis in assembly order (Figure 16.21).

▼ Illustration in Industry

Daniels & Daniels
Newbury Park, California

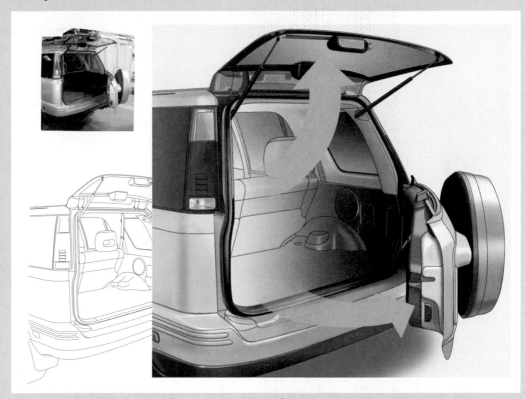

Images Copyright 2003 by Daniels & Daniels.　　　*daniels@beaudaniels.com*　　　*www.beaudaniels.com*

Software

Adobe Illustrator
Adobe PhotoShop

Hardware

Windows PC
Digital Camera

Being able to take your own digital photographs of subject matter results in an image better-suited for illustrating than by relying on what ever is available from the client. In this case, a photograph taken at the desired angle was placed into Adobe Illustrator where an accurate line tracing was executed and an EPS file generated. This tracing was done keeping in mind how it will serve in Photoshop as a template for area selections. The EPS file was rasterized in Photoshop at a resolution appropriate for the final illustration—in this case 600 dpi. Area selections were saved and recalled as necessary. Layers were used extensively to separate the selections as well as to order areas of the illustration for visibility as colors matching the manufacturer's color chips were applied. As a final touch, selective outlines were kept to show edges, panel joints, and details.

Rendering the Illustration

Rendering is the process of applying lighting and material algorithms to the vector geometry in a scene to arrive at a raster picture. Three-dimensional modeling and rendering programs can't render resolution-independent pictures. They get their resolution from rendering tens of millions of pixels at 72 dpi that when scaled to 1200 dpi or higher, produce continuous-tone results—like a photograph.

Of course, the forte of programs like 3ds max is *photorealistic* rendering. You can see from the user view in Figure 16.21 that even with default materials and lighting, very realistic results can be achieved.

By now you understand when color is appropriate and when black-and-white line illustrations are effective. What about using a 3D model not to produce color renderings but to produce black-and-white line renderings?

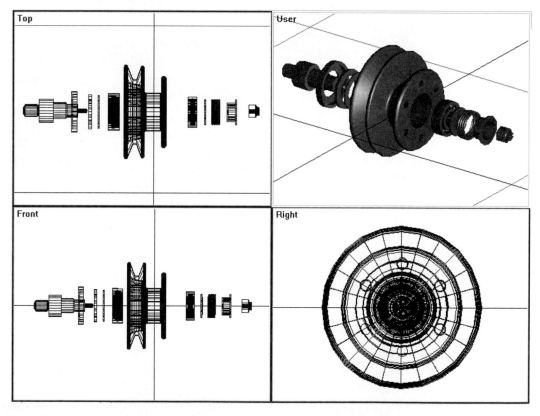

Figure 16.21   Completed pulley assembly with parts displayed along the assembly axis.

Line Rendering the Model

Figure 16.22 shows the result of assigning a white material to all objects and choosing **Edged Faces** by right-clicking on the **User View Name** (upper right corner of the view port). The parts are shown in pure white by setting omni lights above, below, in front, and behind the assembly, and setting their values to be high: a **Multiplier** of **10** or greater. Because it is a surface model, all the surface elements appear. This is a common characteristic of technical illustrations created in 3D modeling programs and points out the need to minimize segments or vertices. This is a display option, not a rendering option. The best you can do in a rendering is to render as a wireframe, and this produces unacceptable results.

Sometimes it's best to use the CBE crude but effective approach. Set your monitor to its highest resolution, make a screen grab (**ALT-Shift-Print Screen**), and paste into Photoshop where you can do selective editing. A screen grab produces a fixed resolution raster illustration that, although not the most desirable for line illustration, can form the basis of an accurate technical illustration when the raster image is placed in Adobe Illustrator and treated like a photo tracing.

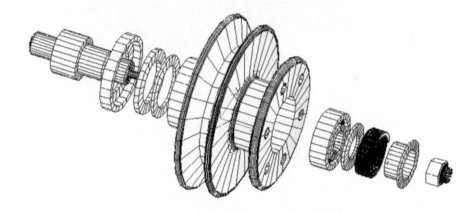

Figure 16.22   Line display achieved with white material applied to all parts.

Continuous Tone Rendering

Building correct 3D geometry is just the first step in making a realistic technical illustration. In Chapter 19, you will learn how to create realistic textures in Adobe Photoshop that can be applied to the objects in 3ds max. Figure 16.23 shows the level of realism that can be achieved with material maps and lighting. A rule of thumb is *"never model what you can map."* For example, threads, knurls, splines, joints, grooves, and other details are mapped onto surfaces using *raster material maps*. To give you a better idea of part detail, Figure 16.24 shows the pulley shaft with *bump maps* applied for the threads, a *material map* for the metal, and a *reflection map* for the highlights. The result is a simply modeled part that is dimensionally correct and visually representative.

▼ **Tech Tip**

3ds max gives you two options for pictorial viewing: a perspective view and an axonometric or *user view*. Both views are infinitely flexible. To get the best pictorial view, work back and forth between the pictorial viewport and the top or front viewports, moving the objects and the position of the viewer until you get just the pictorial view you need.

Figure 16.23 Photorealistic rendering of the assembly with raster materials applied.

Figure 16.24 The pulley shaft with bump, material, and reflection maps applied.

▼ **Tech Tip**

Three-dimensional programs were never intended to produce PostScript line illustrations. However, program additions called *plug-ins* are available for this task. David Gould's Illustrate 5.2 (http://www.davidgould.com) is a 3ds max plug-in used mostly for cartoon animation, but it has the flexibility to produce both raster and vector line illustrations from 3D data.

Review

Technical illustrators often are presented with shape and size information in the form of paper engineering drawings. When these drawings are made "to scale," they can be scanned and used to create shapes or profiles that form the basis of a 3D technical illustration.

A 3D technical illustration starts as a computer model. These models are extremely powerful and flexible in that they can provide an infinite number of illustration views. However, the output from these programs is not resolution independent like PostScript.

In this chapter you have learned how to use the geometric information in engineering drawings to create a 3D technical illustration. Part profiles are traced in Adobe Illustrator and imported into 3ds max where operations such as **Lathe** and **Loft** turn the 2D profiles into 3D models. By modeling a subject, any and all axonometric and perspective views can be generated, and by applying raster textures you create in Adobe Photoshop, high levels of realism are possible.

This technique results in accuracy appropriate for technical illustrations. You shouldn't be overly concerned with absolute dimensions, but rather with scale proportionality. In Adobe Illustrator you should create profiles with the fewest vertices necessary to accurately describe each shape. In 3ds max you should specify the minimum number of primitive, loft, and lathe segments both to simplify the geometry for line rendering and to assure a small file size.

Text Resources

Boardman, Ted. *3ds max 5 Fundamentals*. New Riders Publishing. Indianapolis. 2002.

Duff, Jon M. *Autodesk VIZ in Manufacturing Design*. Thompson Learning. Albany. 2002.

Gooch, Bruce. et al. *Interactive Technical Illustration*. ACM. Symposium on Interactive Graphics. Washington, D.C. 1999.

Murdock, Kelly L. *3Ds Max 4 Bible*. John Wiley & Sons. New York. 2001.

Internet Resources

3D Clearing House. Retrieved from: *http://www.3dluvr.com/*. December 2002.

3ds max Links. Retrieved from: *http://www.upbeat.net/studiomax*. December 2002.

How to Modernize Your Paper Engineering Drawings. Retrieved from: *http://www.gtx.com/documents/whpaper.pdf.* December 2002.

Illustrate 5.2 (Software Program). Retrieved from: *http://www.davidgould.com.* December 2002.

Illustrator Message Board. Retrieved from: *http://www.desktoppublishing.com/illustrator/illustratortalk.html.* September 2002.

Importing Adobe Illustrator Files. Retrieved from: *http://www.maxhelp.com/content/tutorials/illus/illustrator.htm.* December 2002.

Programs Convert Engineering Drawings from Paper to Computer. Retrieved from: *http://www.psu.edu/news/News/1995%20Press%20Releases/January/drawings.html.* December 2002.

On the CD-ROM

For those of you who would like to try out the procedures described in this chapter, several resources have been provided on the CD-ROM. Look in *exercises/ch16* for the following files:

pulley_assy.bmp Import this raster file into Adobe Illustrator to follow the steps in the chapter.

assy_1.bmp through assy_5.bmp These scans are technical assemblies for additional practice.

tracover.ai If you want to get right to importing the profiles into 3ds max, this file is ready to use.

Look in *exercises/ch16/max* for the following files:

pulley.max This file contains the finished assembly with default colors assigned to the part. You can practice moving the parts into exploded assembly position.

nut.max If you want to use a more detailed slotted nut, merge this geometry into your *pulley.max* file and uniformly scale so that the hole diameter of the nut is the same as the corresponding end of the shaft.

Chapter 17

Using AutoCAD Data in Illustrations

In This Chapter

If you are an illustrator working in the manufacturing, processing, construction, or architectural industry, AutoCAD will probably be part of the mix of tools available to you for technical illustration. This is not so much because of AutoCAD's prowess with illustration tasks, but rather because of its ubiquitous nature in creating 2D drawings and 3D models. AutoCAD's DWG native and DXF intermediate formats encourage passing geometric data between illustration and layout programs.

Chapter Objectives

In this chapter you will understand how to:

▼ Make use of intermediate CAD formats such as DXF and IGES

▼ Select AutoCAD's DWG, DWF, DXF, and PLT file formats for illustration purposes

▼ Use AxonHelper to provide both viewpoint and rotational data for arriving at standard views in AutoCAD.

▼ Manipulate geometry into standard axonometric views in AutoCAD using the **ROTATE3D** command

▼ Manipulate the camera into standard axonometric views in AutoCAD using the **VPOINT** command

369

- ▼ Change 3D data into usable 2D data using **HIDE** and **SOLDRAW** commands

- ▼ Choose appropriate output file formats for specific illustration tasks including the EPS and TIF formats

- ▼ Bring AutoCAD geometry into Adobe Illustrator using **Paste**, **Place**, and **Open**

- ▼ Edit AutoCAD geometry once in Illustrator

A Note on CAD and CAD Data

It would be impossible to cover all CAD tools currently on the market. Although many of the upstarts have fallen by the wayside, a number of powerful parametric CAD tools have gained wide acceptance including Autodesk's Inventor, SolidWorks, Pro Engineer, and Solid Edge. There remain strong pockets of Catia, and Ideas, in large manufacturing companies. These tools share many standard 3D features: extrusion, lofting, sweeping, Boolean, and feature-based operations. They also share the most important aspect of design: data.

These CAD programs are used for multiple purposes. They define geometry, relationships, and constraints that are passed on to prototyping or machine tools so that parts can be built. In these cases, drawings are never made and data for technical illustrations is hard to come by. Other CAD tools pass their data to *postprocessing* programs that perform tests such as finite element analysis (FEA). Engineering drawings, if they are made at all, are made at the end of the design process and only if needed.

But the common thread that runs through these dissimilar CAD tools is the ability to translate their proprietary data into *intermediate formats* and as luck would have it (or not luck, rather careful planning), two intermediate formats are of particular interest to technical illustrators: DXF and IGES.

DXF and IGES Formats

AutoCAD's DXF (Drawing Exchange Format) and the American National Standards Institute's IGES (Initial Graphics Exchange Specification) provide intermediate data formats for the exchange of CAD data between dissimilar systems. Adobe Illustrator accepts data in AutoCAD's DXF format; 3ds max imports DXF and IGES files and exports DXF. When 3D data is exported in DXF format from Max and placed into Adobe Illustrator, the Z world is suppressed and the XY world is matched to Illustrator's XY paper coordinates. This requires that the geometry be rotated into correct position in max before the data is exported and placed in Illustrator.

An Example of Translating CAD Data

Before we delve into the particulars of how to use AutoCAD drawing data in technical illustrations it may be helpful to see the path geometry created elsewhere takes to end up in Adobe Illustrator.

Figure 17.1 shows an assembly in Autodesk's Inventor. Inventor has a native format of its own (as does Catia, Ideas, and the others). However, Inventor can save its data in IGES format, and that's our path to a technical illustration. Note that although the view port shows the assembly in pictorial position, it is the viewpoint, and not the object, that is rotated. You can tell this by the position of the axis icon in the lower left corner of the view port.

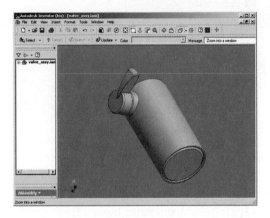

Figure 17.1   Model in Autodesk's Inventor.

This file is saved in IGES format. When imported into 3ds max (Figure 17.2), Inventor's world axis system is matched to max's world axis system.

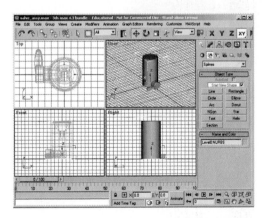

Figure 17.2   IGES model from Inventor brought into 3ds max.

AxonHelper's rotation data is predicated on viewing the rotated object in the front view. But because Adobe Illustrator will match its 2D XY paper coordinates with max's 3D XY world coordinates (top view), pretend that max's top view is really the front view. Rotate the assembly −90 degrees about the X axis in the top view (Figure 17.3).

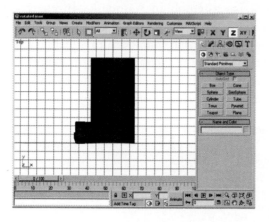

Figure 17.3   Geometry rotated in top view.

Once in this position, use AxonHelper's **Rotate** function to determine object rotation about the world axes to arrive at a given view. We have chosen a 55-25-25 dimetric view. Figure 17.4 shows the result of performing these rotations. At this time, we can render the top view port and have a fixed-resolution illustration.

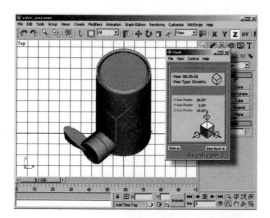

Figure 17.4   Geometry rotated in top view using AxonHelper's rotation data.

But to get the geometry into Adobe Illustrator where a resolution-independent 2D illustration can be created, this file is exported in DXF format. When **Placed** in Adobe Illustrator (Figure 17.5a), the wireframe is translated to Illustrator's XY dimensions. This is unusable in this condition, but with a little editing and overdrawing, an accurate illustration is achievable (Figure 17.5b).

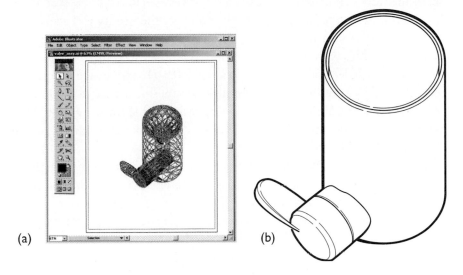

(a)　　　　　　　　　　　　　　　　　　　(b)

Figure 17.5　　　The DXF geometry is placed into Adobe Illustrator.

The Theory Behind Using AutoCAD

Autodesk has a suite of products available for the design of engineered and constructed products. These include AutoCAD, 3ds max, Autodesk VIZ, Mechanical Desktop, AutoCAD Designer, and Inventor. (Some of these products may have been combined or dropped since this book was written, but that's the CAD game.) These products share one thing in common: the ability to create highly accurate geometric descriptions. These descriptions are suitable for making standard engineering drawings. But of greater importance to designers is the ability to communicate this data to analysis programs and to machine tools capable of using the data directly to create products. We will concentrate on the most recent releases of AutoCAD—AutoCAD 2000 and AutoCAD 2002—though the same tasks outlined in this chapter can be completed with any of the popular 3D CAD tools such as SolidWorks, CADKEY, ProEngineer, Catia, or Auto-Trol. Likewise, the PostScript illustration, though described in Adobe Illustrator, could just as easily be completed in CorelDraw or Freehand.

For all this power, CAD programs, like AutoCAD, generally have a serious limitation when it comes to producing publication-ready technical illustrations: these programs are not optimized to produce PostScript output from their accurate data. They generally do not make use of PostScript text fonts, so any text that comes from a CAD program is de facto less than publication quality. They utilize only a small portion of the library of PostScript routines available in a printing device's PostScript interpreter.

Utility programs such as Hijaak Pro can be used to convert AutoCAD PLT data into PostScript, but with less than optimal results. Plot files from AutoCAD are a series of strokes and arcs are actually a set of very small cords. So even if you translate PLT into PostScript, you probably won't get the quality you want.

This doesn't mean that you *can't* make technical illustrations using AutoCAD alone. It just means that you will be severely limited in your ability to make use of PostScript routines available in publication RIPs. (Raster image processors are devices that translate resolution-independent PostScript and fixed-resolution bitmaps into the desired resolution for a given output device.) If you use AutoCAD to make your technical illustrations, your illustrations will look, to be honest, like that were done in AutoCAD.

The solution is to use AutoCAD for what it does best: creating highly accurate 2D and 3D geometry that can be brought into PostScript illustration programs. If data already exists in DWG or DXF formats, and that data is three-dimensional, there is no reason to start over in Illustrator to create the view you want. With the aid of AxonHelper, as was demonstrated in the first section of this chapter, you can either assume a viewing position or rotate the geometry into a position that results in a standard axonometric view. That way, geometry from AutoCAD can be accurately matched with constructions, projections, or sheared views created in PostScript illustration programs by more traditional 2D methods.

This chapter highlights the fundamental process of using 3D data created in AutoCAD to form the basis of highly accurate publication-quality illustrations in Adobe Illustrator. For more information concerning construction procedures using AutoCAD, see the list of text and on-line resources at the end of this chapter.

Getting AutoCAD Data into Adobe Illustrator

In order to determine whether or not it is advisable to use AutoCAD data in technical illustrations, it is best to understand how the two programs share data. AutoCAD holds both geometric and associative data in its native drawing (DWG) format. This 3D database contains a significant amount of information that is not needed by an illustrator but is important in AutoCAD. The DWG file contains no image header; the screen image is generated in real time from the data. Because of AutoCAD's intended engineering market, this geometric data is optimized for screen display, vector stroke plotting, and postprocessing by dedicated programs that act on the AutoCAD data. AutoCAD's data can be 3D vectors, splines, Bezier curves, surfaces, and solids; it also includes vector stroke and TrueType text.

Adobe Illustrator, on the other hand, can use only 2D curves and plane surfaces. Though Illustrator can display TrueType fonts, most prepress operations, as has been mentioned several times, discourage their use, preferring PostScript fonts. (TrueType was created to solve monitor display problems, not as an improvement to PostScript fonts for printing.) Illustrator's native AI data is optimized, not for plotting or display, but for interpretation by a PostScript printing device. To use AutoCAD successfully in technical illustrations, its data must be usable in Adobe Illustrator.

Result of COPY and PASTE

Probably the easiest way to bring AutoCAD data into Illustrator is via the clipboard. If you work on the Macintosh, your only solution is to run an emulator such as Virtual PC where the two operating systems can share the same clipboard.

The Windows clipboard passes data between programs using the Windows Metafile (WMF) format. However, between Illustrator and AutoCAD, the resolution of this data depends on the level of display at which the copy is initiated in AutoCAD. AutoCAD makes decisions as to the accuracy of the display data based on the current level of zoom. So when you select objects in AutoCAD and **Copy**, you are copying the display data and not the actual AutoCAD geometric data.

Figure 17.6 shows a 3D housing in AutoCAD. What is the result of copying from AutoCAD and pasting this image into Illustrator? First, the **HIDE** command was executed to remove shading and to determine visibility. Figure 17.7 shows an enlarged detail created by **ZOOM|EXTENTS** and then **Copy** and **Paste**. Compare this to Figure 17.8 where the same detail was zoomed larger in AutoCAD first before selecting objects, copying, and pasting.

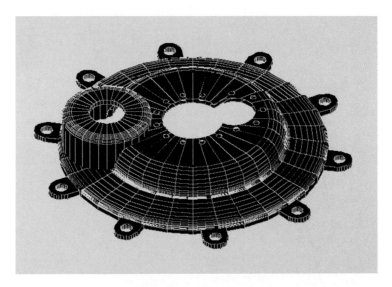

Figure 17.6   3D model in AutoCAD with **HIDE** option.

The following conclusions can be reached about using **Copy** and **Paste**:

▼   Vectors are pasted into Illustrator as a single grouped object, including a background panel.

▼   Curves are represented as multisided polygons.

▼   The resolution of the vectors is dependent on the display resolution in AutoCAD at the time of the **Copy**.

This technique should be reserved for quickly bringing AutoCAD data into Illustrator to determine position, sizes, etc., or as a background image for tracing in Illustrator (see Chapter 11).

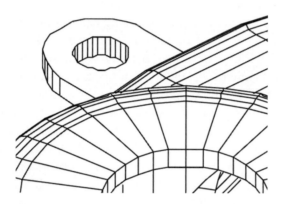

Figure 17.7   Zoom out results in low resolution.

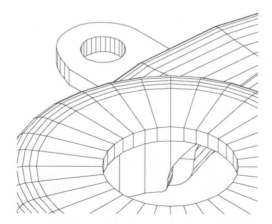

Figure 17.8   Zoom in results in higher resolution.

Result of PSOUT

Another option is to write out a PostScript file in AutoCAD. Figure 17.9 shows the result of opening a PostScript file in Illustrator that was created in AutoCAD using **PSOUT**. The usability of the geometry is independent (as you might imagine, because PostScript is resolution independent) of the zoom level in AutoCAD. The following conclusions can be reached about using **PSOUT**:

▼ Vectors are brought into Illustrator as a single grouped object, including a background panel. Ungroup and **Delete** this background panel.

▼ Vectors are resolution-independent PostScript lines.

▼ Curves are represented as multisided polygons.

▼ All vectors are written regardless of the status of **HIDE** in AutoCAD. This is also true in both model and paper space.

By writing PostScript from AutoCAD you can avoid the resolution display dependence of the Windows clipboard. Additionally, the EPS file can be used across platforms. However, because all geometry is written to the file without regard to visibility, a large amount of editing is required in Illustrator.

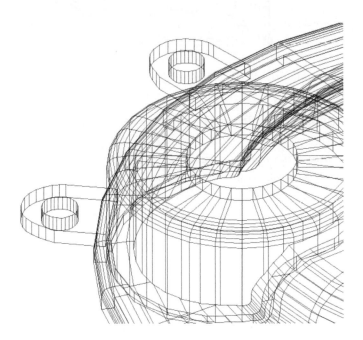

Figure 17.9   The result in Illustrator of creating a PostScript file in AutoCAD using **PSOUT**.

Result of EXPORT

The **EXPORT** command in AutoCAD provides two additional options for creating data that can be used in Illustrator. The WMF option produces the same results, as does **Copy** and **Paste**. The Windows Bitmap (BMP) option produces a low-resolution screen image (Figure 17.10) of limited use in technical illustration.

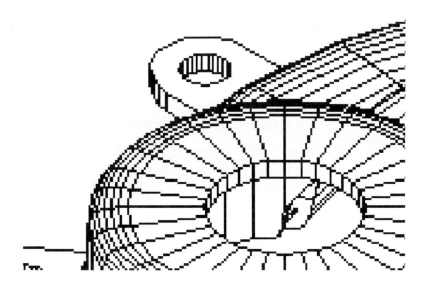

Figure 17.10   The result in Illustrator of creating a Windows Bitmap file in AutoCAD.

Result of Writing to a Plot File

One option is to write AutoCAD data to a file optimized for output on an X-Y plotter. This PLT file can be converted by Hijaak Pro into EPS format and then opened or placed into Adobe Illustrator.

Use the **PLOT** command and specify an HP 7475 plotter as the desired output device; make sure the **Write File** box is checked. The results of this process are:

▼ The **PLOT** option creates smaller XY steps around circles, ellipses, and curves than does **Copy** and **Paste**.

▼ Hijaak converts these vectors into PostScript polylines.

▼ The converted PLT file will generate a superior print on PostScript devices when compared to AutoCAD's **PSOUT** option.

▼ The file will be very large because of the increase in the number of vectors used to describe the shapes.

Converting PLT data to EPS data with Hijaak Pro is not a solution if the desired result is editable Bezier curves in Illustrator. But if the intention is to produce better printing EPS artwork directly out of AutoCAD (with no improvement in Illustrator) then this process has promise.

Result of SAVEIMG

AutoCAD can render a scene if geometry is comprised of valid surfaces or solids. The scene can be saved in up to 32-bit Tagged Image File (TIF) format. The TIF format can be considered publication quality if it has high enough resolution and color bit depth because TIF contains tags that optimize halftone screen translation by a PostScript RIP. If the geometry is a wire frame (comprised only of vertices and connections), it cannot be rendered and subsequently output in TIF format. This option has illustration possibilities because both resolution and bit depth can be specified in AutoCAD. However, the output is raster, fixed resolution, and of limited use in Illustrator.

File Formats in AutoCAD

AutoCAD's native DWG format is supported in Illustrator 10 PC, through either the **Open** or **Place** commands. AutoCAD's intermediate DXF format is supported through both the **Open** and **Place** commands in Illustrator. However, not all AutoCAD elements are translatable—3D polylines, for example, do not translate. Because you probably will have little or no control over which AutoCAD elements have been used in AutoCAD models except for those thst you have actually created yourself, directly using DXF data in Illustrator usually requires extensive editing. For display on the Web, AutoCAD provides the DWF file format, as well as the *Whip!* viewer plug-in. The result is a vector graphic that can be zoomed and panned but not edited. DWF format graphics are not readable in Illustrator.

Isometric Pictorials in AutoCAD

AutoCAD includes an automated 2D isometric drawing feature, similar to more dedicated products such as IsoDraw and InterCap. The result of this automated process is a 2D vector file that can be brought into Illustrator by any of the methods previously discussed. It may be advantageous to complete constructions in AutoCAD because of the parametric nature of its database and because AutoCAD contains more sophisticated construction and editing tools than does Illustrator—trim to the extension of a line, for example, and tangent constructions.

Another method of arriving at automated isometric views in AutoCAD is to use the **VIEW|3D VIEWPOINT** menu selection to access isometric views of 3D data from compass directions. This is aimed more at architectural designers where north-south-east-west makes sense. Still, these automated isometric views of 3D data aid in quickly choosing the most descriptive isometric view.

2D Drawings from 3D Data in AutoCAD

Unless you are using AutoCAD as your primary illustration tool, you probably will never concern yourself with the differences between *model* and *paper space*. To make use of AxonHelper's viewpoint and rotation methods described in the next section you'll always be working in AutoCAD's default model space. For instructions on using paper space to arrange multiple views of the same data refer to one of the references listed at the end of this chapter.

Viewpoint and Rotation Data from AxonHelper

AxonHelper has been described in previous chapters, but here we will use it in an entirely different way. In addition to scale and shear data for standard axonometric views, AxonHelper provides data for changing viewpoint and object rotation in 3D CAD programs such as AutoCAD. These two operations (viewpoint and object rotation) yield the same results, though through different means. One rotates the viewer while the object remains stationary, and the other rotates the object while the viewer remains stationary.

AxonHelper

Standard axonometric views in AxonHelper have been achieved previously by scale construction and shearing. The exact same views can be achieved by rotating 3D geometry about the world axis system (the circular arrow button) or by changing the position of the 3D camera as it looks at the world origin (the eyeball button). You do one or the other; you don't do both. *Viewpoint* and *rotation* data are found by choosing the isometric, dimetric, or trimetric view and the clicking on the viewpoint (eyeball) or rotate option, and in the cases of dimetric or trimetric, the desired view (Figure 17.11).

Viewpoint

The *viewpoint* is the position of the observer (you) in a given view. Because an axonometric view is an orthogonal view (just not a principal view) the viewer is assumed to be an infinite distance from the plane of projection. This means that the magnitude of the distance is unimportant. Figure 17.12 shows the result of translating the viewpoint to a position X = 1.0, Y = 1.0, Z = 1.0. This establishes a vector, a direction parallel to which all projections will be made. Specifying X = 100.0, Y = 100.0, Z = 100.0 will not change the view (the viewing vector) nor will it increase the accuracy of the view. The relationships, not the magnitudes, of the offsets are what are important.

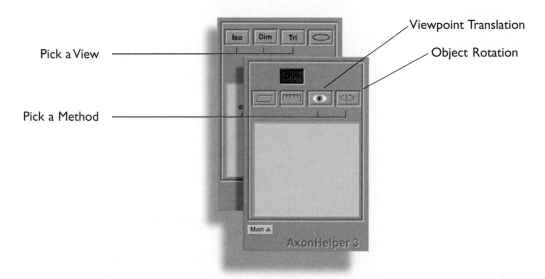

Figure 17.11   Viewpoint and rotation data in AxonHelper.

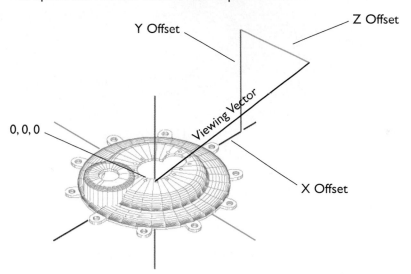

Figure 17.12   The method of achieving an axonometric view by translating the viewpoint.

Viewpoint data is represented in AxonHelper by an XYZ position in space. To achieve this standard view in AutoCAD, type **VPOINT** in the command line and enter the XYZ viewpoint data found in AxonHelper. AutoCAD then displays the 3D data in standard position. The view can be brought into Illustrator by any of the methods previously mentioned in this chapter. Figure 17.13 shows data for a 55-25-25 dimetric in AxonHelper, the necessary instructions in AutoCAD, and the resulting view. By changing which values are assigned to the three axes, a wide range of views can be generated (Figure 17.14).

▼ Note

Each standard view is actually 48 different views, depending on which value you assign to a particular axis and whether the number is positive or negative. For example, Figure 17.14 shows thumbnails of five views for the 55-25-25 view created in Figure 17.13.

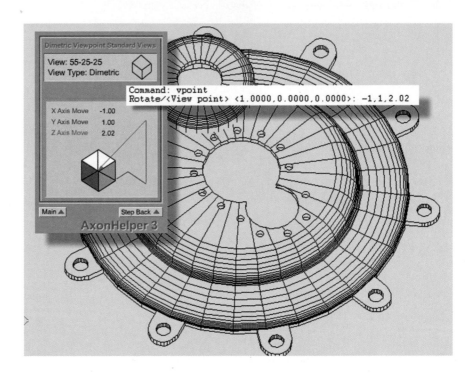

Figure 17.13   A 55-25-25 dimetric created in AutoCAD by translating the viewpoint.

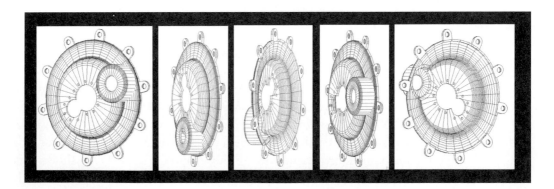

Figure 17.14   A matrix of 55-25-25 views achieved by assigning different values to the XYZ viewpoint position.

As was shown for viewpoint data in Figure 17.14, 48 separate views can be formed from the same standard view position data. Simply substitute one value for another, changing positive and negative values. Remember that in all but one view (usually the front view), world and view axes are not coincidental. You have to be in the correct view, and using the correct axes, to arrive at the standard view you want.

Rotation

Changing the viewpoint moves the observer while objects remain stationary. There are times, however, when the observer must remain in the same position. Standard axonometric views are created when the observer remains stationary and the selected geometry is rotated about three mutually perpendicular axes.

AxonHelper assumes that objects are in principal position and aligned with the world X, Y, and Z axes. The right hand rule is employed to specify positive or negative rotation. AutoCAD performs 3D rotations relative to the current user coordinate system (UCS) or in the absence of an active UCS, the view axes of the view in which the rotations are performed. Use the front view to perform this operation (view and world axes are coincidental), keeping in mind that in AutoCAD, the Y axis is vertical, and the Z axis is in and out of the screen.

To achieve a standard view by rotation in AutoCAD , type **ROTATE3D** in the command line and enter the X, Y, and Z rotation data from AxonHelper. Figure 17.15 shows AxonHelper rotation data for a 50-15-35 trimetric view, the necessary instructions in AutoCAD, and the result of the rotation after **ZOOM|EXTENTS**.

The rotations are executed in order. Note in Figure 17.15 that the negative value of the Z-axis rotation was used to arrive at a view from the right. Afer a while, you will be able to predict what the view will look like ahead of time. For example, in Figure 17.15 were an X axis rotation of –50 degrees to be used (rotation away from you, right-hand rule), a view from below will be created.

One reason many illustrators prefer to keep all geometry stable in relation to the world axes is that it's difficult to return rotated geometry to its original position…especially after 6 months have passed. For this reason, always keep an *unrotated* version of the CAD file for future reference. If you need to use the geometry for other rotation of viewpoint illustrations, you need to start with all objects in principal position, otherwise the values from AxonHelper don't have any meaning.

Kevin Hulsey
Carmel, California

Images Copyright 2003 by Kevin Hulsey.   *khi@hulsey.com*   *www.khulsey.com*

Because vector and raster illustration programs have individual strengths, challenging technical illustration tasks often require the use of multiple programs. Adobe Illustrator provides the flexible and resolution-independent vectors for accurate construction, while Adobe Photoshop contributes the creativity and freedom of raster graphics.

The cutaway illustration of the Radiance of the Seas cruise ship began with individual deck plans supplied by the client. These ship drawings were redrawn for each deck in Adobe Illustrator. A perspective grid was built whereby the deck plans, section views, and ship profile (side view) were distorted into the grid.

All walls and furniture were constructed using the floor plans for placement. The finished line art was exported from Illustrator to a layered CMYK Photoshop file, and as you might imagine, careful layer maintenance was an important contributor to accomplishing the high level of detail in this illustration. Floor and carpet patterns were created from carpet samples in Illustrator, placed into deck floor plans, exported to Photoshop, and distorted into perspective. Colors and textures were then painted on multiple layers using the line art as a mask. The size of the final CMYK illustration file was in excess of 600 MBytes.

Software

Adobe Illustrator
Adobe Photoshop

Hardware

Macintosh G4
Scanner
CD RW Drive

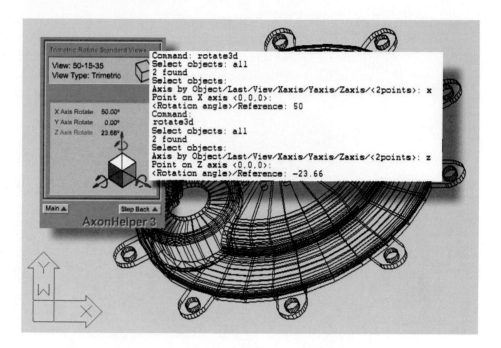

Figure 17.15   A 50-15-35 trimetric projection created in AutoCAD by rotating the geometry relative to the X, Y, and Z axes.

Rendering in AutoCAD

As was previously mentioned, AutoCAD, beginning with Release 14, has incorporated many 3D rendering features normally reserved for dedicated products such as 3ds max. These features include lights, materials—including bump, opacity, and reflection maps—atmospheric effects, and ray traced rendering.

The resulting rendering is a TIF or TGA (Targa) raster file. Use the TIF for insertion into page layout programs such as Quark Express, PageMaker, FrameMaker, or InDesign. Use the TGA file for video compositing. These raster files are resolution dependent and can be used for custom textures or backgrounds in Illustrator. For editing, these rendered views must be brought into raster programs such as Photoshop.

Perspective Camera Views in AutoCAD

Perspective in AutoCAD makes use of the dynamic view (**DVIEW**) command. This allows a camera to be set a known distance from a target (the **DISTANCE** option) with a specified focal length (the **ZOOM** option). By manipulating these two options a perspective view like that in Figure 17.16 can be achieved.

By using the **POINT** option, you can enter viewpoint information from AxonHelper just as you did for axonometric views. Unless you pay attention to the relative size and units of the object, the camera will probably be too close to the origin (0, 0, 0) for a usable view. Enter **DVIEW** again and choose the **DISTANCE** option. This will move the camera back along the vector leading to the target.

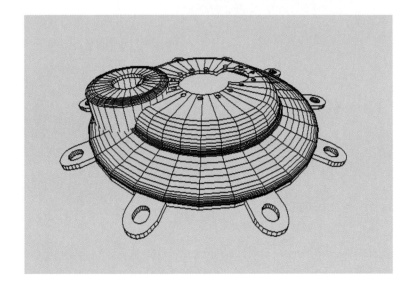

Figure 17.16   The housing viewed from a 15-mm focal length camera from an isometric-like viewpoint.

▼ Note

Technical illustrations of large architecture, machinery, or process equipment are appropriately done in perspective. Avoid perspective for any subject you can hold in your hands. Figure 17.16 shows the distortion that can occur when nonarchitectural objects are viewed in wide-angle perspective. Additionally, it is nearly impossible to match perspective images at a later time. Be careful not to sacrifice the technical communication quality of an illustration by utilizing extreme perspective effects.

Editing in Illustrator

Once you have the model from AutoCAD in Illustrator you have several questions to answer:

▼   Is the quality of line work from AutoCAD appropriate for the illustration?

▼   Is it worth the effort to edit the wireframe for visibility?

▼   If editing is too involved, can the AutoCAD drawing be used to guide a tracing?

Figure 17.17 shows how the housing is used in Illustrator when a **Copy** and **Paste** from AutoCAD is overdrawn with curves and subsequently stroked. The raster template is shown in the background to give you an idea of the amount of editing that would be required were **PSOUT** used to generate an EPS file. And in addition to the editing that would have to be done, the file would be vector segments and not curves.

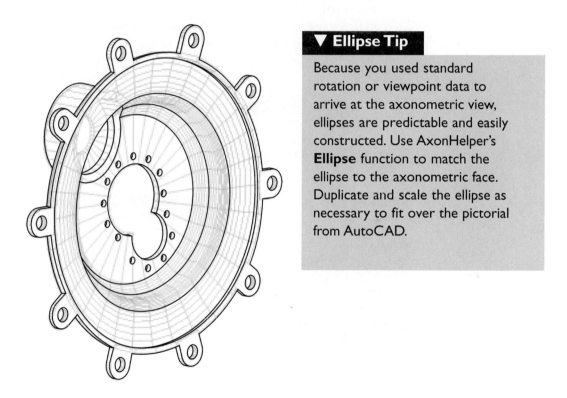

▼ Ellipse Tip

Because you used standard rotation or viewpoint data to arrive at the axonometric view, ellipses are predictable and easily constructed. Use AxonHelper's **Ellipse** function to match the ellipse to the axonometric face. Duplicate and scale the ellipse as necessary to fit over the pictorial from AutoCAD.

Figure 17.17   An edited AutoCAD vector drawing (rear) and a PostScript Illustrator overdrawing (front).

Standard Views in 3ds max

Standard axonometric views can be created in 3ds max by using the view information from AxonHelper. Max creates axonometric views any time the viewpoint is rotated. Unfortunately, this is a free-form technique, and predefined standard views are not achievable in this manner. To arrive at standard axonometric views, you have to establish a camera and force the camera to make a parallel projection.

▼ Orient the subject as desired to the world X, Y, and Z axes. This usually means that the object is centered on the world origin.

▼ Select the eyeball (viewpoint) method and desired standard view from AxonHelper. For our example we have chosen a 70-10-10 dimetric, a view looking down on the object. Write down the XYZ viewpoint position from AxonHelper.

▼ Create a camera in 3ds max. Select the camera.

▼ Right-click on the **Select and Move** command and enter the XYZ camera position from AxonHelper. You may have to scale this value up depending on the size of your subject. For example, 0.53 may need to be 53.0, 530, or 5300 (Figure 17.18). Note that AxonHelper values have been translated into feet and inches of the same relationship.

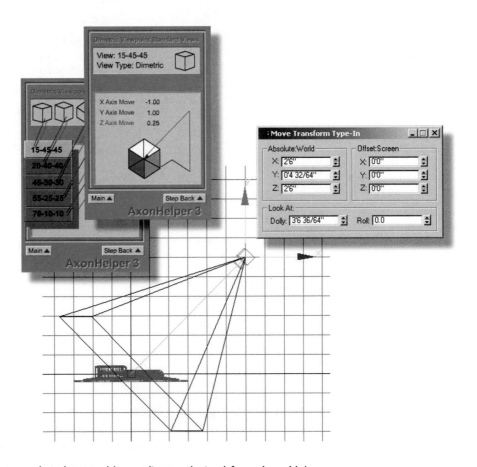

Figure 17.18   Camera placed at world coordinates derived from AxonHelper.

▼ Select the camera's target. Right-click on **Select and Move** and enter 0, 0, 0 for the camera's target. Note: The target must be at 0, 0, 0 or you won't have the desired view (Figure 17.19).

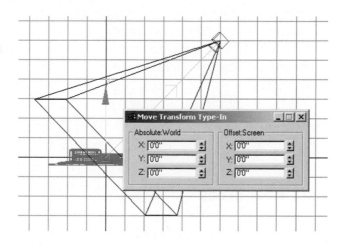

Figure 17.19   Camera target is moved to the world origin.

▼ Select the camera and choose the modify option. Check the **Orthogonal Projection** box. This forces the view from the camera's position to be the desired standard axonometric view.

▼ Type **C** to switch the current view port to the camera, and you'll see the standard axonometric view (Figure 17.20).

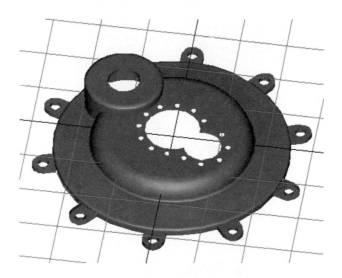

Figure 17.20   Standard 70-10-10 axonometric view in 3ds max.

Review

CAD systems are excellent sources of geometric data for technical illustrations. However, because CAD models are created to define geometry in ways amenable to manufacturing and not for high-quality printing, the output from CAD systems may need to be edited or translated to be usable in technical illustration.

The heart of CAD is the data format such systems input and output; to effectively use CAD data, you must understand the strengths and limitations of CAD file formats.

AxonHelper provides both viewpoint and rotation information for arriving at standard axonometric views in AutoCAD and 3ds max. CAD geometry, once in Illustrator, can be edited into an illustration or used as an accurate template over which PostScript illustration is drawn.

Text Resources

Leach, James A. *AutoCAD 14 Instructor*. WCG/McGraw-Hill. New York. 1998.

Leach, James A. *AutoCAD 2000 Instructor*. WCG/McGraw-Hill. New York. 1999.

Internet Resources

AutoCAD Drawings as Computer Files. Retrieved from: *http://csanet.org/newsletter/winter02/nlw0206.html*. December 2002.

Printing With AutoCAD. Retrieved from: *http://gccprinters.com/support/doc/autocad.html*. December 2002.

On the CD-ROM

For those of you who would like to follow along with the viewpoint and rotation procedures in this chapter, several resources have been provided on the CD-ROM. Look in *exercises/ch17* for the following files:

housing.dwg   Open this file in AutoCAD to follow the viewpoint and rotation steps presented in this chapter.

housing.bmp   Open this file in Illustrator to practice overdrawing.

housing.eps   Open this file in Illustrator to practice editing.

model01.dwg through model5.dwg   Use these AutoCAD models as the bases for additional viewpoint and rotation exercises

Perspective Drawing

Axonometric Projection

Technical Animation

Technical Illustration

Rendering

3D Modeling

Technical

Chapter 18

Modeling for Illustration

In This Chapter

In Chapter 16 you discovered how traditional scale paper engineering drawings are turned into 3D technical illustrations. Paper drawings were scanned and subsequently placed in Adobe Illustrator where they were used to guide creation of the profiles. The primary technique involved creating spline profiles that could be used to sweep (lathe), loft, or extrude.

In chapter 17, you were shown how existing 3D CAD geometry is brought into raster, vector, and modeling programs so that publication-quality illustrations can be made. You also gained knowledge as to what types of technical illustration might be created directly in AutoCAD.

Now, you are ready to learn how to analyze geometry and plan appropriate modeling strategies. You'll soon discover that there are many approaches to modeling, possibly even an infinite number of strategies. The approach you use will be determined by the capabilities of the software that you are using and your own abilities and preferences. However, some strategies may lead you down an unproductive path, resulting in less than acceptable final results.

If you have paper drawings of the parts you are going to illustrate in three dimensions, you will probably want to scan them and create profiles in Adobe Illustrator as we did in Chapter 16. If you have 3D engineering geometry in digital form, you will want to edit the data down to what you need to make the illustrations, and either import the geometry into 3ds max or write the geometry to a 2D PostScript file to create profiles in Illustrator.

But if you have photographs or drawings not to scale, or if you have no drawings at all but are in possession of the physical parts, and if you don't have 3D digital data, you'll have to create the illustration directly by modeling. In this chapter we will use Discreet Logic's 3ds max to model parts for an illustration.

Chapter Objectives

In this chapter you will understand how to:

▼ Analyze the fundamental geometry of a part and choose efficient and effective modeling methods

▼ Understand the level of precision required for an illustration

▼ Know when to create models from primitives

▼ Lathe a profile about an axis

▼ Extrude a profile along an axis

▼ Loft a profile along a path

▼ Create stamped shapes using patch grids

▼ Create a NURBS surface from NURBS curves

▼ Use Boolean operations to create compound shapes

When to Model from Scratch

We will be using an aircraft illustration as the subject for our modeling discussion. In this case, Humbrol's Airfix 1/72 Bristol Beaufighter from World War II. This satisfies three conditions for modeling from scratch: a lack of detailed scale engineering drawings, no 3D geometric CAD data, and possession of actual parts that can be measured. Figure 18.1a shows the aircraft and its parts. Figure 18.1b shows the type of graphical views that support the assembly. These views come from the assembly documentation included with the kit and can be scanned to form the basis for profiles.

▼ Note

The authors have found it particularly effective to "model from a model." You can find kits to create scale models of cars, trucks, planes, ships, construction equipment, buildings, locomotives, the human body, and animals. Because these models need to be assembled, they have already been broken down into component parts. Eventually, you will be able to analyze products and break them down yourself.

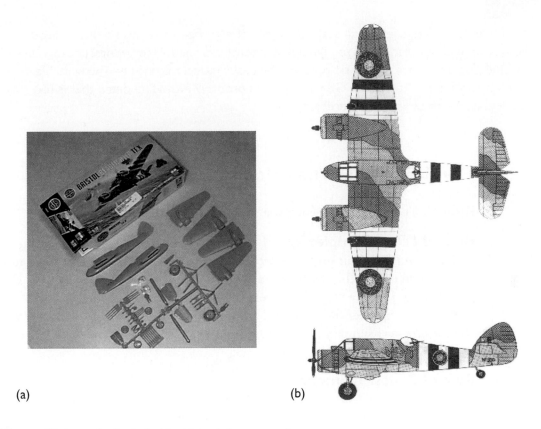

(a)   (b)

Figure 18.1   A plastic hobby kit and documentation.

In this chapter we will concern ourselves only with creating representative geometry. In Chapter 19, you will be shown how to create realistic raster texture maps so that the model can be rendered realistically. When modeling is complete, the airplane will look like Figure 18.2.

Figure 18.2   A computer model created from the parts shown in Figure 18.1a.

Considerations When Modeling

Before you commence modeling you should be aware of the following rules:

Rule of Least Complexity

Always create the most simple geometry possible. Use deformations and raster texture maps to modify the shape or add detail rather than additional geometry. Use primitives and texture maps whenever possible.

This rule keeps your file sizes small and the geometry flexible. In Chapter 19 you will see how very simple geometry can appear complex with the addition of raster texture maps.

Rule of Fewest Elements

Always specify the fewest number of elements necessary to produce smooth shapes based on the viewing distance to the illustration. Even simple geometry can become overly complex if, for example, you specify too many sides to a cylinder, cone, or sphere, or if you use too many vertices to define a profile. If you don't know how many elements will be enough, start with 16 to 20 elements in a proxy object. Then, when the final view is determined, a finished, more complex object can be merged and substituted for the proxy before final rendering and printing. You can always increase (**Tessellate**) or decrease (**Optimize**) the complexity later.

Rule of Smallest Printer Dot

Know your output resolution and viewing distance. Never model detail having edges that when viewed from a final distance are smaller than the target printer's resolution. If you spend time modeling detail (nuts, bolts, screws, rivets, panel joints, etc.) that at rendering scale and resolution are smaller than a printer dot, they will simply drop out. An illustration that looks fine at 2450 dpi will probably look horrible when displayed on screen as part of a multimedia program at 160 x 120 and 72 dpi.

Rule of Viewer Field

Never model geometry that will be unseen or covered up by other parts. Imagine the sinking feeling (in your bank account) when you find out that much of the geometry you spent hundreds of hours modeling will never be seen in the illustration. It is a mistake *not* to determine viewpoint and scale early in an illustration project.

Because 3D modeling gives you access to *any* view, the temptation is to make viewpoint decisions later in the process. With traditional manual methods the viewpoint was the *first* thing you determined, and then only after careful analysis. Why? Because if you were wrong, you had to start over. If you know the intended use of the technical illustration and intended standard view and scale, you can make intelligent decisions about what must be modeled and what can be left out.

Rule of the Rendering Engine

If your rendering engine does a good job of determining visibility, keep Boolean operations to a minimum. A Boolean (as you'll learn later in this chapter) is a way of joining, subtracting, or intersecting objects. The result is a new compound object with increased complexity when compared to the original parts. Sometimes it is just better to let the rendering engine determine the intersections.

Geometric Analysis

Just about all parts can be made from simple geometry after careful analysis. The manner in which these primitives are arranged in space gives the illusion of detail when they are rendered with lights and textures. Still, some parts (if you need greater accuracy) are conducive to more sophisticated modeling techniques. Figure 18.3 shows our Beaufighter in axonometric wireframe mode. Several of the modeling techniques have been identified. Note the difference in geometric complexity of the parts. Some components are regular and defined by a few vertices and elements. Others are more irregular, requiring a greater amount of data to describe their surfaces.

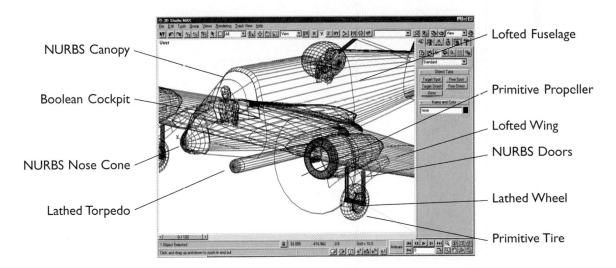

NURBS Canopy

Boolean Cockpit

NURBS Nose Cone

Lathed Torpedo

Lofted Fuselage

Primitive Propeller

Lofted Wing

NURBS Doors

Lathed Wheel

Primitive Tire

Figure 18.3   A model is usually comprised of parts making use of several modeling techniques.

You'll soon discover other program's to help you make illustrations. The pilot and gunner in Figure 18.3 were created in Fractal Poser, saved in DXF format, imported into 3ds max, and scaled and edited as necessary.

Your Modeling Environment

There are several settings and options you can use to make your modeling more efficient.

Work on Paper

The default 3ds max view port color is a medium gray. You may assign different grays to all parts during the modeling phase; it is more effective to change this color to white. Choose **Customize|Customize User Interface|Colors** and pull the **Viewports** menu down to **Viewport Background**. Click on the color chip and drag the slider to white. This will be your default view port color the next time you open Max. To render with a white background, choose **Rendering|Environment|Background Color** and move the slider to white.

Rendering to the screen with a black background carries no resource overhead. The monitor could care less how it lights up the pixels. However, *printing* a rendering with a black background will rapidly use up ink or toner. Avoid making prints of black view port backgrounds unless absolutely necessary.

Work in Views

Max doesn't save view sets but it does remember active views. First choose **Customize|Viewport Configuration|Layout** and select the four-view port layout. Right-click on each view and select view ports like those shown in Figure 18.4. When you get a view you want to go back to, make the view active: right click on the view. Choose **Views|Save Active...View** to save the view orientation (zoom, pan, camera orbit).

Set Model Scale Units

Scale is much less important in modeling than when drawing. Still, you may want to establish a scale so that grid lines make sense. Choose **Customize|Units Setup** and set appropriate units.

Save Start-up Files

You will probably want to save a number of start-up files with different view port configurations and scales. Create a directory especially for this.

Work with Appropriate Axes

Max allows you to work in world, view, or local axes, among others. Generally it is easier to work with view axes unless an object has been rotated, in which case view or world axes make little sense and you will want to switch to local (object) axes.

Work in Separate Files

Unless you are blessed with workstation level equipment, performance will suffer as model complexity increases. You may want to split the illustration model into several subassemblies and merge them together for final rendering. If you work with common axes and scale, merging is a simple operation.

Modeling With Primitives

Various 3D modeling programs have different sets of standard primitives. Some programs let you add to the set of standard primitives. A primitive is economical because it is based on a standard mathematical description. Two identical primitives share the same description and, depending on the program, may require less processing or storage (and you know how that helps). Figure 18.4 shows how two primitives (torus and cylinder) can be used in combination to create the final parts shown in Figure 18.5.

An airplane tire is adequately described by a torus. The torus has the minimum number of divisions necessary to define the surface. If you go back and look at Figure 18.2, you can see that with a view from the front and above, little of the actual landing gear can be seen. You just need to hint at it with primitive geometry.

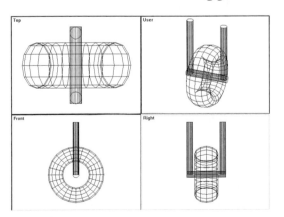

The intersection of the two cylinders is created by the rendering engine and is not the result of joining the two parts. Compare the wireframe intersection of the axle and support in Figure 18.4 with its final rendered intersection in Figure 18.5. This is an example of the *rule of the rendering engine.*

Figure 18.4   Torus and cylinder primitives used to represent landing gear geometry.

Figure 18.5 Intersection of axle and strut determined at rendering time.

Lathe a Profile about an Axis

The wheel could probably be made from cylinders and tubes, but to demonstrate a lathe application, we will create the wheel from a profile. The step-by-step description of how this is done was covered in detail in Chapter 16 and will not be repeated here.

In the top view, create the profile of the wheel as a continuous spline line. You can see in Figure 18.6 that this profile describes the rim and hub and uses the wheel and axle for size and position. In the user view you can see that the profile is positioned at the center of the tire and axle. Note how the largest tire cross section determines the size and shape of the wheel. You can also see that we are going to make only half a wheel. That is because we will be looking at the wheels only from one side.

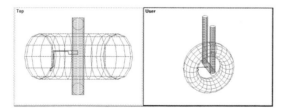

Figure 18.6 Wheel profile aligned with axle and tire ready for **Lathe** operation.

After performing the **Lathe** operation, and orienting the lathe and axis correctly, the finished wheel appears in Figure 18.7. When rendered (Figure 18.8), the wheel has a bead, a rim, and a hub. Just as the rendering engine determined the intersection of the axle and strut, it has determined the intersection of the axle and wheel hub. Figure 18.9 shows two other parts that were lathed, the torpedo and engine cowl.

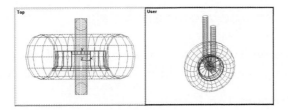

Figure 18.7   Final wheel created by the **Lathe** operation.

Figure 18.8   Rendered landing gear assembly showing intersections created at render time.

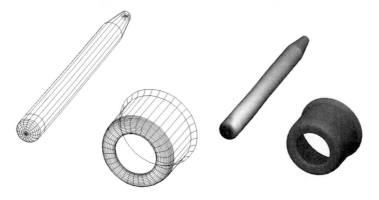

Figure 18.9   Other parts created by the **Lathe** operation.

Extrude a Profile along an Axis

The shape of the airplane's aileron isn't appropriate for lathing, though it might be created by editing a cylinder primitive. It is better modeled by a thin *extrusion*.

Create or Import the Profile

Because we have orthographic information on this part, we can create a profile in Adobe Illustrator by drawing over a scan of the top view. Import this profile into 3ds max (as we did with lathe profiles). Figure 18.10 shows the profile (with control points and handles) in Illustrator and after the profile has been imported into 3ds max and scaled.

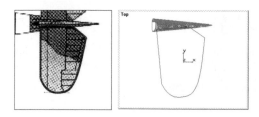

Figure 18.10   Use Illustrator to create accurate profiles when you have drawing information.

Extrude the Thickness

This profile, of course, has no thickness. With the profile selected, choose **Modify|Extrude** and specify a thickness. Note in Figure 18.11 that the aileron has been rotated into position before extrusion and that the object's coordinate (local) axes have been rotated along with the geometry. Extrusions always create thickness in the profile's Z-axis direction.

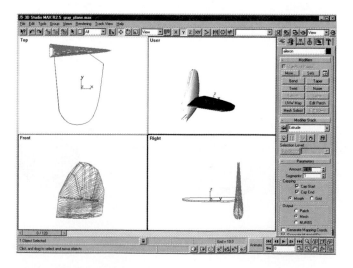

Figure 18.11   Extrusion is done along the profile's Z axis.

Make the Other Aileron

Rather than go through the same steps for the other aileron, use the **Mirror** function to copy, align, and offset the identical geometry. Figure 18.12 shows the **Mirror** dialog box with the appropriate options selected. The copy must be offset so that it is separated from the original aileron by the thickness of the tail. (You don't really have to calculate this offset distance. Just start entering numbers and within two or three tries you will have it.) Figure 18.13 shows the final tail assembly.

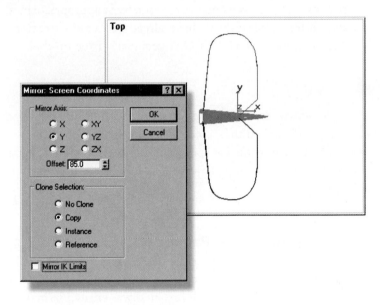

Figure 18.12   Copy, align, and offset the extruded aileron geometry.

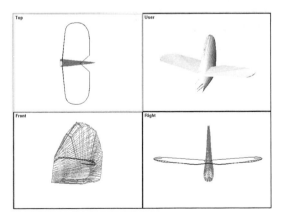

Figure 18.13   The completed tail assembly.

Loft a Profile along a Path

The fuselage of our Beaufighter can be modeled in any number of ways. For example, a profile could be lathed about the plane's axis. That shape could then be selectively edited to produce the desired shape. Or, the shape could be created with NURBS curves (as is covered later with the nose cone), but because the fuselage is essentially a ruled, single-curved surface, NURBS probably are not the best choice. The cross-sectional shape of the fuselage is essentially the same along its length, narrower at the front and back, so the most appropriate methods would be *extrusion and lofting*. Because we have already covered extrusion, we will loft. The file *fuselage.max* on the CD-ROM gets you started on this.

Create Shape and Path

Because we have only top and front view drawings, the beginning cross-sectional shape was created visually by inspecting the individual halves of the model fuselage. The cross section (the red loft shape) was created in max's right side view while the length of the fuselage (the loft path) was created in the front view (Figure 18.14).

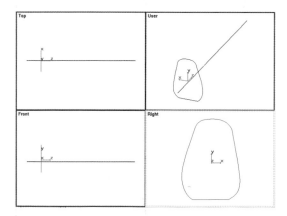

Figure 18.14   The cross-sectional shape and fuselage path ready for lofting.

The position of the path and shape one to another is unimportant. The shape's **Pivot** (local origin) will be matched to the paths **Starting Vertex** (vertex 1). If you pick the shape first, the path is moved to the shape. If you pick the path first, the shape is moved to the path.

Loft the Shape

Select the shape. Its object axis system is displayed. Choose **Create| Geometry|Compound Objects|Loft** and click on the **Get Path** button. When the cursor finds a valid path, it turns into a cross and an ellipse appears next to it (Figure 18.15). If you wait long enough without clicking the mouse, a tag reading path will appear. Click on the path. 3ds max creates a lofted surface that is the length of the path with a consistent shape cross-section. (If you specify **Move**, the shape and path will be used for the loft. If you specify **Copy,** the original shape and path will be left alone (Figure 18.16). Use the **Copy** feature so you can delete the loft if you need to edit the shape or path.

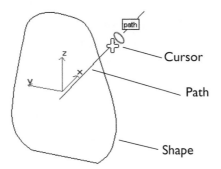

Figure 18.15   Select the shape and then the path to complete the lofted surface.

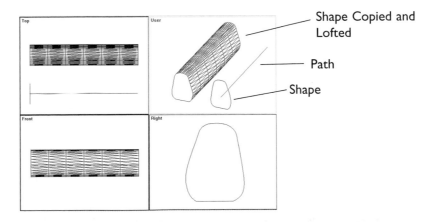

Figure 18.16   The lofted surface after the shape and path have been copied.

Convert the Loft to a Mesh

Vertices, faces, or elements can be moved or scaled once the loft is completed. But to do this you will need to convert the loft (first represented simply by a shape and path) to a mesh. Select the loft object and choose **Modify|Edit Mesh**. The loft is converted to a mesh for editing.

Edit the Loft Elements

With the mesh selected, choose **Modify** and the red vertex icons in **Modify Roll-out**.

▼　In the front view select the last set of vertices at the tail of the fuselage.

▼　Press the spacebar to lock the selection.

▼　Switch to the Right view and choose the icon for **Scale and Non-Uni-form Scale**. This will allow you to scale the tail profile differently in the X and Y directions.

▼　Toggle the X axis lock button and scale the tail profile width down considerably. Toggle the Y axis lock and scale the tail profile height down as well.

Move along the fuselage, scaling the profiles to interpolate between the previous shape and the next, going back and forth to fine-tune both shape and position, until all are completed. The user view of Figure 18.17 shows the transition between the nose and the tail. When rendered, the fuselage takes on the smooth appearance of Figure 18.18.

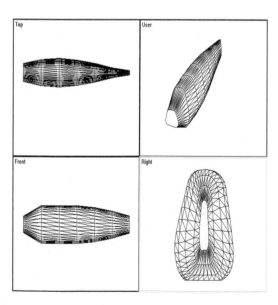

Figure 18.17　Vertices at fuselage sections are scaled differently in the X and Y directions to achieve the desired shape.

Figure 18.18   The rendered fuselage after editing lofted shapes.

Patch Surfaces

Although there were not any patch surfaces on the Beaufighter, modeling by using patches is a common method for representing objects that started out as rectangular sheets of metal or plastic and whose form is the result of stamping or molding. Essentially, a patch surface is a plane divided into either triangular or rectangular subdivisions. Much like we edited the lofted fuselage in the previous example, the sheet can be edited (vertices, elements, or faces) to achieve a desired shape. Look forward to Figure 18.21 where a completed stamping has been created from a patch surface.

Create the Patch Surface

   Choose **Create|Geometry|Patch Grids**, check the **Quad Patch** button in the **Command Roll-out**, and draw out a rectangular sheet in the view in which you will be looking directly down on the object. In our case that is the top view (Figure 18.19).

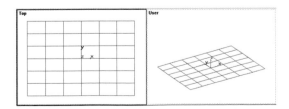

Figure 18.19   A quad patch surface.

André Cantarel
Eppelheim, Federal Republic of Germany

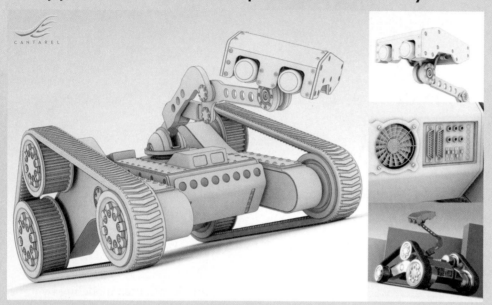

Images Copyright 2002 by Andre' Cantarel. *andre@cantarel.de*   *www.cantarel.de*

The power and flexibility of 3D modeling programs make them an obvious choice for creating highly detailed technical illustrations. However, such programs are limited in the options for rendering—especially if black-and-white line illustrations are desired. The solution I chose was finalToon from CEBAS Visual Technologies, with global illumination shading done with finalRender Stage-1

Final Toon is a true line renderer and outputs vectors in Adobe Illustrator format. Technical illustrations require nonphotorealistic rendering techniques, decisions as to what is hidden and visible, and the selective use of line weights to show form and function. A true line renderer makes such transformations between two and three dimensions possible. For example, technical illustrations require that thick outlines stay clear and visible against the shading of the surface. Extensive shading isn't needed in a line illustration, so as you can see from the example in this case study, a low range of shading is used. Such shading is contrary to realistic surface shading models, like the color rendering of the robot, but is more than acceptable in technical illustration renderings.

Software

3ds max 4.2
CEBAS
  finalToon 1.0
  finalRender–1
Adobe Illustrator

Hardware

Dual AMD 1.8

Edit the Patch

You can consider the patch surface to be a sheet of pliable material that can be stretched over a form. The more **Length** and **Width Segments** you specify on the surface, the more intricate the shapes that can be formed.

▼ With the patch selected, choose **Modify| Edit Mesh** and select the vertex icon.

▼ Select the vertices that define the shape of the stamping.

▼ Press the spacebar to lock the selection. Click on the Y axis lock button and pull the material up. The patch will look like the user view of Figure 18.20. (The vertices have been pulled up along the Y axis, so they can't be seen in the top view.)

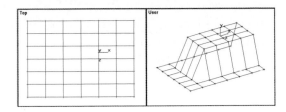

Figure 18.20   Vertices of the patch object are pulled along the Y axis.

▼ Selecting another group of vertices and pulling them higher results in the final stamping shown in Figure 18.21.

▼ The extra flashing around the base of the stamping can be trimmed away by selecting and deleting vertices.

Figure 18.21   Finished patch surface stamping ready for trimming.

NURBS Surfaces

Up until now we have been modeling by defining polygonal surfaces that 3ds max smoothes at rendering time. To get really smooth shapes, you must have a lot of these polygons. Go back and look at Figure 18.8, our landing gear assembly. The tire was defined by a torus of 24 divisions. You can actually see the chords of the ellipses. To make them disappear (at this close viewing distance) would probably require 36 or 48 divisions, increasing the complexity of the data considerably. To make smooth polygonal surfaces you need lots of polygons.

A solution to this is NURBS. NURBS (*non-rational uniform B-splines*) use mathematical splines to describe curved surfaces. If you look at Figure 18.3, you'll see three examples of NURBS surfaces: the landing gear doors, the pilot's canopy, and the fuselage nose. The geometry appears simplistic because the surfaces are defined by curves, not polygons. Max defines the actual NURBS surfaces at render time, without the heavy overhead (and approximation) of polygons. For our NURBS example, we will model the airplane's nose cone. Look ahead to Figure 18.27 to see the final shape of the nose cone. To follow along, load the file *nose.max* from the CD-ROM.

▼ Note

NURBS should not be used for every modeling task. They are particularly well suited for irregular, double-curved, or warped surfaces. Additionally, because Max must define the surface at render time, rendering times for NURBS surfaces will be longer.

Create the NURBS Curves

We will use the front of the fuselage as a template for the nose. That way, we know that the nose cone and fuselage will match nearly perfectly. If you had carefully analyzed the shapes and determined that the back of the nose cone would match the front of the fuselage, you could have started the fuselage with that profile. The you could use a copy as the basis of the nose cone.

Begin by looking at the front of the fuselage in the left view port. Choose **Create|Shapes|NURBS Curve** and click on the **CV Curve** button. Carefully draw around the nose of the fuselage (Figure 18.22). This will be the base curve for the NURBS nose cone.

Create the Side Profile

In the front view, create a spline curve. (It doesn't have to be a NURBS curve because we are only using it as a guide for scaling our NURBS profiles.)

Create Copies of the Original NURBS Curve

In the front view and with the X-axis lock on, hold the **Shift** key down and make copies of the original NURBS curve. Place the last one at the tip of the nose cone. This is shown in the front view of Figure 18.23.

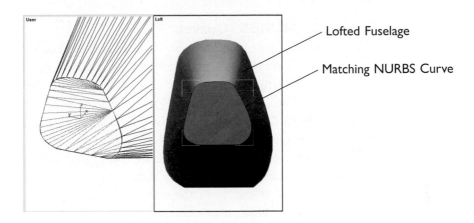

Lofted Fuselage

Matching NURBS Curve

Figure 18.22   The nose of the fuselage becomes the pattern for the large end of the nose cone.

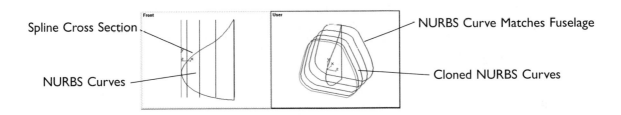

Spline Cross Section

NURBS Curves

NURBS Curve Matches Fuselage

Cloned NURBS Curves

Figure 18.23   Create a profile in the front view for the shape of the nose cone.

▼ NURBS Note

Life is easier when making NURBS surfaces from profiles based on the same original curve. For one thing, you have less editing to do when you reshape each curve. For another, you guarantee that the surface is defined by curves that were all drawn in the same direction. When curves are drawn in opposite directions, or when the starting vertex is not in the same relative position, the lofted surface is twisted as it tries to connect successive NURBS profiles. This can be corrected by rotating the profile until the vertices do, in fact, line up. But in doing so, you lose the orientation of the shape and it must be edited into correct final form.

Scale the NURBS Curves

In the front view, uniformly scale the profiles, moving them in the Y direction until they coincide with the spline curve you drew as a guide (Figure 18.23). Make sure the profiles are centered on the guide as demonstrated in the left view port of Figure 18.24.

Shape the NURBS Curves

The nose cone becomes decidedly rounder as it approaches the tip, so the curves must be edited, making the transition between the end of the fuselage and a very small circle at the tip. In turn, select each of the NURBS curve copies (the original stays unedited because it has to meet with the fuselage). Select a curve and choose **Modify|NURBS Curve|Point**, and tweak the shapes, making them rounder as they approach the front. Make sure you keep the top and bottom of the curve aligned with the guide. See the user view in Figure 18.25. Each curve profile passes through the top and bottom of the front view shape.

The end curve at the tip of the nose cone is very small. Scale this curve even further to close the nose cone off at the front.

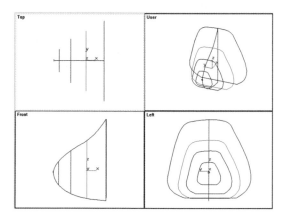

Figure 18.24   Scale each of the copied curves until they conform to the guide.

▼ Note on Attaching

When you *attach* two objects such as NURBS curves, you create a structural relationship so the geometry engine can do something with them—like build a surface. This is different from *grouping* where a logical relationship allows objects to be moved, scaled, or rotated together.

Attach the Curves

In order for the individual curves to be considered elements of a surface they must be attached. This associates the curves with a single NURBS surface.

▼   Select the NURBS curve where the nose cone attaches to the fuselage.

▼   Select **NURBS Curve|Attach** from the **Modify Roll-out**.

▼ Click on the selected curve and drag the cursor to the next profile. A
   dotted line is shown, and each curve turns blue as it is found.

▼ Continue with each curve in order until all are red and attached. The
   curves now comprise a single NURBS object.

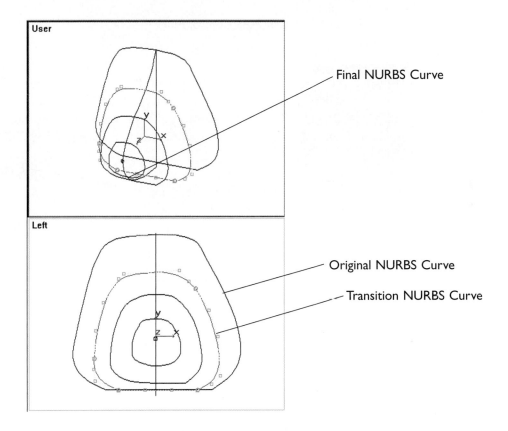

Figure 18.25   Transition NURBS curves are made rounder toward the tip.

Create the NURBS Surface from the Curves

Once the curves are attached, instruct Max how you want the surface defined. In
this case, we are going to loft a surface in the "U" or depth direction using the
attached curves.

▼ Click on the small green icon in the **NURBS Curve|General Roll-out**. A
   NURBS control panel like that shown in Figure 18.26 will be displayed.

▼ Click on the **U-Loft** icon and click on the original NURBS curve. As you
   move the cursor over the next curve, a dashed line will again appear and
   the next curve will turn blue.

▼   While it's blue, click on it and continue from curve to curve until all have been identified. You'll see the NURBS surface being defined as you move from curve to curve (Figure 18.26). The complete nose cone is shown in Figure 18.27.

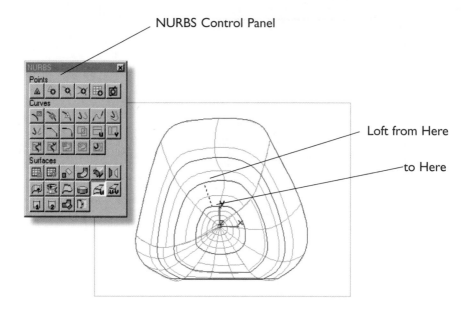

NURBS Control Panel

Loft from Here

to Here

Figure 18.26   The nose cone defined as a NURBS surface from a series of NURBS curves based on the same original profile.

▼ Note

If segments (and not the entire profile) of the curves turn blue, make sure you connect to the similar segment on the next curve. You may have to piece halves, quarters, or eighths of the NURBS surface together. Figure 18.26 shows where two segments of the same quadrant of the curves are connected.

Boolean Operations

Even after forming geometry by combining primitives, or by lathing, extruding, or lofting profiles, it may still be necessary to add or remove portions to arrive at a final shape. Boolean operations provide tools to join two separate objects, remove selected geometry, or intersect parts.

There are several opportunities to perform Boolean additions in our aircraft geometry. For example, the engine nacelles could be joined to the wings to form a single object. And the wings themselves could be joined to the fuselage. But because 3ds max handles these intersections easily when rendering, they are left as separate objects.

Figure 18.27   The finished rendered nose cone that was modeled as a NURBS surface.

Although none of the parts in this example were formed by Boolean intersection, this operation remains the most powerful of the three. For example, had more detailed drawings been available, front, top, and side extrusions could have been intersected to create nearly finished geometry. This technique is shown in simplified form in Figures 18.32 and 18.33. The file *intersect.max* is on the CD-ROM for you to try your hand at this operation.

For our Beaufighter, three Boolean subtractions are needed: the pilot's cockpit, the gunner's station, and the landing gear wells on the underside of the engine nacelles. The key to performing subtractions is to create the appropriate *tool* to cut away the desired material. Removing a portion of the fuselage as shown in Figure 18.28 starts the cockpit. The file *fuselage.max* is provided on the CD-ROM for you to practice this subtraction operation.

When you perform Boolean operations on compound geometry (geometry that is the result of previous Boolean additions, subtractions, or intersections), unpredictable results can occur. This is minimized by converting the result of a Boolean to an *editable mesh* before the next Boolean operation.

Always choose **File|Hold** to place the current scene in memory before performing a Boolean operation. You then can choose **File|Fetch** to restore. The order of the operation can impact the results, so you may want to change the order of the operation. Make sure you pay attention to A minus B or B minus A for subtractions.

Create the Tool

The tool is a solid block that extends beyond the fuselage, cutting flat edges horizontally and vertically. Inspect the alignment of the red tool in the view ports of Figure 18.28. It is helpful to render the user view port so that the result of the subtraction can be observed.

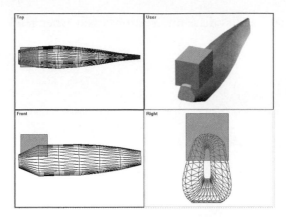

Figure 18.28   Tool in place for the first Boolean subtraction.

Perform the First Subtraction

Select the fuselage. This becomes part A in the A minus B Boolean subtraction. Choose **Create|Geometry|Compound Objects|Boolean** and make sure the A minus B subtraction is selected. Choose **Pick B** and **Move**; click on the red cutting tool. The result is shown in Figure 18.29.

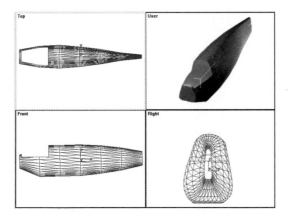

Figure 18.29   Result of the first Boolean subtraction.

Perform the Second Subtraction

The cockpit needs material removed to form its depth. A vertical tool is created that closely follows the intersection of the first tool and the fuselage (Figure 18.30). This was done by creating a closed spline shape and then extruding that shape to an appropriate depth. Select the fuselage and perform an A minus B subtraction. You should have a finished part that looks like Figure 18.31. It's finally starting to look like an airplane!

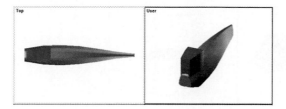

Figure 18.30   Second cockpit tool in place.

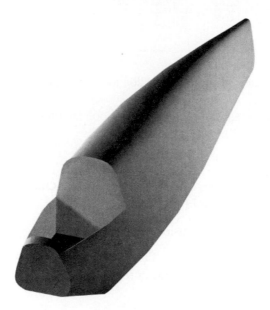

Figure 18.31   The completed fuselage after the cockpit has been formed by two Boolean subtractions.

Boolean Intersections

A *Boolean addition* keeps all material of the two Boolean operands or objects. A *Boolean subtraction* removes the common material (like drilling a hole). A *Boolean intersection* keeps only that material common to the two shapes. If addition is like welding two shapes together, and subtraction is like cutting or drilling, intersection is like casting material into a cavity.

What if your illustration is populated with aircraft at some distance? Certainly you wouldn't need the level of geometric detail of a craft in the foreground. The solution is to create a primitive shape that can be edited as necessary and provides sufficient detail at an extended viewing distance.

Figure 18.32 shows three airplane extrusions aligned in space. These profiles were created by tracing the views of Figure 18.1 in Adobe Illustrator and importing those shapes into 3ds max. The profiles were then extruded.

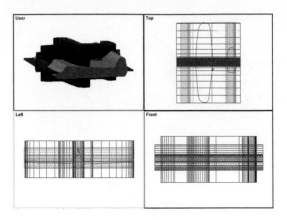

Figure 18.32   Top, front, and left-side geometry in position for Boolean intersections.

This method requires two successive Boolean intersections. It doesn't matter which two you pick to start. For example:

▼ Select the front extrusion.

▼ Choose **Create|Geometry|Compound Objects|Boolean** and then **Intersection** and **Pick Operand B**.

▼ Select the top extrusion. Max keeps only the geometry common to the two extrusions.

▼ Right-click on **Boolean** in the **Stack Roll-out**. Select **Convert to: Editable Mesh**.

▼ Repeat the intersection for the editable mesh and the side extrusion. The result of the two intersections is shown in Figure 18.33.

Figure 18.33   The result of intersecting top and front geometry and subsequently intersecting left-side geometry.

Finishing Up

We'll let you complete the remainder of the parts. Carefully analyze each component so that the most efficient modeling method can be utilized.

When all the parts have been modeled and a gray color assigned to each, choose **Create|Lights|Omni** and place an omni light above and in front of your scene, about 50 feet above the plane. Set the multiplier to 3.0 and render the user view port. Compare the results in Figure 18.34 with Figure 18.1b. We have credible results for working from a small-scale model with little additional information.

Figure 18.34   Views of the completed aircraft model.

But what about all the insignia, camouflage, and other details that make the plane more realistic? These will be applied as raster textures in Chapter 19.

Modeling from Life

Up until now you have created technical illustrations from scale drawings, CAD data, photographs, and in this chapter, parts that can be easily measured. But what about technical subjects that are not easily measured or may possibly be inaccessible? This requires a combination of methods and, to be quite honest, a fair amount of research, interpolation, and intelligent guessing on your part.

Figure 18.35 shows a dust collector that is a little large to be easily measured. It is possible to climb upon it and take some measurements, but looking at the age of the structure, I wouldn't chance it. So we are left to gather our data from the ground. Additionally, it is a permanent structure, meaning that it can't be disassembled or brought inside where it might more easily be modeled. What would be your strategy so that it might be accurately modeled and so later technical illustrations could be created from the model?

Figure 18.35   A dust collector to be modeled.

▼   The first step is to secure photographic documentation. Take a digital camera and record the views that are available to you (Figure 18.36). Take other photographs of details and textures that can be used later for materials. Because the camera's lens will introduce a measure of perspective, the photographs should be taken with the center of vision near the geometric center of the subject.

▼   Create scale profiles in Adobe Illustrator to guide construction, normalizing the perspective where necessary. This is where you would need to make use of obvious symmetry and the alignment of components. These profiles may be simple shapes to guide overall height, width, and depth (Figure 18.36) because the modeling program will take care of many of these symmetrical sizes.

▼ Note on Photographs

Models created from photographs are sensitive to the focal length of the camera. The shorter the focal length, the more convergence is introduced into the image. This greater convergence makes orthogonalizing the photograph more difficult and introduces a measure of error into shapes. If you have a choice, always choose long (telephoto) length to minimize this convergence.

▼ Import the profiles into 3ds max to guide your modeling. You know that they will be imported into the top view, matching Illustrator's X Y paper coordinates with max's X Y world axes. This helps you plan your modeling strategies (Figure 18.37). You can see from Figure 18.36 that not all features have been located. In fact, not all features *have* to be located, only their approximate locations based on known coordinates.

Left Side View   Front View

Figure 18.36   Profiles to guide construction.

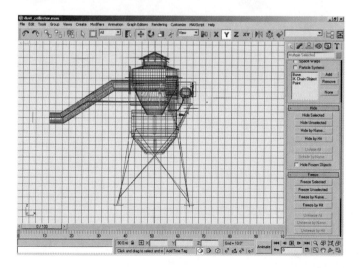

Figure 18.37   Wireframe model based on imported profiles.

The final model is shown in Figure 18.38. Only basic material colors have been assigned to the various parts. In Chapter 19 we will explore how materials can be added to increase the realism of the illustration. Digital photographs for the dust collector can be found on the CD-ROM in *exercises/chapter18*.

Figure 18.38   Modeled dust collector (John Davis).

Line Illustrations from Models

Getting quality PostScript line illustrations from 3D computer models has always been problematic. In Chapter 17, several strategies for generate line illustrations were explored using AutoCAD and 3ds max. Earlier in this chapter a case study was presented featuring a tool (CEBAS finalToon and finalRender) dedicated to postprocessing a 3D model into editable PostScript vectors.

This process isn't absolutely automatic, and anywhere from some to considerable editing must be done in Adobe Illustrator to arrive at the final results. Two-dimensional spline objects (circles, lines, gengons) from 3ds max are each assigned to their own layer. In fact, Andre's robot consumed over *950 layers* in Adobe Illustrator! However, the results can be startling. Refer back to the case study on page 406. Along with the color raster rendering are examples of monochromatic raster techniques. Figure 18.39 shows the result of translating 3D vectors into 2D vectors.

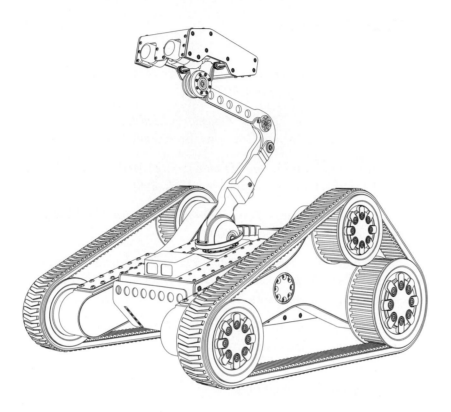

Figure 18.39   Model vectors translated into PostScript vectors (Andre Cantarel).

Review

The advent of powerful PC-based modeling programs such as 3ds max has changed the face of technical illustration. Where 2D drawings were made, 3D models now provide an unlimited number of illustration views. The complete technical illustrator must be able to create models from which static and dynamic illustrations can be made.

In previous chapters you learned how to scan technical documents in order to create profiles for 3D modeling operations. In this chapter you learned how to analyze components of a model so that the most effective modeling technique might be applied.

Every illustrator must be able to use primitives when appropriate or to lathe or sweep, extrude, loft, modify grids, create NURBS surfaces, and perform Boolean operations. Together, these methods form the illustration tool kit needed by today's technical illustrator.

Text Resources

Bartlet, Brandon (Ed.). *3D Studio Architectural Rendering*. New Riders Publishing, Indianapolis. 1996.

Covert, Pat. *Basics of Scale Automotive Modeling: Getting Started in the Hobby*. Kalmbach Publishing Company. Brookfield, Wisconsin. 1999.

Khemlani, Lachmi. *Into 3D with form.Z: Modeling, Rendering, Animation*. McGraw-Hill Professional. New York. 1999.

Luebke, David, et al. *Level of Detail for 3D Graphics*. Morgan Kaufmann Publishers. San Francisco. 2002.

O'Roark, Michael. *Principles of Three-Dimensional Computer Animation: Modeling, Rendering, and Animating With 3d Computer Graphics*. W.W. Norton & Company. New York. 1998.

Smith, Peter H., and Siulva, Walter. *Introduction to Solid Modeling*. Distance Engineering, Inc.. Ann Arbor. 1999.

Internet Resources

3D model library. Retrieved from: *http://www.3d02.com/*. January 2003.

Accustudio models. Retrieved from: *http://www.accustudio.com/exchange/prnindx.htm*. January 2003.

CAD modeling and rendering forum. Retrieved from: *http://www.gavilan.edu/cgd/main/cadmr.htm*. January 2003.

Index of United Kingdom CAD Modelers. Retrieved from: *http://www.applegate2.co.uk/products/psa/4114.htm*. January 2003.

Medical modeling from Graphic Pulse. Retrieved from: *http://www.graphicpulse.com*. January 2003.

Modeled and mapped examples. Retrieved from: *http://www.3-d-models.com*. January 2003.

Southern Model Builders. Retrieved from: *http://www.southernmodel builders.com*. January 2003.

The modeler's bookstore. Retrieved from: *http://scalemodel.net/bookstore/ main.asp*. January 2003.

On the CD-ROM

For those of you who would like to try out the procedures described in this chapter, several resources have been provided on the CD-ROM. Look in *exercises/ch18* for the following files:

cockpit.max	This file contains a finished shaped fuselage and the two tools needed to subtract the cockpit.
dust1.jpg through **dust3.jpg**	Use these digital photographs to model the dust collector.
fuselage.max	This is a lofted shape ready for you to begin editing. Pattern shapes have been provided and appear in the right view port.
intersect.max	This file contains three orthogonal extrusions to intersect. Start with the front and top extrusions. Once this is completed, intersect the left-side extrusion with the result of the first intersection.
nose.max	Use this file to practice NURBS surfacing by copying the base curve, editing the copies to conform to a profile shape, and creating a U-loft surface.

Chapter 19

Raster Materials

In This Chapter

In Chapters 16 to 18 you have seen how computer models can be turned into effective illustrations. However, up until now you have had to accept the most simplistic materials—usually shades of gray. This is fine for gray plastic, but what about all the other materials in the world?

Three-dimensional modelers create vector, surface, and solid geometry. It is interesting that materials are applied to this geometry using an entirely different type of graphic data: raster graphics. So you model in vector, but you render in raster. This is significant because modeling and rendering skills come from different parts of your brain. Modelers are pragmatic, analytical, and structured. Material mappers are creative, intuitive, and visual. Because of this, many large illustration shops have individuals who specialize in materials and others who do the modeling. It's even not uncommon to see light and camera specialists.

The heart of rendering in 3ds max is its **Material Editor.** This is the part of the program that coordinates the various raster maps (like reflections and bumps). But in order to get raster textures that max can use, you have to be familiar with the workings of a raster editor, like Photoshop.

Once materials have been assigned to 3D objects, the scene must be rendered. To be rendered effectively, lights must be carefully positioned—much as a stage technician positions lights for a play or musical. Many readers have already made the leap: effective technical illustration starts with accurate and efficient geometry, adds descriptive raster textures, and finishes with effective lighting. The final piece of the equation is to assign appropriate rendering algorithms to the different textures so that when the illustration is finally rendered, materials look realistic.

Because raster textures are fixed resolution, the raster dimensions of a material and the target resolution of the final rendering must be carefully coordinated. If the material resolution is too coarse, the material won't look realistic. A material map with a resolution finer than the rendering will be wasted.

So that you can see the complete process, we will be using the aircraft example from Chapter 18 (Figure 19.1) for the majority of the examples. A separate wheel model is used to demonstrate rendering resolution.

Figure 19.1   The airplane model that will be the subject of our raster texture exercises.

Chapter Objectives

In this chapter you will understand:

- ▼ Raster graphics and its typical file formats

- ▼ How simple materials can be created in 3ds max's material editor

- ▼ How materials can be enhanced by textures

- ▼ The various mapping channels such as diffuse, environment, bump, opacity, and reflection

- ▼ How to use displacement maps as a way of editing geometry

- ▼ How omni lights and spotlights impact the realism of a scene

- ▼ Rendering formats, color depth, and bitmap size

Raster Graphics

A raster graphic consists of a rectangular array of data points. Each point is a picture element, or *pixel*, and can be addressed independently of any other pixel in the array. Everything in a raster graphic is a pattern of these pixels. A circle is a pattern of pixels that when viewed as a whole, appears as a circle. A raster letter "A" is simply a pattern of dots that is interpreted as the beginning letter in the alphabet.

Vector graphics (like you make in Illustrator) are not comprised of pixels at all, but rather with objects such as lines, rectangles, circles, and text.

This distinction is important because by using raster graphics subtle nuances in surfaces can be approximated. Adjacent pixels can be displayed differently to show minute changes in surface texture. This is more problematic in vector graphics. It's easier to illustrate organic nongeometric shapes in high, but fixed, resolution raster format. It's easier to illustrate linear geometric objects in resolution-independent vector format.

This raster power and flexibility comes at a price, however. Because the condition of every element in a raster array must be stored, raster textures require greater computing resources than do vector textures.

To better understand raster textures, you should become familiar with common terminology.

Bitmap

A *bitmap* is a rectangular array of pixels intended for display on a monitor or printing on a printer. The *dimensions* of the bitmap describe horizontal or vertical pixels. A bitmap of 800 x 600 has 800 pixels horizontally and 600 pixels vertically, resulting in almost a half million total data points.

Bit Depth

Each pixel in a bitmap can be described by one or more pieces (or bits) of information. This is referred to as a bitmap's *depth* and adds even more information about the image. A bit depth of 1 means it is impossible to display grays. Grays must be approximated by arranging white and black pixels in patterns that can be interpreted as gray by a viewer. This is referred to as *dithering*.

▼ Note

Some illustrators don't like dithering, but it is the fundamental way that rendering engines represent colors that can't be printed (that are out of the printer's gamut). Great dithering is an illustrator's friend. Lousy dithering makes materials look unrealistic.

Color Palette

The more colors in a particular palette, the more realistic the rendering. The number of colors is directly related to the amount of information (bit depth). Table

19.1 displays the relationship between bit depth and the colors available in a given palette.

▼ Digital Color		
Color Depth, bits	Name	Colors Available
1	Bitmap	Black and white
8	Indexed	256
16	High	64,000
24	True	16.7 Million

Table 19.1   Rendering bit depth impacts the color palette.

Some of you, no doubt, are aware that formats and printers tout "32-bit color" or even "64-bit color." But understand that even 16.7 million colors may be overkill. William Horton, in his definitive text *Illustrating Computer Documentation*, states that there exist only 7,500,000 measurable colors. Trained colorists (people formulating colors for paint companies, for example), even under ideal conditions, can only distinguish 1,000,000 colors. And even people trained to recognize colors (like air traffic controllers) can recognize only 50 colors.

What does this mean for raster textures and rendering? If most people can distinguish as few as 50 colors, what are we doing rendering everything in 24 bits? The penalty in terms of file size and rendering time is horrific with little difference in perceived realism (Figure 19.2). The solution? Carefully match bit depth with audience, use, and available reproduction technologies.

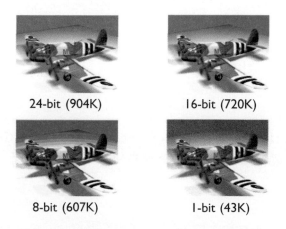

24-bit (904K)   16-bit (720K)

8-bit (607K)   1-bit (43K)

Figure 19.2   Reproduction in 1, 8, 16, and 24 bits and each image's corresponding file size.

Rendering Methods

Once a raster texture is assigned to a surface, the mathematics applied to the rendering determines, to a large part, the realism of your final print. The following discussion concerns the strengths and limitations of the basic rendering algorithms. Three-dimensional modeling and rendering programs implement these rendering algorithms differently, adding unique controls in an attempt to differentiate their products. But no matter what the method is called, it all boils down to how the value for each pixel in the rendering is calculated. To illustrate the differences, we will be using the engine cowl from our aircraft example.

Flat Shading

This technique assigns the same value (color, brightness) to every pixel of a surface. On polygonal models (Figure 19.3) this results in characteristic facets between adjacent planes (Figure 19.4). Objects made out of flat planes and dull materials (paper, cardboard, flat painted metal) can be rendered this way with excellent results. The benefit of using flat shading is the speed of the rendering because only

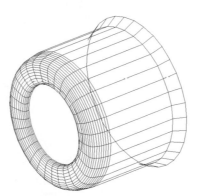

one value needs to be calculated for every pixel on the surface. Figure 19.5 shows a surface flat shaded using a value represented by the vector normal to the surface. For this reason, many illustrators use flat shading as they build their models, reserving more time-intensive rendering methods for final images.

Figure 19.3   A polygonal wireframe model showing flat plane intersections.

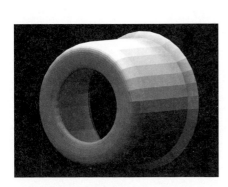

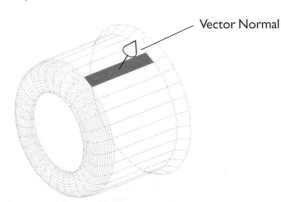

Vector Normal

Figure 19.4   Flat shading produces facets.   **Figure 19.5**   All pixels are assigned the same value.

Gouraud Shading

This technique averages pixel values between vertices and produces smooth, but predictable, transitions across surfaces. Surfaces seem to blend together (Figure 19.6). Because adjacent surfaces share vertices, all plane intersections begin with identical values, assuring smooth transitions between adjacent polygonal surfaces. In Figure 19.7, note that adjacent planes share the same vector values.

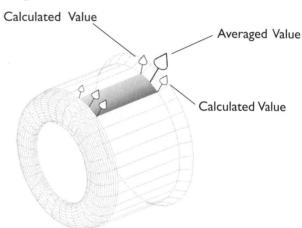

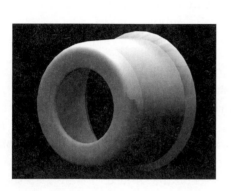

Figure 19.6   Gouraud shading produces smooth transitions between planes.

Figure 19.7   Gouraud shading averages pixels between vertices.

Phong Shading

Phong and Gouraud shading differ in that Phong shading calculates a unique value for every pixel on every visible surface. This is necessary if there are deviations between vertices that are not accurately represented by averaging. It is necessary to calculate the value of every pixel if reflection, transparency, or bump maps are applied to the surface. Figure 19.8 shows a surface distressed by applying a bump map to a Phong-rendered material.

Figure 19.8   Phong shading calculates the value of every pixel on a surface.

Ray Tracing

Ray tracing produces the most realistic results by tracing every light ray in a scene from the pixel it falls on, back to all light sources. In that way, highly detailed reflections can be rendered. This realism, however, comes with a price. Fully ray traced illustrations increase rendering times by as much as 100:1 over Phong rendered materials. Reserve ray tracing for scenes that have subtle lighting, reflections, and materials (Figure 19.9).

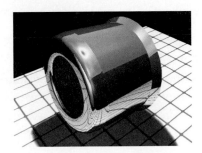

Figure 19.9   Ray tracing calculates the effect of every light source at every pixel.

▼ **Note**

Entire scenes are not ray traced, only desired materials. You can have flat, Gouraud, Phong, and ray traced materials in the same scene. So it's important to use a material that gives you the results you desire. Flat shading chrome is nearly impossible. Ray tracing cardboard is a waste of time.

Raster File Formats for Materials

We will begin creating raster materials for our airplane cowl. But first, a discussion of raster file formats and their applicability in making materials is in order. For a more in-depth discussion of raster file formats see the Technical Illustration File Types section in Chapter 1.

3ds max has an impressive array of import filters for bringing raster images into its material editor. However, some are more appropriate than others. If you want to control colors as well as file sizes, it is preferable to create raster material maps in only one or two file formats, each using the same color palette. Photoshop is excellent for translating and managing the colors from various sources.

Windows Bitmap (BMP)

This raster format is preferred because of its lossless RLE compression and 24-bit color depth.

Graphic Interchange Format (GIF)

This format also uses lossless compression but is appropriate for images having large areas of solid colors and is limited to 8 bits of color information (256 colors). Because opacity and bump maps often have large areas of white and black, this format is best suited for these applications.

Joint Photographic Experts Group (JPG)

This format uses lossy compression, so artifacts may be a consideration under extreme compression. The 24-bit color space is appropriate for continuous-tone materials.

Targa Graphics Adapter (TGA)

This file format is device dependent, relying on Truevision's display hardware. It is capable of 24-bit color, appropriate for continuous tone. Because recent versions of hardware support 64 bits, TGA files can contain a large amount of non-color information that is unusable in 3ds max's material editor.

Tag Image File Format (TIF)

This file format is specifically designed for printing to PostScript printers. Like the TGA format, TIF contains information unnecessary for creating material maps for illustration. However, the TIF format is ideal for scenes rendered out of max.

What are the risks in making inappropriate file format choices when making materials?

▼ Imperfections or artifacts may become obvious when scaling a texture (JPG).

▼ Greater storage space is required because files contain more information than is needed. This is true in terms of bit depth (TGA) as well as header and footer information (TIF).

▼ Colors may shift as max tries to match colors to the current palette. The solution is to use a consistent raster file format, such as BMP or JPG.

The Material Editor

The heart of creating materials in 3ds max is its **Material Editor**. In this part of max, you can create elaborate materials and apply them to geometry in the scene. To help understand the capabilities of this part of 3ds max, we will start with the most simple material and progress to what should be a very realistic rendering of the cowl.

Basic Materials

The most rudimentary materials contain only color and surface characteristics and do not make use of a raster texture. Figure 19.10 shows the **Basic Parameters** rollout from the **Material Editor**. Two versions of a material have been applied to the cowl—one flat and dull, and one with a more polished surface. Both materials are based on the same ambient, diffuse, and specular colors and Phong rendering. The difference lies in the dull material having *low shininess* and *shininess strength* while the reflective material has much greater strength. By keeping the shininess curve broad, the material looks more like brushed metal, rather than a hard, polished surface. Already, the engine cowl has a realistic look.

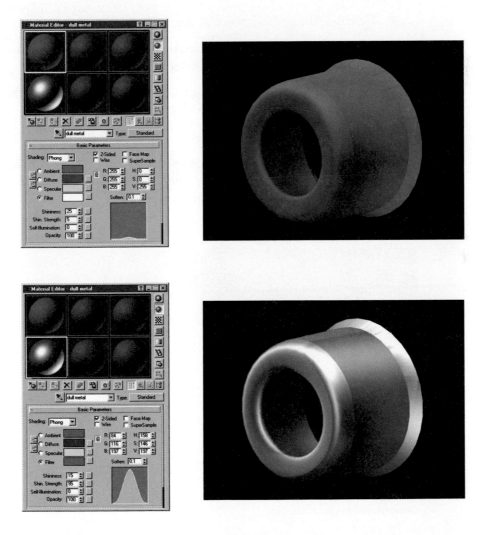

Figure 19.10   Flat default material and a simple raster map.

Textures

But engine cowls don't look like perfect metal spinnings. The color isn't uniform, the surface is comprised of riveted or glued sections, and there are dents, scratches, and probably, oil streaks and dirt. To get these textures, we will create raster maps that when applied to the engine cowl will yield the desired results.

Diffuse Maps

A diffuse map establishes the overall texture color for our engine cowl and is built on the shiny material developed for Figure 19.10. Figure 19.11 shows the flat raster map that when wrapped around the cowl will produce a brass front, blue top, and gray bottom. A little calculation here will keep you from having to scale the map later.

▼   Calculate the circumference of the cowl and note its length.

▼   Determine the desired resolution. If your final rendering will be at 300 ppi, use an even multiple of this number; start at 150 ppi.

▼   Create the Photoshop file using the same dimensions and at the desired resolution.

▼   Save the file in JPG format, medium compression.

Figure 19.12 shows the **Maps Rollout** section of the **Material Editor** with the file *diffuse.jpg* loaded into the **Diffuse Channel**. The preview has been changed from a sphere to a cylinder to better represent the final result.

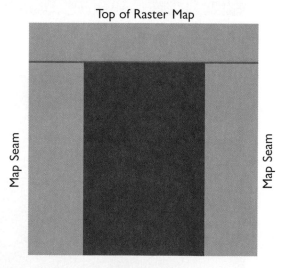

Figure 19.11   Two-dimensional raster map.

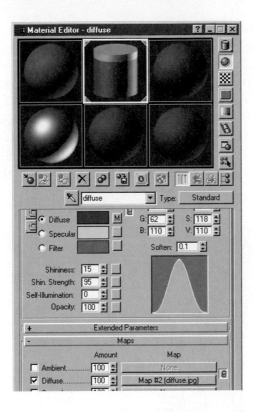

Figure 19.12   Raster map loaded into the diffuse channel.

Any raster texture must be mapped to the part geometry using one of several UVW mapping methods. The dimensions of mapping coordinates are called UVW so not to be confused with XYZ world, view, or local space coordinates. The rendering enging knows how to apply the raster map to the selected geometry only after assigning these UVW mapping coordinates. Because our engine cowl is essentially cylindrical, we will choose a cylindrical material map. To apply mapping coordinates to the engine cowl follow these steps:

▼   Select the engine cowl.

▼   Select **UVW Mapping** from the **Modify** drop-down menu.

▼   A representation of the mapping coordinates called a *gizmo* is superimposed over the cowl.

▼   Select **Cylindrical**, **Alignment Y**, and **Fit**.

▼   The gizmo can be scaled or rotated, changing how the raster map is applied to the object.

A mapping coordinate representation (gizmo) is overlaid on the selected engine cowl as shown in Figure 19.13. When the material is applied to the engine cowl, an even more realistic illustration is created (Figure 19.14). The material map may be upside down or backwards when first applied. Rather than going back and editing the material map in Photoshop, you can change the position of the mapping gizmo relative to the object (see Tech tip below). Select **UVW Mapping|Gizmo** from the history hierarchy, and rotate the gizmo using the rotate tool positioned on the gizmo's axes (you don't have to click). When the axis is selected, it changes color. If you are rendering the view port, you'll immediately see the results of the rotated mapping.

UVW Mapping Gizmo

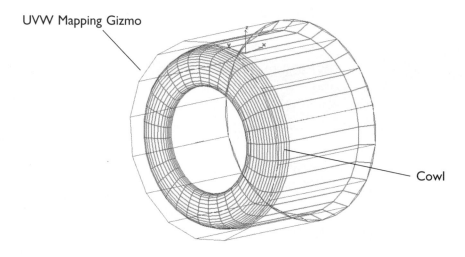

Cowl

Figure 19.13   Cylindrical mapping coordinates aligned with cowl geometry.

▼ Tech Tip

The top of the mapping gizmo is marked with a small line (shown in red in the figure to the right). This will correspond to the positive Y axis of the bitmap as you created it in Photoshop. The green line on the gizmo shows you the seam where the map is wrapped together. The rotation tool is positioned over the Y axis, allowing the gray portion of the map to be revolved until it is on the bottom. Look back at Figure 19.11 and compare the top and seam of the raster map with their positions on the object as controlled by the UVW map gizmo.

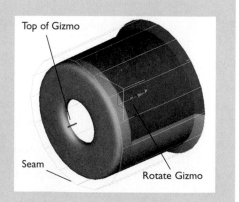

Top of Gizmo

Seam

Rotate Gizmo

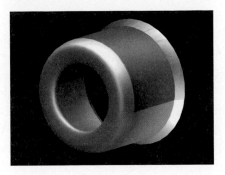

Figure 19.14   Final cowl material applied with diffuse shininess.

But of course, we want much more realism in our cowl diffuse map. Figure 19.15 shows a more detailed raster map, including oil streaks, panel joints, and wedges at the rear that will eventually be made transparent. The result of the more detailed diffuse map is shown in Figure 19.16.

▼ Note

It is possible to assign different materials to a single object through the use of a **Multi-Sub Object Material**. Essentially, you create a master material that has other materials assigned to it. Because an object can have only one material at a time, the master material is assigned to the object and the subobject materials are assigned to parts of the object.

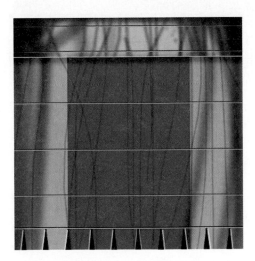

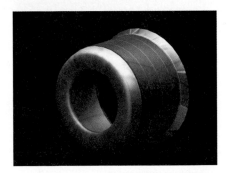

Figure 19.15   Realistic raster material map.   **Figure 19.16**   Map applied to cowl.

Bump Maps

A bump map is a gray-scale raster image. It is overlaid on a diffuse map and causes the pixels of the diffuse map to be altered at rendering time relative to the surface that the diffuse map's pixels are mapped upon. Bump mapping doesn't impact geometry at all, only the pixels mapped to the geometry. Very fine detail can be created by combining diffuse and bump maps.

It's important to match diffuse and bump map resolutions. You want a bumping pixel to match a diffuse pixel. And you want at least one rendering pixel to match both. The effect of a bump map depends on the strength of the bump and lights. The bump's strength tells the rendering engine how far to move the diffuse pixel above or below the base surface. Lights are required to reveal the change in the surface caused by the bump.

To give the surfaces even more realism we want to make the cowl look as if it has been through (naturally) a war. Figure 19.17 shows a gray-scale image created from the detailed diffuse map in Figure 19.15. The map uses white as a baseline for no surface bump. The black pixels represent pixels that are treated as indented at render time. Grays, as you might expect, make transitions between flat and depressed areas. The triangular wedges at the bottom of the map will eventually be turned transparent by an opacity map. Figure 19.18 shows the **Map Rollout** with the diffuse and bump channels being used and a new material in the fourth slot.

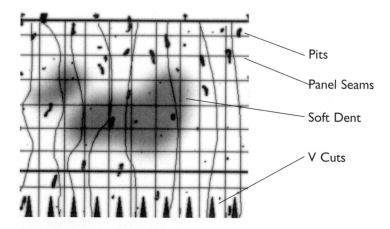

Pits

Panel Seams

Soft Dent

V Cuts

Figure 19.17   Bump map based on material map.

▼ Note on Bump Maps

By basing all material maps on the same base file, you are assured that the maps will line up. In Photoshop, do this by creating a master .psd file with the various maps on different layers. You then can selectively discard layers and save the maps individually.

▼ Illustration in Industry

Vector Scientific, Inc.
Los Angeles, California

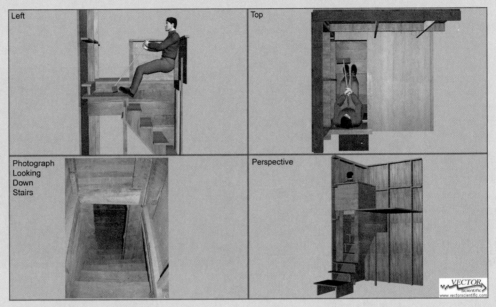

Images Copyright Vector Scientific.

Software

3ds max
Photoshop
Poser

Hardware

Windows PC
Digital cameras

Vector Scientific was hired to reconstruct an event involving a man falling down the stairs of a barn while working. The barn was over 100 years old and had not been subject to safety standards when it was built. As a result the barn staircase was narrow with an even narrower door about halfway down. In essence, the problem to solve was: Could the man have fallen down the stairs while he was working?

Photographs of the scene were taken with a digital camera, and the scene was reconstructed using Photoshop and 3ds max. The anthropometric data for the man was known and a proportional model was constructed in Poser and imported into 3ds max. Once in 3ds max, the model of the man was placed at various locations along the stairs to give a visual representation of why the man could not have fallen the way he had claimed. The advantage of using three-dimensional software to model the scene was the capability to view the scene from angles not available in the physical world (such as removal of the left wall to facilitate viewing of the stairs from the side). The matter was resolved before going to trial.

By using the same file as the basis of both diffuse and bump maps you assure that painted features (such as seams and joints) align with the bump pixels that either lift or depress the pixels. By having the painting of a seam include the highlight and the seam opening, for example, you can bump the opening down. The combination of the painted seam and the opening being depressed with a bump map increases the bump map's effect.

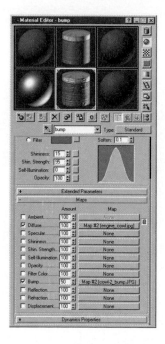 

Figure 19.18   Result of bump map application.

Opacity Maps

An opacity map tells the rendering engine the state of transparency of every pixel in the diffuse map. White pixels on the opacity map are 100 percent opaque. Black pixels are 100 percent transparent. Gray pixels are interpreted as being translucent, depending on their values.

In Chapter 18 we stressed that it is important to keep geometry as simple as possible. One could create a wedge-shaped tool and Boolean (cut out) the shapes at the rear of the cowl but that would make the cowl geometry (which up to now has been a simple lathed profile) much more complex. The alternative is to create an opacity map and treat the wedges as transparent *at rendering time,* leaving the geometry unaltered.

The opacity map (Figure 19.19) is simply the edited bump map from Figure 19.17. By using the same base map, you can be assured that the opacity pixels will correctly align with the diffuse and bump maps. The one change is to sharpen the edges of the wedges so that the transparency begins at a sharp edge. Figure 19.20 show the rollout, new material, and the result of combining diffuse, bump, and transparency maps.

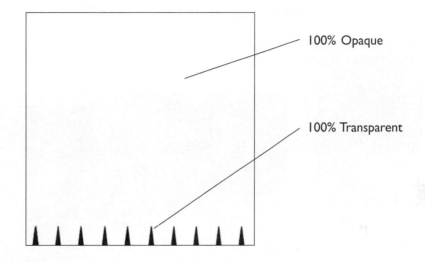

100% Opaque

100% Transparent

Figure 19.19 Opacity map.

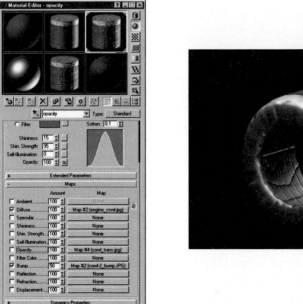

Figure 19.20 Result of applying opacity map.

Environment Maps

As an example of an environment map, consider the bubble canopy for the rear gunner. The glass material can be represented without a raster material by a basic material (color only) and a very low opacity setting. But since one would expect the sky to reflect into the canopy, the file *sky.jpg* has been loaded into the reflection channel having a strength of 30. See Figure 19.21.

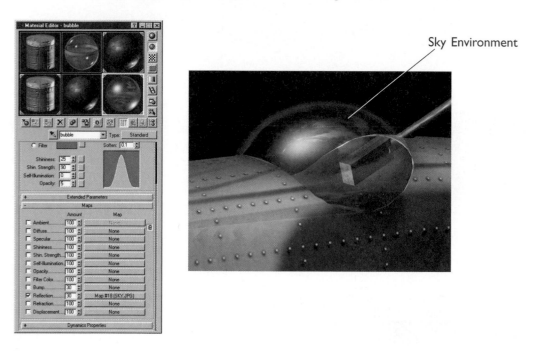

Sky Environment

Figure 19.21   Environment map applied to canopy.

Complete the Mapping

Various mapping techniques were used to complete our Beaufighter. In each case, the emphasis was on keeping the geometry as simple as possible and achieving realist effects through raster maps. For example, you'll notice rivets on the fuselage skin in Figure 19.21. These are part of a realistic material map. Material maps are very flexible in that they can be selectively applied to objects in your illustration as needed. If you need greater detail, you can simply apply a more detailed raster material map to the same basic geometry.

The Tires

Figure 19.22 shows a completed tire and wheel assembly with the tire realistically rendered. The rollout shows that a diffuse "dirt map" has been used to give the tire the appearance of having visited unpaved runways. Additionally, a bump map adds realism to the side wall with raised lettering.

You would, of course, add this level of detail only if the tires were to be featured in the illustration. Otherwise, just use a basic color.

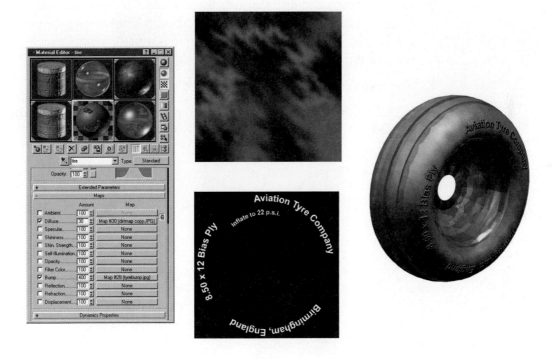

Figure 19.22   Dirt and bump maps applied to the aircraft tire.

Previously, you have seen how UVW mapping coordinates are represented by the *mapping gizmo* and how that gizmo can be rotated to realign the map to the object. The gizmo can also be scaled or moved around the object like a decal. Additionally, you can tell the rendering engine whether or not you want the texture *tiled*. That is, whether you want the texture applied once or repeated across the geometry, and how many times in which U, V, or W directions. You can set tiling in the **Material Editor** but this isn't advisable because you may want the material not to tile somewhere in the scene. Instead, set tiling in the **UVW Map Rollout** for a specific application of the texture.

The Propellers

We are going to show the propellers as if they are spinning. Figure 19.23 shows the propeller geometry, simply a thin cylinder with a swept hub, along with the engine cowl just completed. So how is the thin cylinder turned into the realistic spinning propeller shown in Figure 19.25?

▼ Create a diffuse map showing the blurred propeller. Load this raster map into the **Diffuse Channel**.

▼ Make a copy of the blurred propeller.

▼ Invert the copy and load that into the **Opacity Channel** (the black will now be treated as transparent). That way, everything around the propellers will be clear at rendering time, allowing the engine cowl to be seen through the spinning propellers.

▼ Play with the opacity strength to get the effect you want.

Figure 19.24 shows the **Material Editor** and the two maps, along with the finished propeller.

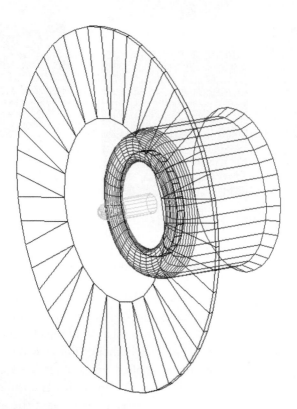

Figure 19.23   Rotating propeller geometry.

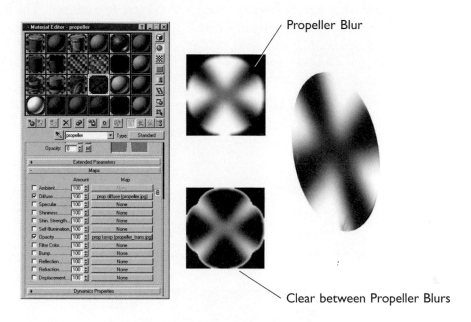

Propeller Blur

Clear between Propeller Blurs

Figure 19.24 Blurred propeller raster material map.

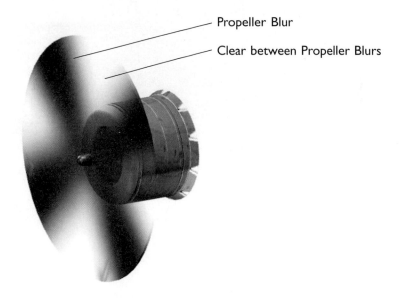

Propeller Blur

Clear between Propeller Blurs

Figure 19.25 Spinning propeller in place.

The Engine Nacelles

Figure 19.26 shows the nacelle geometry, a cylinder that has been edited into a taper. The materials used (Figure 19.27) include a diffuse map, showing blue on top and white on the bottom, along with oil streaks and dirt. The bump map shows large, soft dents so that the nacelle isn't so geometrically perfect.

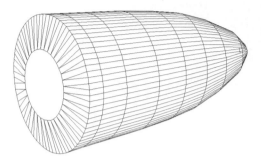

Figure 19.26   Nacelle geometry.

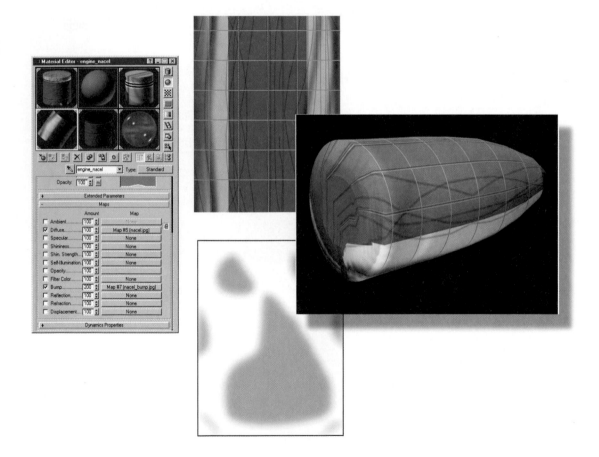

Figure 19.27   Nacelle material and bump maps and their effect.

The Wings

The wings began life as lofted shapes. The cross-sectional profiles were edited until the final geometry shown in Figure 19.28 was arrived at. Like the cowl and nacelles, the wing is blue on top and white underneath, but this time in an irregular pattern. The diffuse map is divided into a top and bottom, and because rivets and their shadows are included in the map, no bump is used (Figure 19.29). Because the wing essentially has a top and bottom and no sides, **Planar Mapping** wraps the wing material around the geometry.

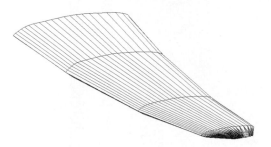

Figure 19.28   Wing geometry.

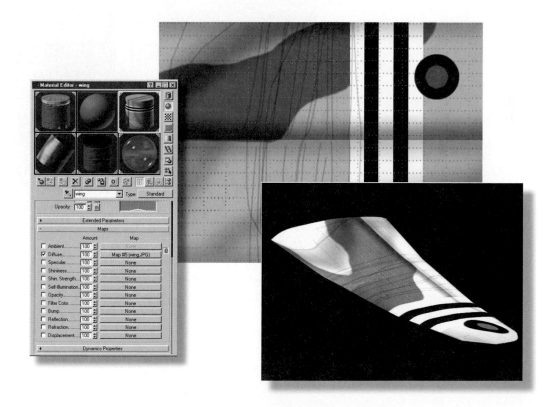

Figure 19.29   Wing material map and its application.

The Fuselage

Because the fuselage is essentially a cylinder, cylindrical mapping is appropriate. The geometry was previously shown in Figure 18.31. The material is a single diffuse map, with joints and rivets painted on the surface (Figure 19.30).

Figure 19.30   Fuselage material map and its application.

The Tail

The tail is somewhat problematic because it doesn't adhere to a regular geometric shape. After some experimentation, the best results were achieved by using a **Planar Map** on the mostly planar geometry (Figure 19.31). The map tiles over the top and down the other side of the tail (Figure 19.32). A caveat: this results in an inverted image on one side, but because the insignia is so small, it is not noticeable. If you want to correct this, copy the material, double the width of the map's canvas, and then flip the copy over. This forms a mirror of the map for the other side.

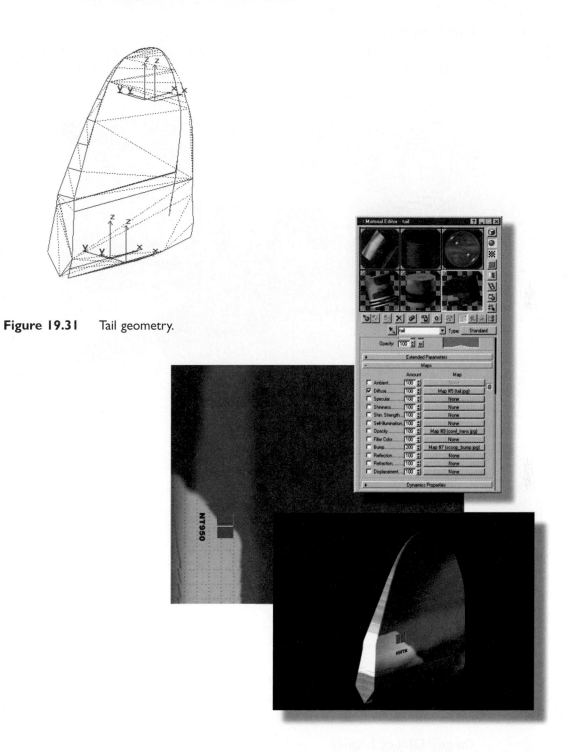

Figure 19.31   Tail geometry.

Figure 19.32   Tail material map and its application.

The Completed Illustration

The final illustration is shown in Figure 19.33. In the next sections, we will discuss how the scene was lighted and what needs to be considered when rendering the final output.

Figure 19.33   Final model with material maps in place.

Materials and Lights

A simple suggestion covers any discussion concerning lighting: keep lights to a minimum. The easiest way to ruin an illustration is to overilluminate it, washing out all the detail you worked so hard to create.

Several lighting objects are available to illuminate your illustration. By selecting the appropriate lights, their placement, and strength, the geometric feature of your subject can be shown crisply and with detail.

Omni Lights

These lights illuminate evenly in all directions from a point, like a lightbulb without a shade.

Target Spotlights

These lights project a cone of illumination much like a theatrical spotlight. The edges of the spotlight can be adjusted to yield a hard or soft edge.

Target Direct Lights

These lights project a cylinder of illumination much like the sun projects parallel light rays from a great distance. An important aspect of lights is the ability to exclude geometry from illumination. By carefully excluding objects, you can have lights impact exactly what, where, and how much light you want. Lights can also cast or not cast shadows. Without shadows your illustration will appear flat. With shadows too strong, all detail in the shadow will be lost. So you can see that effective lighting can often spell the difference between successful and unsuccessful 3D illustration.

Follow these guidelines when setting up lights:

▼ Create a strong overhead omni or target direct light to act as general illumination. Outdoors, set the light above the scene. In a room, place it as if it were a ceiling light.

▼ Set a target spotlight behind and above the viewer to illuminate the focal point of the scene.

▼ Set another omni light or spotlight behind the scene and off to the side. This provides the back lighting necessary for cylindrical or spherical geometry. This light can be light blue to represent sky light or orange to represent the sun.

▼ Render the scene and see how it looks. Adjust the brightness of the lights to get the detail you desire.

▼ Now, set additional lights to reveal detail that would normally be hidden. Exclude objects that you don't want additional lighting on.

Figure 19.34 shows our airplane scene with lights set and identified. Figure 19.35 shows the landing gear detail with an extra omni light set specifically to illuminate the gear and tire. As you might imagine, the strong omni light and spotlights directed from above would keep any of the detail under the wings in deep shadow. The detail omni light placed under the wing (Figure 19.34) solves this problem.

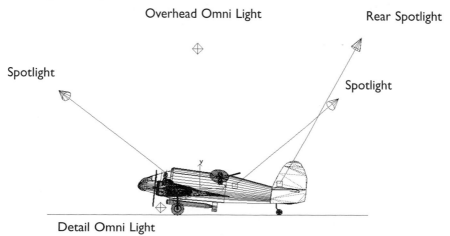

Figure 19.34   Lighting scheme for the airplane model.

Figure 19.35   Lights should be positioned to reveal specific features.

Rendering Raster Images

Rendering quality is a function of several variables: resolution of material maps, effectiveness of lighting, resolution of the rendered frame, and rendering method. To say the least, the higher the resolution of both material maps and final output, as well as the number of lights and sophistication of material rendering method (flat shading vs. ray tracing, for example), the longer the rendering time. Remember that the geometry in the 3D scene is resolution independent. You can always re-render at a higher resolution.

Also keep in mind that resolution isn't the same (though it's directly related to) pixel density. Every rendering is essentially completed at 72 dpi. If you later scale the rendering to 50 percent, you'll have a 144-dpi image. Unfortunately, the physical size of the image will be cut in half. So to plan your rendering you need to know:

▼ The physical size of the final illustration (say, 10 x 5 in).

▼ The resolution (pixel density) of the final rendering (say, 300 dpi)

Multiply the final size by the final resolution and you get the rendering dimensions (10 x 300 = 3000; 5 x 300 = 1500). So to get a 10 x 5 in rendering at 300 dpi you eventually have to render your 3D illustration at 3000 x 1500 pixels.

Quick Render First

A quick rendering at 160 x 120 should tell you everything you need to know about the materials, shades, and shadows of the scene. You can hide a number of lights to minimize calculations. The quick rendering command takes its parameters from the full render scene command.

Set Individual Cameras

To experiment with small changes in viewpoint, establish several cameras to produce different views. This is preferable to freely moving the camera because it is difficult to return to an exact vantage point (unless you note the world XYZ location of camera and target for each position).

Animate the Camera

If you animate the camera, you can render out each frame in the camera's path, as if you have a motor drive on your camera.

Final Rendering

The default aspect ratio for rendering is 1.333:1 (640 x 480, 800 x 600, 1024 x 768, etc.). Additionally, other formats can be chosen as needed (Figure 19.36).

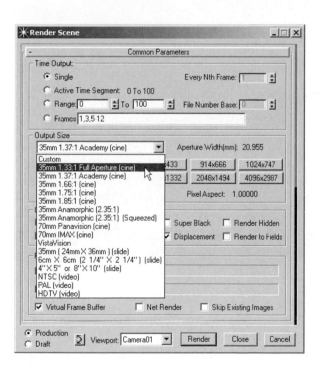

Figure 19.36   Choose the appropriate output aspect ratio.

Matching the appropriate file format (Figure 19.37) to the desired aspect ratio is important to minimize unnecessary work later. For example, NTSC video may expect frames in TGA format while publications accustomed to reviewing 35-mm slides (1.333:1) would appreciate TIF for final high-quality printing.

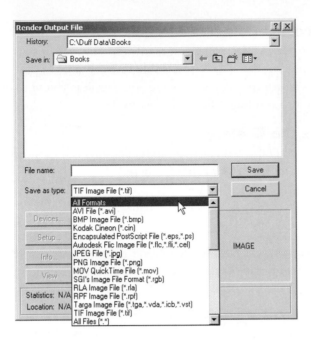

Figure 19.37   Match the data file format to the output size.

Because the raster output is of fixed resolution, the relationship between output size and final image resolution is inverse: the larger the output size, the lower the resolution. Let's look at an example that demonstrates this relationship.

Figure 19.38 displays four renderings, the first done at 160 x 120, and then three at double the output size: 320 x 240, 640 x 480, and 1280 x 960. By scaling each to the same overall dimensions, effective resolution has been increased. Remember that as scale is halved, resolution doubles. The question is: At what point does increasing resolution have no effect on final output quality? Consider the file sizes in Table 19.2.

▼ Bitmap and File Size

Bitmap size	160 x 120	320 x 240	640 x 480	1280 x 960
File size	76Kbyte	301Kbyte	1.2MByte	4.8MByte

Table 19.2   File size increases exponentially as resolution increases.

All files in Table 19.2 are of 24-bit CMYK TIF format. Notice the dramatic increase in size as resolution is doubled from 144 to 288 ppi. Can you see a visual difference in Figure 19.38 that justifies the increase in storage, rendering time, and printing?

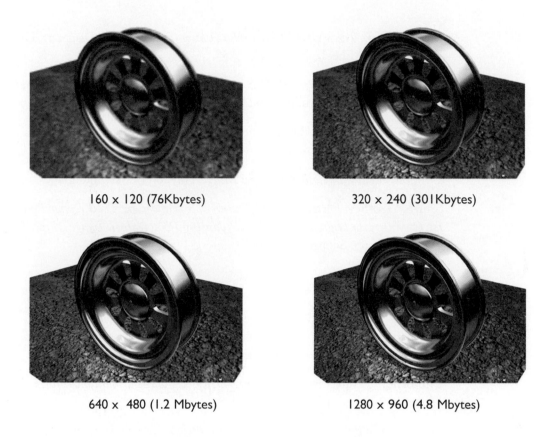

160 x 120 (76Kbytes)   320 x 240 (301Kbytes)

640 x 480 (1.2 Mbytes)   1280 x 960 (4.8 Mbytes)

Figure 19.38   Scaling the rendering increases the output resolution.

Review

Raster materials, when intelligently applied to accurate geometry, result in effective technical illustrations. To create detailed raster materials you need to develop skills in a raster editor, such as Photoshop, but you may find that some of the best textures are available from nature. A digital camera or scanner and some experience in Photoshop will open an unlimited source of raster textures. When diffuse, environment, bump, and opacity maps are combined, a high level of realism can be achieved. Placing appropriate lights in a 3D scene is crucial to achieving realistic results. Final rendering output size and data formats must be matched for the intended use of the illustration for optimal results.

Text Resources

Brodatz, Phil. *Textures: A Photographic Album for Artists and Designers.* Dover Publications. Mineola, New York. 1999.

Flenning, Bill. *3D Texture Workshop: Painting Hollywood Creature Textures* ,Volume One. DAZ Productions. Draper, Utah. 2002

Horton, William. *Illustrating Computer Information.* John Wiley & Sons. New York. 1991.

London, Sherry. *Photoshop Textures Magic.* Hayden Books. Indianapolis. 1997.

Smith, Geoffrey. *Photoshop 5 3D Textures f/x and design: The Premier Resource for Creating 3D Digital Realities by Producing Photorealistic Image Maps.* The Coriolis Group. Scottsdale. 1998.

Internet Resources

Instruction on making textures. Retrieved from: *http://www.wellesley.edu/CS/courses/CS215/handouts/12-textures.html.* January 2003.

Photoshop textures. Retrieved from: *http://www.siteformat.com/tutorials/photoshop/textures/.* January 2003.

Seamless tiling textures. Retrieved from: *http://www.indiaeye.com/free.htm.* January 2003.

Texture resources. Retrieved from: *http://www.allfreelance.com/textures.html.* January 2003.

Texture tutorials. Retrieved from: *http://www.photoshoproadmap.com/photoshop-tutorials-tips/textures-backgrounds.html.* January 2003.

On the CD-ROM

For those of you who would like to try out the procedures described in this chapter, several resources have been provided on the CD-ROM. The max files for the Beaufighter are in *exercises/ch18.* Look in *exercises/ch19* for the following directory of texture maps:

/texture_maps   This directory contains the JPG texture maps for the exercises in this chapter.

Axonometric Projection

Perspective Drawing

Technical Animation

Technical Illustration

Rendering

3D Modeling

Technical

Chapter 20

Animation and Technical Illustration

In This Chapter

The advent of computer modeling and rendering has changed the way illustrators think about illustration. Previously (in the manual-analog world), illustrations required extensive planning because once you committed to a certain view (such as isometric, dimetric, trimetric, or perspective) or rendering technique (like stipple, cross-hatch, or airbrush), it was prohibitively expensive to change. Digital illustration has changed that.

Simply moving the camera to a new location and setting the camera's target at an appropriate position in space can generate *any view* of 3D digital geometry. *Any material* can be applied to geometry and the scene re-rendered. If a bearing changes from sintered bronze to high silicone aluminum, no problem.

Likewise, many traditional technical illustrations were static. Assembly of parts were shown by spatial overlap and flow lines. Sometimes the flow lines became so convoluted that the illustration looked more like a wiring diagram. At best, illustration elements could be copied and moved along the flow line, with a movie frame exposed at each position, creating a crude animated assembly. But with modeled geometry, movable in space, true technical animations are simply a logical outcome of the modeling-rendering-illustrating-animating process.

We will use the two projects from Chapters 16 through 19 (the pulley assembly and Beaufighter) as examples of how 3D technical illustrations can be effectively animated. In the first example, the pulley assembly (Figure 20.10) will be animated so that its assembly can be observed. In the second example, the Beaufighter from Chapter 18 will be placed into an environment so that its operation (taxi and takeoff) can be animated.

Chapter Objectives

In this chapter you will understand:

▼   How an effective environment can add to the success of an animated illustration

▼   How animation enhances existing static 2D technical illustrations

▼   Animation file formats and how technical illustration animations can be integrated into multimedia training and Web documentation

▼   Camera parameters in 3D animation

▼   Lighting parameters that make understanding technical animations easier

▼   How to plan an animation using a *frame planning sheet*

▼   How to render the animation into an appropriate format

Effective Technical Illustration Environments

Most traditional technical illustrations pretend that a white object is hung in space in front of a white curtain. This focuses attention on the parts themselves, which is good. But without an environment, it is often difficult to determine scale and spatial positioning (Figure 20.1). Parts that overlap (because they are white and the background is also white) might be interpreted as being the same part. Parts whose edges unfortunately align can be confused as being the same part. Chapter 10 covered techniques that increase spatial discrimination in 2D drawings. If you would like to see more examples of 2D techniques for creating spatial cues, review that chapter.

An environment might be as simple as a grid or colored panel behind the objects, building depth into the scene. The gray panel behind the parts in Figure 20.2 builds another layer of depth and helps visualize object limits. Notice that some parts hang out over the panel, further bringing the parts to the front. Even a small

▼ Tip

Include only that portion of the *entourage* (the surrounding environment) necessary to place the subject of your illustration in context. Remember what you are trying to accomplish with the illustration. Too much environment takes away from the impact of the subject.

portion of the actual environment can aid both in depth and scale perception. The architectural illustration in Figure 20.3 includes a portion of the *entourage*, the environment surrounding the structure.

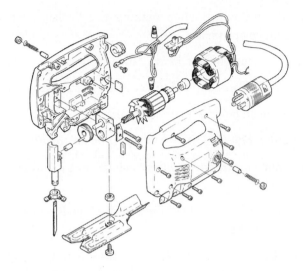

Figure 20.1   Traditional illustrations omitted the environment in which the parts were viewed (Tim Asher).

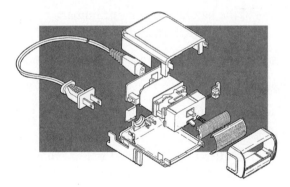

Figure 20.2   Simple environmental techniques help in depth perception.

Figure 20.3   Placing the illustration in its natural environment can aid in spatial perception (Greg Maxson).

Some illustrations don't require a realistic environment while others do. For example, trucks, trains, cars, planes, and boats are hard to visualize outside the environment in which they operate, while small electronic parts would get lost if you put them *in situ*.

Consider the following when making the decision whether to place your illustration in an environment:

▼ Will the environment add or detract from the function of the illustration?

▼ Does an environment add unnecessary complexity to the illustration?

▼ Is creating an environment economically viable?

▼ If an environment is necessary, can it be abstracted?

▼ How much of the environment will be seen in either a static or animated illustration?

▼ In the case of an animated illustration, how long will any portion of the environment be visible?

In a modeled and rendered technical illustration, the environment can be thought of as a diorama or theatrical stage and the viewer of the illustration, the audience. This means that you can employ all the techniques available to these practices:

▼ Make use of the natural tendency of humans to read perspective into a scene. Objects that are further away are smaller, paler, and bluer.

▼ Store scene elements off stage (out of the camera's view) until needed.

▼ Model or render only those parts of objects that will be seen.

▼ Simulate objects and depth by using 2D mattes (paintings or digital photographs).

▼ Hide parts of the scene that you don't want to model or render in shadows or behind foreground geometry.

Animation's Function in Technical Illustration

Animations were avoided when technical illustration was done manually because an additional skill set was required and the process was not (usually) economically justifiable. The same can be said of 2D digital illustrations. By their training, technical illustrators have not been cinematographers or videographers because these two functions are totally unrelated to making technical illustrations. However, with 3D digital illustration, the tendency (unfortunately) is to animate everything that is modeled. Everything is chrome (ray traced) and moving, exploding, and on fire. Why? Because you can set it up, turn it on, go to lunch, and when you come back (hopefully), it's done! But embrace this tenant:

Animation's function is to illustrate what can't be effectively shown by a static image.

There is still a considerable overhead with animated illustrations. Animations require more exacting modeling because you'll be seeing more of the objects as they or the camera move. You'll need more realistic material mapping because you'll be zooming in and looking at details not normally visible. You'll be spending more time actually rendering the product (because you'll be rendering multiple frames), and you'll be spending more capital—time and money—because you'll want to render the animations in the shortest time and store them on acceptable media.

Animation File Formats

Chapter 1 covered raster and vector file formats in detail. Animation formats stem from these static formats, while several animation formats are specifically designed for time-based events. Many proprietary animation formats exist but you should always match the file format to the illustration's use, and your client's needs.

▼ Note

Always capture your illustration (static or dynamic) at the greatest resolution, bit depth, and in the case of animations, frame rate. Consider this the baseline data from which less robust illustrations and animations can be sampled.

If you need greater technical information than is provided here, see the previously referenced Murray and VanRyper's *Encyclopedia of Graphic File Formats*, published by O'Reilly & Associates. This book is out of print but is available used at *Amazon.com*.

Graphic Interchange Format (GIF)

This format allows multiple images to be stored in a single file. These images can then be displayed in sequence, creating an animation (animated GIF). GIF is limited to 8-bit color and automatically uses LZW lossless compression. Additionally, GIF animations contain a separate data channel that can be used to signify a single color to be transparent. GIF does not store audio data.

Uses: Small animations that have large areas of solid colors or values (like the white background of a traditional line illustration). Because of its automatic compression, GIF animations are appropriate for distribution over the Internet and viewing in all graphical Web browsers. Most multimedia software accept GIF animations.

Audio /Video Interleave (AVI)

Microsoft's multimedia format uses a single track and interleaved "chunks" of audio and video data. AVI is capable of holding 24-bit color and 8-bit audio. It implements either lossless [(run-length encoding (RLE)] or lossy (video) compression, but many applications implement only the RLE method.

Uses: Production-quality visual animations with less than production-quality sound control. Appropriate for any animation delivered on Windows platforms or over the Internet.

QuickTime (MOV)

QuickTime is Apple's format for storing audio and motion video data. It features multiple lossless and lossy compression options. QuickTime records up to 24-bit true color and 8-bit sound.

Uses: Production-quality animations requiring separate audio and video tracks. Windows machines must have *QuickTime for Windows* installed. Web browsers must have the *QuickTime Plug-in*.

Motion Picture Expert's Group (MPEG)

A robust audio and video format capable of 24-bit color, MPEG compresses both audio and video in a single synchronized data stream. Data is compressed within each frame and between frames using predictive encoding. MPEG records only the change from one frame to the next. Either hardware or software CODECs are required to create or view the animation.

Uses: Production-quality animations delivered on CD-ROM or streamed over networks from video servers.

FLASH

This is not a generalized animation format, but rather a format used by Macromedia's Flash animation application. Flash creates resolution-independent vector graphic frames that can be delivered over the Web or packaged with the FLASH engine for stand-alone playing.

Uses: Resolution-independent stand-alone animations, over the Web, or within multimedia such as Macromedia's Director or Authorware.

Animation Cameras

If you are familiar with traditional 35-mm cameras, you will understand how computer cameras respond during animation. However, because you can control any and all parameters associated with cameras, you may be tempted to manipulate too many of the parameters, resulting in a video with unnatural movement and display.

Keep the following in mind when using cameras during animations:

▼   For normal human vision, use focal lengths of 50 to 75 mm.

▼ Avoid short focal lengths (10 to 20 mm), especially for small objects. This tends to introduce excessive perspective distortion into the scene and will make small objects appear large and out of scale.

▼ Keep camera movement to a minimum. Let objects move within the scene.

▼ Move the camera's target first, and then the camera itself. Moving the target is like moving your eyes. Moving the camera is like moving your head.

For example, Figure 20.4 shows the shaft from the pulley assembly. You can hold this part in your hands. See how the appearance of the object changes dramatically as the camera's focal length is changed from telephoto (axonometric) to wide-angle (extreme perspective).

In Figure 20.4 the **Orthographic Projection** feature in 3ds max's **Modify|Camera Rollout** is unchecked. The 50-mm focal length results in a field of view (FOV) of roughly 39 degrees, about that of human vision. In Figure 20.5 this orthographic feature has been checked. By doing so, all perspective has been removed, essentially creating a camera of infinite focal length. The orthographic image is slightly larger than the perspective image (because the slight convergence of the 50-mm lens was removed). The reasons that the two images appear almost identical are that:

▼ The object is fully contained in the FOV.

▼ The FOV approximates that of natural vision.

▼ The object is small, reducing the potential of perspective convergence.

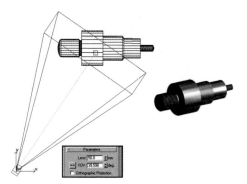

Figure 20.4   Perspective camera.

Figure 20.6 uses a wider-angle lens (25-mm focal length and 71-degree FOV) making the object appear much larger because of increased perspective convergence. Notice that the object now has significant perspective convergence and that the camera has been moved much closer to the subject.

In Figure 20.7 the focal length has been further reduced to 15 mm. The FOV is now over 100 degrees, introducing unacceptable perspective distortion, making the object appear curved. Only really large objects (like buildings) require such a wide angle to be totally in the FOV at close range.

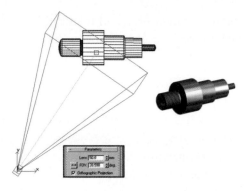

Figure 20.5   Axonometric camera.

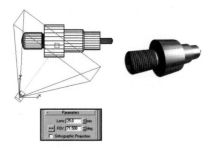

Figure 20.6   Wider angle (FOV) creates greater perspective convergence.

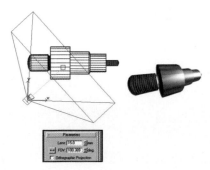

Figure 20.7   Extremely wide FOV creates unacceptable distortion.

Light the Animation

Lighting an animation is more difficult than lighting a single scene illustration because objects in the scene may move out of one illumination source and into another. Several hints:

▼ Start with a single source of illumination. If you want the scene generally illuminated, use a strong omni light with a multiplier set at 1.5 or greater. If you want a particular part of the scene illuminated, use a target spotlight.

▼ Add sources of illumination to accomplish specific visual results, like casting shadows to show contour or to illuminate certain features.

▼ Don't add lights that conflict with your overall lighting scheme. For example, don't add spotlights that cast shadows in directions different from your main lighting source.

▼ Make extensive use of excluding objects from illumination or shadows. Make your lights illuminate only what you want them to.

▼ Attach lighting sources to objects (and exclude all other objects) so they remain constantly illuminated no matter where they are in the animation.

Figure 20.8 displays an object lit by a single overhead omni light that's in front and to the left. Light radiates in all directions evenly from an omni light, so it is appropriate for lighting a scene where the camera's target moves around.

▼ Note on Lighting

In Figure 20.8 the shadow is cast below and to the rear of the pulley. You want to avoid placing the light behind the subject and looking into the shadow because little detail can be seen. Of course, this is a problem in animation because your point of view may change. The solution is to illuminate the back side of the subject with a light that excludes all else, so that the shadows from the primary light source are not confounded.

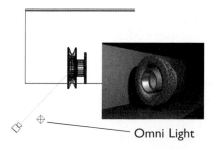

Omni Light

Figure 20.8   An omni light radiates light equally in all directions.

Figure 20.9 shows the same scene illuminated by a target spotlight. Note the difference in shadows cast on the ground and wall and the level of illumination outside the falloff area of the spotlight.

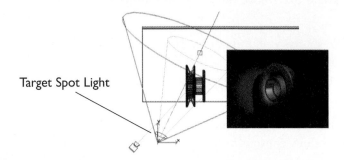

Target Spot Light

Figure 20.9   A target spotlight creates a hot spot and illumination falloff.

Plan the Animation

Most young illustrators are amazed when they realize the amount of planning expected before animations are actually begun. In fact, it is advisable to plan the animation *before* beginning *modeling* and *mapping*. It would be a shame to spend hours or days modeling sections that won't appear in the animation because they are hidden by other parts, are always out of the FOV, or are at a distance where they wouldn't be distinguished. Likewise, developing intricate material maps for those parts would also be a waste of time.

Planning animation is a three-step process. The more significant the animation, the more important are the first two steps.

Step 1: Key Frame the Action

A *key frame* is a place in an animation where the motion (of objects, cameras, or lights) is completed. A storyboard records, in sketch form, the appearance of the animation at each key frame. Every animation is different, so there is no set number of key frames or storyboards.

The animation program will interpolate movement (determine in-between positions) from one key frame to another. Movement between key frames can be edited to fine-tune the action.

Step 2: Plan the Action

A frame-planning sheet (on the CD-ROM in PDF format) refines the animation one step further. On this sheet, the position of every object in the scene (geometry, lights, and cameras) is planned. Because this is done on a time line, beginning at the first frame and ending at the last, the status of every object can be established at any frame in the animation.

Step 3: Execute the Animation

After the animation has been key framed and the behavior of all elements in the animation planned frame-by-frame, the actual animation can begin. Because you have a preview of the animation with the storyboards, you now know which geometry can be abstracted or omitted all together. You also know for how many frames certain features will be in the camera's field of view and at what distance. This provides information for the level of model detail (low polygon count) as well as the required detail of material mapping.

Finish the Animation

You may be animating a scene that already formed the basis of a static technical illustration, and in that case, you probably have over-modeled and over-mapped the objects. In this case, after planning cameras and lights, the animation becomes a value-added bonus of the original 3D illustration.

Or, you may be starting from scratch. Consider the following suggestions to increase your efficiency:

▼ Build the scene from large to small, from gross to specific. This follows a theatrical model of stage design. For example, if your animation takes place in a room, build the room first and set the camera to match the storyboard for frame 1. Then populate the room.

▼ Use *proxies* to plan the animation. If you are working as part of a team, develop a naming convention for parts in the scene. The animation can proceed while objects are actually being built and the final objects substituted (merged) into the animation when completed.

▼ Render out test frames that correspond to your key frames. This allows you to check the appropriateness of your initial planning, as well as the effectiveness of your modeling, lighting, and mapping.

▼ Test render the animation at small size (160 x 120), sampling the frames. If you have a 120-frame animation, render every tenth frame at 160 x 120 to get an idea of how things are going.

▼ Run the animation in a player. This let's you control the action frame by frame, forward and backward.

Render the Animation

The final animation is rendered frame by frame at the desired resolution (bitmap dimension) and in an appropriate format. In 3ds max this essentially means AVI or MOV formats.

Employ a rendering algorithm and material type appropriate for the scene. For example, if the scene contains only matte, dull materials (concrete, fabric, and soil), there is no reason to use ray tracing for materials or scene rendering. If objects only appear in the distance, there may not be any reason for them to be material mapped at all—a simple default color may suffice.

▼ **Note**

Other options exist for rendering the animation, based on your needs and capabilities. For example, the animation can be rendered as a series of Targa (TGA) frames and edited into an animation by Truevision-compatible hardware. The TGA format has the greatest color depth of the supported file types as well as access to 16 bits of alpha channel (for things like visual effects).

Use the Animation

Animated technical illustrations can be used in many ways. A perusal of the Web sites listed at the end of this chapter reveals that animated technical illustrations are used across business, industry, the legal system, science, medicine, and the military. These companies use technical animations for:

▼ *Enterprise visualization.* Mock-up and analysis of manufacturing and systems solutions

▼ *Digital prototyping.* Design and testing of manufactured and constructed products before committing to production

▼ *Manufacturing process simulation.* Simulation of production activities for optimization

▼ *Collaborative design.* Shared design data that encourages group design activities

▼ *3D animated evidence in litigation.* Creation or re-creation of visual displays used in trials

▼ **Tip**

If you're not ready for a product such as 3ds max, consider Micrografx's Simply 3D. (Micrografx is now subsumed under Corel and is no longer directly supported. However, you can get Simply 3D from *Qualitylinks.com* or *Amazon.com*.) This tool will get you into the majority of lights, cameras, motion, and rendering for under $30.

▼ Illustration in Industry

Richard Tsai / EDAW, Inc.

Irvine, California

Images Copyright 2002 by EDAW, Inc.   TsaiR@edaw.com   www.edaw.com

Software

3ds max 5.0
AutoCAD 2002
Forest plug-in
Adobe Photoshop
QFX

Hardware

Pentium x 2 1000
Wacom Tablet

EDAW Inc. was asked to produce visual simulations to aid decision-makers in the widening of a major residential thoroughfare in the City of Irvine, California. Engineering CAD data, architectural sketches, verbal descriptions, and 3ds max were combined to produce models that provided a high degree of precision and realism in representing design details. Additionally, the combination of modeling and photography allowed several design alternatives to be entertained. The accuracy of compositing rendered models over digital photographs was critical in producing reliable, defensible public exhibits. The portrayal of correct tree species and maturation were particularly important elements considered in the landscape design. Photoshop and QFX were used for blending and retouching the final images. As tools for decision making in the public sector, visual simulations such as these have proven invaluable for adjacent residential communities.

The Pulley Example

The pulley (Figure 20.10) is comprised of parts on either side of the pulley itself. The pulley file (*pully.max*) can be found on the CD-ROM in *exercises/ch20*. Some of the parts are made from a light reflective material, making it difficult to distinguish them from the light background. Likewise, were the camera to remain in the current position throughout the animation, the parts on the right side of the assembly would disappear behind the pulley body and not be seen. To solve these two problems, the following strategy is used:

▼ Choose the background wisely so that individual parts can easily be seen.

▼ Perform the animation in two stages. Assemble the left-hand parts first, and then the right-hand parts.

▼ Move the camera from left to right as the parts are assembled.

Figure 20.10   The pulley assembly contains parts of various materials, making an environment choice difficult.

Create an Environment

Place the pulley assembly in an environment that makes it easier to distinguish the individual parts. Figure 20.8 shows four possibilities. In Figure 20.11, shadows are used to bring the parts off a background. In Figure 20.12, a grid is applied to the background to further distinguish the parts, and in Figure 20.13 the shadows are recast. In Figure 20.14, a darker background is used to better contrast with the lighter-hued parts.

Shadows alone are marginally effective in creating greater contrast between the parts and the background because shadow shapes and values may be similar to the parts. In Figure 20.11, the shadow is distorted by moving the omni light source off to the side. This has the added benefit of *chiaroscuro*, a technique used to place areas of light next to areas of shadow, and areas of shadow next to areas of light. Note how the small pieces on the right have better contrast with the background where they are in front of the shadow.

A grid background might be effective because a grid is geometrically different from the parts (Figure 20.12). Adding shadows to the grid (Figure 20.13) further increases depth perception. Finally a dark background with a white grid, like that in

Figure 20.14, can be used. Unfortunately, the darker background fights for attention with parts in the assembly that are darker than aluminum. The best solution is to combine the grid and shadows as shown in Figure 20.13.

Figure 20.11   Shadow on wall.

Figure 20.12   Background grid.

Figure 20.13   Shadow and grid.

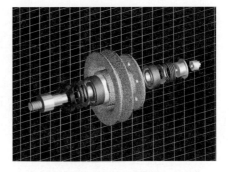

Figure 20.14   Dark wall and light grid.

Design the Animation

For the purpose of this example, we are going to design the animation using the following parameters:

▼ Frame size: 320 x 240

▼ Frame rate: 15 frames per second

▼ Length: 5 seconds (75 frames total)

Using a frame planning sheet (Figure 20.15), the motion of the assembly is designed. Note how the pulley body itself remains stable while parts move along the assembly axis. Parts on the left side assemble over frames 4 to 35 while the camera and its target traverses left to right. Parts on the right side assemble over frames 40 to 69, while the camera remains in position on the left side of the pulley. The five frames at the beginning and end of the animation provide a rest before the motion starts and stops. Let's see how to do this.

Figure 20.15   Frame planning sheet for the pulley animation.

Set Up the Animation

Load the file *pulley.max* from the CD-ROM. In 3ds max, click on the **Animate** button in the bottom menu (it will turn red). Then right-click on the animation controls. You should see the **Time Configuration** dialog box (Figure 20.16). Click on **Custom** to set the frame rate and animation length. Note that the **Animation Slider** (Figure 20.17) now reads 0/74, signifying that the animation is sitting on frame 0 of 75 frames.

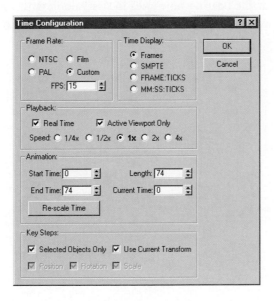

Figure 20.16   **Time Configuration** dialog box allows you set the frame rate and animation length.

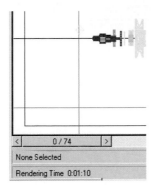

Figure 20.17   **Animation Slider** reflects current and end frames.

Key Frame the Parts

Using information from the frame planning sheet, click the **Animate** button and move the **Animation Slider** to the frames at which movement for each part is completed; move the parts individually to their final positions at that frame position. Do this first for parts on the left side and then the right.

Figure 20.18 shows the left-side parts in their assembled position at frame 35. You can have the parts all move together (the easy way and not too realistic) or stagger the parts so that the last parts wait on the first before they move. To do this, go to the **Track View** and move the beginning and end points of each part to the frames where you want the movement to start and stop (Figure 20.19). Refer to the frame planning sheet.

You can see that the bearing now goes in first, along with the shaft, followed by the spacer, felt washer, and cap—just like the parts would be assembled at the factory!

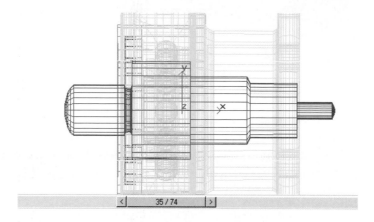

Figure 20.18   At frame 35 all left-side parts have moved to their assembled position.

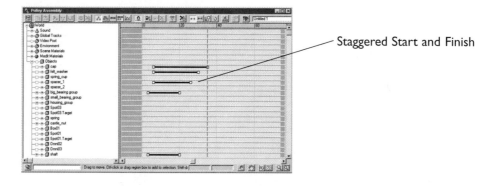

Staggered Start and Finish

Figure 20.19   Parts in the **Track View** with their starting points staggered so that the last parts wait on the first parts before starting assembly.

▼ **Note**

Work in the front view and constrain movement to the X axis. This way, you can assure that the parts move along the assembly axis.

Continue with parts on the right side by moving the frame slider to frame 69 and positioning the small bearing, spacer, spring, spring cup, and hex nut into their final assembled positions (Figure 20.20). Adjust their start and stop frames in the track view as you did with the left-side parts. From Figure 20.20 you can see that all the right-hand parts start their movement together but only the nut completes its movement at frame 69.

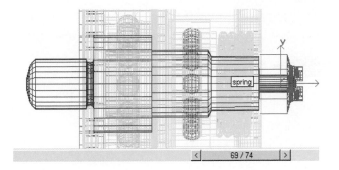

Figure 20.20   At frame 69 all right-side parts have moved to their assembled position.

Figure 20.21 shows all the parts in the **Track View**. You can see that all the right-side parts begin moving together at frame 35 but complete their movement in order, with the castle nut coming to rest at frame 69.

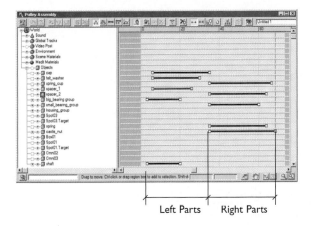

Figure 20.21   At frame 69 all right-side parts have moved to their assembled position.

The final step is to animate the camera, from its starting position on the left side of the assembly to the middle of the assembly at frame 35, staying there until frame 70.

Work in the Top View

By working in the top view, no up and down movement (Y axis) is introduced into either the camera or its target. Click on the **Animate** button and move the frame slider to frame 35. Move the camera to a position slightly to the right of the pulley body. Move the camera's target to the center of the assembly (Figure 20.22). Keep the camera the same distance from the parts. Leave the camera at the same position for the second half of the assembly. Change to the camera view and play the animation. Adjust the starting and ending frames of individual parts as necessary.

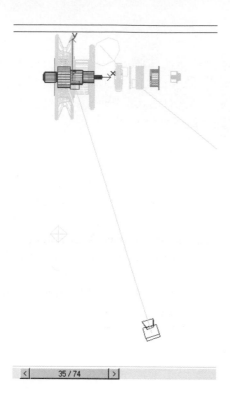

Figure 20.22   Camera and its target have moved into position to view right-side parts by frame 35.

Render the Animation

Once the animation performs, as you want, you may want to render the key frames (4, 35, and 69). If these look right, render all the frames to an AVI file. By default, the renderer will dither the images in true color, appropriate for our animation because of the smooth transitions of the metallic surfaces. When you choose **Rendering|Render** from the menu, set frame size to 320 x 240 and check **Active Time Segment (0-74)**. Check **Save File** and click the **Files** button to get the file into AVI format and into an appropriate directory.

You will be presented with choices for compression. Refer back to the section on file compression in Chapter 2 for in-depth discussions on compression techniques. If you have time, you may want to render out the pulley animation at high, medium, and low compression settings to see the differences in quality and file size.

Depending on your computer, it may take from 20 to 120 minutes to render 75 frames. The final animation (*pulley.avi*) can be found in on the CD-ROM in *exercises/ch20*.

The Beaufighter Example

The airplane from Chapters 18 and 19 required more extensive modeling and mapping than did the pulley we just animated, so as you might imagine, its environment would also be more difficult to make. In a way it is, but because the Beaufighter (Figure 19.33) is the star of the animated illustration, its environment can contain less detail than you might imagine and still look realistic.

Make the Physical World

The airplane will operate in a nearly 360-degree world, so we want to be able to animate and illustrate the plane during its taxiing, run-up, and takeoff.

Start by creating a hemisphere (Figure 20.23) that will serve as the sky for our world. Check the scale of the plane so that the hemisphere is large enough. (This isn't really that important because your world can always be scaled later as necessary.)

Below the dome of the sky is a thin cylinder the same diameter as the dome that will serve as the ground for our world. It has been moved away from the dome so you can see it but will be moved back for the remainder of the example.

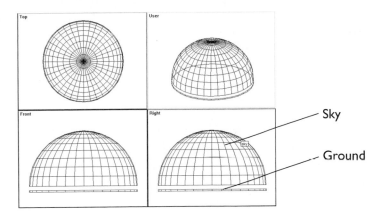

Figure 20.23   Hemisphere and ground for the airplane environment.

Set Natural Lights

Because this is an outdoor daylight scene, you'll need fairly strong lighting. Start with two strong omni lights (**Strength=2.0**) and an overhead spotlight (Figure 20.24) at the top of the dome (inside, or it won't shine on our world!).

▼ **Note**

You will want to experiment with lighting combinations in order to achieve realistic outside lighting. This is difficult because with computer illustration and modeling tools, you can't get your light source as far away as the sun. You want to light the entire scene evenly, but with a strong shadow-casting light source overhead.

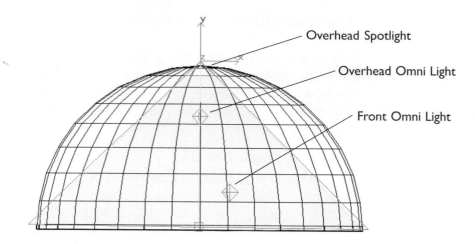

Figure 20.24   Strong overhead lighting provides natural lighting for the scene.

Set a Camera

Inside the dome, create a camera slightly above the ground so that the sky and the ground can be seen at the same time (Figure 20.25).

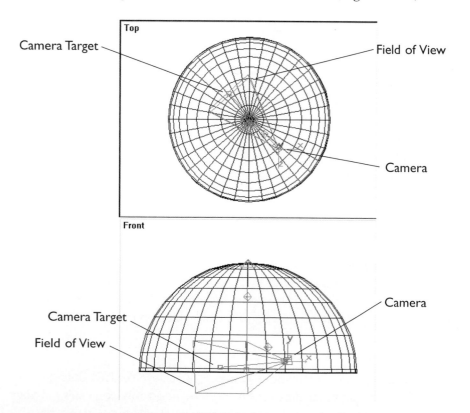

Figure 20.25   With the camera set low, a view of both the ground and the sky is achieved.

Apply Materials

To get an idea what the scene will look like, apply the material *sky.jpg* to the dome (use spherical mapping coordinates), and *grass.jpg* to the ground (use planar mapping coordinates). These materials are in the 3ds max materials library.

▼ Note on Tiling

The grass material is not a seamless texture. However, if you set tiling to repeat a sufficient number of times, the grass texture will be both small enough and rough enough to make the individual tiles indistinguishable. The sky also doesn't tile seamlessly. If the camera will be making a 360-degree pan, this *would* be a problem.

Render the Camera View

As a test, render the camera view port. If you are going to be animating the scene with a fairly static camera (as we are doing), you will want to rotate the spherical gizmo for the dome's UVW map until the most desirable part of the sky is shown in the camera. Figure 20.26 shows a rather uninteresting and confusing part of the sky. In fact, it shows where the edges of the sky map are wrapped around the dome and touch (the seam). After the UVW map gizmo for the sky dome is moved up and rotated, the sky looks more realistic (Figure 20.27).

Figure 20.26   Camera view rendered where the seam in the sky map is visible.

Figure 20.27 Camera view rendered after the sky UVW map gizmo has been rotated into a more favorable position.

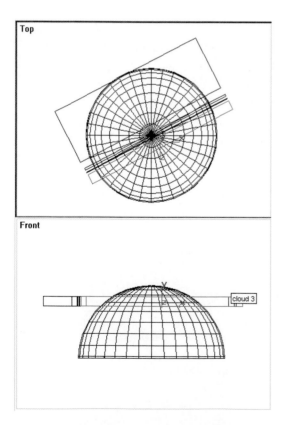

Cloud Shadows

The cloudy sky projected on the dome will surely cause shadows on the ground, so several cloud planes are positioned high above the camera (Figure 20.28). Place these clouds within the dome so that the spotlight representing the sun will cast their shadows across the ground. Figure 20.29 shows the environment with shadows falling almost parallel to an imaginary picture plane. Position the clouds so that a major shadow area will be in front and behind the subject. This frames the subject in sunlight with a darkened background.

Figure 20.28 Clouds positioned inside the dome and below the major light source.

Figure 20.29   Shadows on the ground formed by clouds.

Distance Terrain

Of course, few environments look like Figure 20.29, where the earth is perfectly flat all the way to the horizon. There are several methods of creating terrain. The one chosen here has the benefit of low polygon count and easy editability. By pulling and pushing the vertices, an irregular mountainous object is formed.

Start with a **Tri-Patch Grid** (Figure 20.30) and edit its vertices until it becomes a mountain range that can be introduced into the environment. You can see in Figure 20.31 that the terrain is placed at the back of the hemisphere and in front of the camera. Because our camera isn't going to move much (just follow the plane during takeoff), the terrain is placed in the camera's field of vision. When the *granite.jpg* material is applied to the terrain (and tiled and bump mapped), the world looks much more realistic (Figure 20.32).

Figure 20.30   Tri-patch forms the basis of distant terrain.

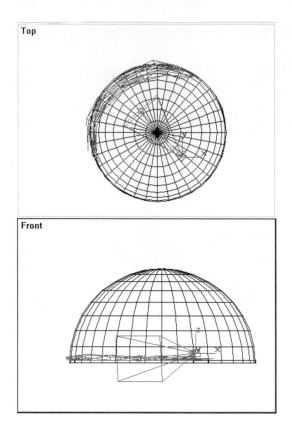

Figure 20.31   Edited tri-patch is positioned at the horizon and within the field of the camera.

Figure 20.32   Rendered distant terrain with *granite.jpg* material and bump maps applied.

Runway and Associated Roads

The runway and taxi strips are modeled as thin boxes to which a concrete raster texture is applied. This texture, created in Photoshop (Figure 20.33) includes joints in the concrete, tire skid marks, and noise to show irregularities. This map can be edited for other concrete features as well—irregular gravel edges, for example. Figure 20.34 shows a top view of the runway in place, while Figure 20.35 shows the camera view. You can see the benefit of using physical clouds since the runway and surrounding terrain consistently share the same shadows.

To get the map to fall on the runway surface exactly where you want, move the planar UVW map gizmo until the runway numbers and skid marks are in the camera's field of view.

Figure 20.33   Raster map used for runways.

Figure 20.34   Top view of rendered runways.

Figure 20.35   Camera view of rendered runways.

Airport Buildings

Because airport buildings are in the distance and won't be in the animation for its entire length, we can get away with fairly crude structures with simple materials. Two or three shapes can be scaled and combined with copies to create a realistic grouping of structures in the distance. Figure 20.36 shows the grouping of buildings, and in Figure 20.37 they are shown in the environment. The buildings are illuminated with their own light sources to increase shadow contrast. The completed environment can be found in the file *world.max* in the *exercises/ch20* on the CD-ROM.

Figure 20.36   Airport buildings in place.

Merge the Beaufighter (*airplane.max*) with the scene. The plane is at the correct scale, but if you didn't personally create both the environment and the plane, one or the other may have to be scaled. Figure 20.38 shows the plane in position ready for an animated taxi and takeoff. Two omni lights are positioned above the plane, one to illuminate the plane itself (multiplier 1.5) and one to cast the plane's shadow on the ground (multiplier 0.5). These omni lights are attached to the plane and will follow the aircraft during the animation.

Figure 20.37   Airport buildings placed in the environment.

Figure 20.38   Airplane merged with the environment.

Set Animation Parameters

As was done in the pulley example, set the frame rate, frame total, and start and stop frames. Use a frame planning sheet to key frame the taxi, run-up, and takeoff. Position the camera's target on the plane at each key frame. Figure 20.39 shows these key frames. The file *take_off.avi* on the CD-ROM shows the completed animation.

(a)

(b)

> ### ▼ Tip
>
> A feature in 3ds max called **Video Post** allows you to switch from one camera to another in the execution of your animation. The **Video Post View** lets you orchestrate exactly when, along the animation timeline, certain cameras are active. As you might imagine, this requires considerable planning. You should include each camera in your storyboard development and record their actions on the frame planning sheet.

(c)

(d)

Figure 20.39   Rendered key frames from the animation.

Review

Animated technical illustrations can provide the most realistic representations possible of proposed or existing objects or situations. With some imagination, and care, you can even create situations that defy natural laws. Animations represent a natural extension of 3D modeling and illustration with the added variables of time and motion.

An effective animator needs knowledge of lights, cameras, and motion, and the same sense for the dramatic found in photographers, cinematographers, and display artists. If you haven't tried your hand at this, it's time to get started!

Text Resources

Hart, John. *The Art of the Storyboard: Storyboarding for Film, TV, and Animation.* Focal Press. Burlington, Massachusetts. 1998.

Miano, John. *Compressed Image File Formats: JPEG, PNG, GIF, XBM, BMP.* Addison-Wesley Pub Co. (ACM Press). Boston. 1999.

Pocock, Lynn and Rosenbush, Judson. *The Computer Animator's Technical Handbook.* Morgan Kaufmann Publishers. San Francisco. 2002.

Reinhardt, Robert and Dowd, Snow. *Flash MX Bible.* John Wiley & Sons. New York. 2002.

Schnyder, Sandy. *The Gif Animator's Guide.* Hungry Minds, Inc. (Wiley). New York. 1997.

Williams, Richard. *The Animator's Survival Kit.* Faber & Faber. Boston. 2002.

Internet Resources

Check out the following Web sites for technical animation products, services, and information.

Animation Technologies. Retrieved from: *http://www.animationtech.com/.* January 2003.

Engineering Arts. Retrieved from: *http://www.engineeringarts.com/.* January 2003.

Flash animation of Cape Hatteras lighthouse relocation. Retrieved from: *http://www.widomaker.com/~litehous/movie.html.* January 2003.

Knot Laboratory. Retrieved from: *http://www.knottlab.com/content/engineering/ e-main.html*. January 2003.

Michael Llewellyn, Illustrator. Retrieved from: *http://www.mikellll.com/*. January 2003.

Systems Integration and Research. Retrieved from: *http://www.sirinc.com/ engmm/engmm.html*. January 2003.

Vector Scientific Biomechanics, Engineering Animation. Retrieved from: *http://www.brcinc.com/*. January 2003.

On the CD-ROM

For those of you who would like to try out the procedures described in this chapter, several resources have been provided on the CD-ROM. Look in exercises/ch23 for the following files:

airplane.max	The airplane file, ready to be merged with the environment.
frame_sheet.pdf	The frame planning sheet used in this chapter to plan camera, light, and object positions throughout the animation.
pulley.avi	The completed 75-frame pulley animation.
pulley.max	The completed pulley assembly.
take_off.avi	The completed airplane animation.
world.max	The environment for the Beaufighter animation.

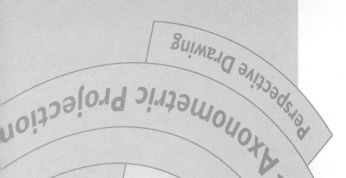

Glossary

3D model   A mathematical representation of spatial geometry displayed on a monitor as lines (vectors), planes, and solids.

3D projection plane   The plane on which the 3D image is projected. The size of this plane is determined by the size of the output raster image. The camera's local Z axis is perpendicular to the projection plane.

Adjacent views   Any two orthographic views separated by 90 degrees, either in viewing direction or object orientation.

Analog data   Data that can take on infinitely variable values. In technical illustration this can be, sketches, engineering drawings, and film-based photographs.

Analog reproduction   Making copies using analog methods; in technical illustration, this usually involves traditional offset lithographic or optical xerographic methods.

Auto tracing   The process of converting a raster image to a vector drawing by automatically detecting sharp value differences in the continuous tone.

Axonometric axes   The world axes in axonometric position along which true measurements can be determined.

Axonometric diagram   The coplanar arrangement of the axonometric plane and two or more orthographic planes rotated into the axonometric plane.

Axonometric drawing   An axonometric view where one or more axis scales has been normalized to full scale.

Axonometric plane   A plane not perpendicular to one of three three world axes.

Axonometric projection   An axonometric view formed by projecting features from orthographic views onto an axonometric plane.

Axonometric shearing   The method of creating an axonometric view by scaling and rotating 2D orthographic views into correct axonometric projection.

489

Axonometric view A view whose direction of sight is not parallel to one of the world axes.

Base color The starting color. The end colors in a blend.

Bit depth The amount of information associated with each position in a bitmap image. The greater the depth of the bitmap, the greater the number of grays or colors that can be displayed at each point.

Bitmap A rectangular array of picture elements (pixels), intended for display on a monitor or printing on a printer.

Black-and-white proof The representation of final artwork in black-and-white or gray form and potentially at a reduced scale or resolution.

Blend A unique object in Adobe Illustrator whereby transition lines or shapes are created between base objects.

Block shading The selective use of areas of black in a line rendering to show shadows.

Brush A pattern or calligraphic brush shape that can be assigned to a path in Adobe Illustrator.

Bump map A grayscale raster material used for impacting surface geometry detail at rendering time.

Camera The XYZ location of the station point (SP).

Camera local axes The central vision line from the camera to the projection plane.

Camera target The center of the field of view, which is the same as the center of vision.

Circle projection A method of creating a scale orthogonal view from an axonometric ellipse so that distances and angles can be constructed and projected back onto the axonometric plane.

Clipping mask A path that is used to limit the display of a background object, gradient, blend, or raster image.

Color key A location along a gradient fill where color can be specified.

Color palette The matrix of colors available for a given image. See also *Bit depth*.

Color proof   The representation of final artwork in color and potentially at a reduced scale, resolution, or color depth.

Color space   The entire gamut of color; the method of specifying color models such as RGB or CMYK.

Compound path   A shape comprised of two or more paths. Treated as a whole, the places where the paths overlap are treated as holes when filled with colors, gradients, or blends.

Constrain angle   A setting in preferences that impacts the angular orientation of all operations in Adobe Illustrator.

Continuous tone   The use of gradient values and colors to represent geometric form.

Depth   A horizontal dimension perpendicular to height and width.

Digital camera   A device that captures images in digital form with depth of field.

Digital reproduction   The creation of copies without analog methods.

Digital scanner   A device that captures images in digital form without depth of field.

Dimensioned views   Orthographic views, drawn either to scale or not to scale, having accurate sizes noted using standard dimensioning practices.

Dimetric view   The axonometric view where the plane of projection is inclined equally to two of the three world axes.

Display color   The red, green, and blue (RGB) color space model used for technical illustrations displayed on color monitors or projectors.

Dithering   The approximation of color or tone outside a color palette by the juxtaposition of colors within the palette.

Dots per inch (dpi)   The measure of density in a raster screen display.

Drawing Exchange Format (DXF)   AutoCAD's neutral data format for exchanging data between dissimilar CAD or graphics applications.

Drawn to scale   Orthographic views with sizes that reflect the true dimensional relationships of the object can be directly compared.

Ellipse Any view of a circle where the line of sight does not form a 90-degree angle and is not parallel to the plane of the circle.

Ellipse (measuring device) Making use of the fact that all diameters of an ellipse represent the same measurement.

End cap The treatment of line ends that simulates lines drawn by ink pens.

Environment map A raster material used for overlaying a stationary environment on a reflective surface.

Field of view (FOV) The conical representation of the range of view (cone of vision) of a camera or a person.

Fill The characteristics of a curved line or closed shape. These characteristics include color and gradient. See also *Clipping mask*.

Flat shading A rendering technique that assigns the same value (color, brightness) to every pixel of a surface.

Font The collection of all typographic characters of a certain style and size.

Front view Also called the front elevation view, the view seen when looking parallel to the horizontal depth axis.

Gamut The range of colors a device is capable of reproducing.

Gizmo The visual representation of the UVW mapping coordinates.

Gouraud shading A rendering technique that averages pixel values between vertices and produces smooth, but predictable, transitions.

Graphic Interchange Format (GIF) An 8-bit, 256-color raster data format with internal lossless compression and alpha channel.

Gradient mesh A method of producing smooth color across a shape by subdividing the shape into a matrix. The intersection of the matrix elements form nodes that can be assigned individual colors.

Gray scale A palette containing 8 bits of value information.

Height A vertical dimension perpendicular to width and depth.

In projection The position of orthographic views such that shared dimensions between adjacent views are in alignment.

Isometric view The axonometric view where the plane of projection is inclined equally to the three world axes.

Job jacket The physical envelop (analog) or virtual directory (digital) that contains all materials associated with a particular job.

Job ticket The control document used to record all activities necessary to accomplish the job.

Joint Photographic Experts Group (JPG) A 24-bit, 16.7 million–color raster data format with internal lossy compression.

Kerning The adjustment of horizontal spacing between characters in a variable spaced font.

Lab XYZ A device-independent color model.

Layer The conceptual organization of graphic elements in an illustration such that they are ordered or viewed selectively.

Leading The spacing from the baseline of one line of type to the base-line of the next line of type expressed in points.

Length A nonorthogonal dimension not parallel to height, width, or depth.

Lens length (mm) This measurement determines whether the camera is wide angle (15–35 mm), normal (35–55 mm), or telephoto (55–300 mm).

Limpel-Ziv-Welch (LZW) A lossless data compression algorithm.

Line blend A smooth rendered transition between base object lines whereby color and stroke are tweened or morphed.

Line rendering The selective use of lines of varying thickness to represent geometric form.

Major axis The axis of an ellipse about which the ellipse rotates as the viewing angle is increased or decreased; the major axis is equal to the circle's diameter.

Meta tool A graphics application that contains both raster and vector tools.

Minor axis The axis of an ellipse, perpendicular to the major axis, that remains concurrent with the circle's local Z or thrust axis.

Mono spaced type Typography where each character commands the same horizontal space.

Node The intersection of lines in a **Gradient Mesh**.

Not to scale Orthographic views without proportional relationships cannot be compared. You *cannot* directly measure these views for technical illustrations.

Offset path A function in Adobe Illustrator that creates a parallel shape a specified distance from a path.

Omni light A light that illuminates evenly in all directions from a point, like a lightbulb without a shade.

Opacity map A raster material used for impacting surface opacity at rendering time.

Optical character recognition (OCR) The process of creating an electronic text file by scanning typographic characters in analog data.

Orthographic view Any view taken from an infinite distance.

Outline stroke A function in Adobe Illustrator that converts a stroke to a closed path shape. The interior of the shape is identical to the original stroke.

Path Any line drawn with the pen tool. Any line surrounding a shape from the toolbox that can be stroked.

Pathfinder The window in Adobe Illustrator containing tools with which to act on paths. These tools include: add, subtract, intersect, exclude, divide, trim, merge, crop, outline, and minus back.

Perspective projection A view formed by the intersection of visual rays from a known vantage or station point with a plane of projection.

Phong shading A rendering technique that calculates a unique value for each pixel on every visible surface.

Photo tracing The process of converting a raster image to a vector drawing by manual methods.

Picture plane (PP) The plane on which the perspective is projected.

Pixel   The smallest addressable unit of a raster graphic.

Plan view   The top view of the object (width and depth) set at a desired angle to the edge view of the picture plane.

Points   The system of typographic measurement where 1 point is equal to 1/72 in.

PostScript   Adobe's page definition language used for creating and printing resolution-independent artwork and printing resolution-dependent raster images.

PostScript drawing tool   A 2D computer application program whose data is resolution-independent vectors and optimized for output to a PostScript printer.

Principal view   Any orthographic view that is taken parallel to the X, Y, or Z world axes.

Printer dots per inch (ppi)   The measure of density of a digital printing device.

Process color   Printing using cyan, magenta, yellow, and black inks; CMYK color.

Profile view   The side view of the object set in the desired relationship to the ground line. The profile view establishes height in the perspective view.

Projectors   Parallel eye sights used for locating features and aligning adjacent orthographic views.

Raster   The technology that displays information as rows of data points. See also *Bitmap*.

Raster editor   A 2D computer application program whose data is a fixed-resolution raster bit map.

Raster graphic   A graphic formed by a matrix of individual pixels.

Raster image processor (RIP)   The engine within a digital printer that translates raster, vector, and outline text data into the pattern of printer dots for a given printer.

Ray traced shading   A rendering technique that traces every light ray in a scene from the pixel it falls on, back to all light sources. In that way, highly detailed reflections can be rendered.

Rendering   The stylistic representation of form, materials, and shadows.

Rotation	The method of creating an axonometric view in a principal view by rotating object geometry relative to the world axes.
Run length encoding (RLE)	A lossless data compression algorithm.
Scanning	The process of converting analog data to digital raster data by sampling the value and/or color using charge couple device technology.
Shape blend	A smooth rendered transition between base object shapes whereby fill and stroke are tweened or morphed.
Side views	Also called profile views or side elevations, the views seen when looking parallel to the horizontal width axis (X axis).
Sphere (measuring device)	Making use of the fact that all elements from a sphere's center to the surface represent the same measurement.
Spot color	A single color used for emphasis.
Standard axonometric views	Axonometric views for which information is known for any of the several methods of creation.
Station point (SP)	In the top or plan view, the point at which all projections (visual rays) converge. In the perspective view, the center of vision.
Stroke	The characteristics of a path or line. These characteristics can be pen, color, stroke width, end cap, and line joins.
Swatch	A custom-mixed color in Adobe Illustrator saved in the swatches palette with a descriptive name.
Symbol	A graphic stored into Adobe Illustrator's symbol palette. Changes in the parent symbol are reflected in the symbol's instances.
Tag Image File (TIF) format	A 24-bit, 16.7 million–color raster data format with internal lossless compression.
Targa Graphics Adapter (TGA) format	A 64-bit, proprietary raster data format used with Truevision display hardware.
Target direct light	A light that projects a cylinder of illumination much like the sun projects parallel light rays from a great distance.
Target spotlight	A light that projects a cone of illumination much like a theatrical spotlight.

Thrust axis The axis that remains perpendicular to a feature as it is revolved in space; the local Z axis; the perpendicular axis of an ellipse is always concurrent with the ellipse's minor axis.

Top view Also called the plan view, the view seen when looking down on a vertical axis from above.

Trimetric view The axonometric view where the plane of projection is unequally inclined to the three world axes.

Unit cube An axonometric cube whose dimensions are unitary (1 in, 1 mm, 1 mile, and so on) used to create accurate axonometric constructions.

UVW map A coordinate system that coordinates the placing of raster material maps onto 3D computer geometry.

Vanishing lines Heights projected to their common vanishing point.

Vanishing point (VP) The apparent point of convergence of parallel elements.

Variable-spaced type Typography where each character commands different horizontal space based on the adjacent characters.

Vector graphic A graphic formed by lines and curves.

Vertical measuring line A line along which true heights can be projected from the profile view.

Viewpoint The method of creating an axonometric view by positioning the viewer's eye (the camera) at a specific location in space.

Visual rays Projectors (eyeball sights) that converge at the station point, forming the perspective drawing by their intersections with the picture plane.

Web-safe An 8-bit color palette matching those colors a Web browser can display without dithering.

Width A horizontal dimension perpendicular to height and depth.

Windows bitmap (BMP) The native Windows raster format using RLE compression and 24-bit color depth.

Index